Improved Crop Quality by Nutrient Management

Reviewed papers from the Improved Crop Quality by Nutrient Management Workshop, Sept. 28 – Oct. 1998, Bornova, Izmir, TURKEY

Reviewed by

D. ANAÇ

Ege University
Faculty of Agriculture
Department of Soil Sciences
35100 Bornova-Izmir-Turkey

P. MARTIN-PRÉVEL

3 rue, st. Hubert,
34820 Teyran, France

D.K.L. MAC KERRON

Scottish Crop Research Institute,
Invergrowrie,
Dundee, DD2, 5DA, United Kingdom

M.L. REILLY

Environmental Resource Management
Department University College,
Belfield, Dublin 4, Ireland

E.J. GALLAGHER

Department of Crop Science,
Horticulture and Forestry,
National University of Ireland,
Dublin, Belfield, Dublin 4, Ireland

A.E. JOHNSTON

Soil Science Department I.A.C.R.,
Rothamsted Harpenden, Herts,
AL5 2JQ United Kingdom

N. ERYUCE

Ege University
Faculty of Agriculture
Department of Soil Sciences
35100 Bornova-Izmir-Turkey

M.M. EL-FOULY

National Research Center
El-Tahrir Str.
Dokki-Cairo, Egypt

Improved Crop Quality by Nutrient Management

Edited by

D. ANAÇ
Ege University,
Faculty of Agriculture,
Department of Soil Sciences,
Izmir, Turkey

and

P. MARTIN-PRÉVEL
Teyran, France

KLUWER ACADEMIC PUBLISHERS
DORDRECHT / BOSTON / LONDON

Library of Congress Cataloging-in-Publication Data

ISBN 0-7923-5850-3

Published by Kluwer Academic Publishers,
P.O. Box 17, 3300 AA Dordrecht, The Netherlands.

Sold and distributed in North, Central and South America
by Kluwer Academic Publishers,
101 Philip Drive, Norwell, MA 02061, U.S.A.

In all other countries, sold and distributed
by Kluwer Academic Publishers,
P.O. Box 322, 3300 AH Dordrecht, The Netherlands.

Printed on acid-free paper

Printed in the Netherlands.

CONTENTS

PART 1: Crop quality – nutrient management by conventional fertilization

PART 2- Crop quality – nutrient management by foliar fertilization

PART 3-Crop quality – nutrient management under stress conditions

PART 4- Crop quality – nutrient management by the diagnosis of crop nutrition

PART 5- Crop quality – nutrient management in soilless culture

PART 6- Crop quality – nutrient management by alternative sources

PART 7- Crop quality – nutrient management in general

PREFACE

Dilek ANAÇ[1] and Pierre MARTIN-PRÉVEL[2]

[1] *Convener of the 1st IAOPN International Workshop on Improved Crop Quality by Nutritient Management – Ege University, Faculty of Agriculture, Department of Soil Sciences, 35100 Bornova-Izmir, Turkey*
E-mail: danac@ziraat.ege.edu.tr

[2] *President IAOPN – 3 rue St.Hubert, 34820 Teyran, France - Phone/fax : +33 4 6770 2295 - E-mail : pfmp@cirad.fr*

Present global population which is roughly 5.7 billion is expected to increase at an annual rate of 80-85 million and over 90% of this increase will be in developing countries. Despite the fact that world food production has grown faster than the population, more than 800 million people in the developing world were undernourished in the early 1990's and millions more suffered from diseases related to micro nutrients and contaminated food and water.

After all, irrational fertilizer use and inefficient use of irrigation water are the main agricultural mismanaged uses. Rapid population growth and continuous urbanization which exhausts the natural resources will result in an increased demand for food in general and for high grade and quality crops as well. In fact, the purchasing power of man in the exporting countries will rise if they will not satisfy the quality requirements of the importing countries.

Practical experience and scientific investigations have shown that of the various cultural measures, balanced fertilization above all exerts a considerable influence on the quality of agricultural products. Balanced crop nutrition increases crop quality, safeguards natural resources and brings benefit to the farmer.

These were the major topics of the **1st IAOPN Workshop on 'Improved Quality by Nutrient Management'**.

The meeting was attended by 99 participants and 15 accompanying persons from 18 different countries. The scientific programme covered a wide range of subjects related to crop quality and nutrient management. These proceedings consist of the 72 refereed oral and poster paper presentations.

In order to elaborate Workshop **conclusions**, the present overall status of quality achievements and possible strategies were debated during a closing session. Three major issues were pointed out :

1. What is quality ? As far as edible produces are concerned, markets tend presently to pay only for *external* quality characters such as:

☐ keeping ability,
☐ resistance to transportation hazards,
☐ absence of visible defects,

□ parameters contributing to the appeal (colour, size, shine, firmness…).

Criteria concerning *internal* quality and especially its expected benefits for the consumer's health start to be also considered by the sellers and buyers. But at the moment they rely mostly on two considerations :

□ a wish for more tasty produces ;

□ an uncontrolled fear for toxicities by pollutants, heavy metals, nitrates, etc.

Nobody takes care that for instance, as reported to the audience by Dr. J. Jonhston, hospitals are sometimes facing new diseases due to nitrate deficiency in the diet. Nor anybody at the moment would accept to pay more for a produce of which the composition (e.g. in essential amino-acids) would be certified as highly superior from a dietetic point of view. However, in developed countries a fraction of the consumers agrees to pay more for produces qualified as 'natural' (originated from the so-called 'organic' or 'biological' farming), because they consider them as more tasty and/or better for their health.

2. How to deal with the 'organic farming' ideology ? Although scientists cannot agree to any external postulate, statements about a claimed better quality cannot be either rejected or accepted without careful check. The recommended general policy was :

□ to *comparatively assess the objective quality* of organically vs. conventionally grown produces under the same soil and climate conditions ;

□ in that view, due to the very strict regulations adopted by the organic farmers unions, to urge developed countries to set up *experimental farms specially designed* for the organic/conventional comparisons, i.e. with large fully comparable sectors permanently devoted to one production system ;

□

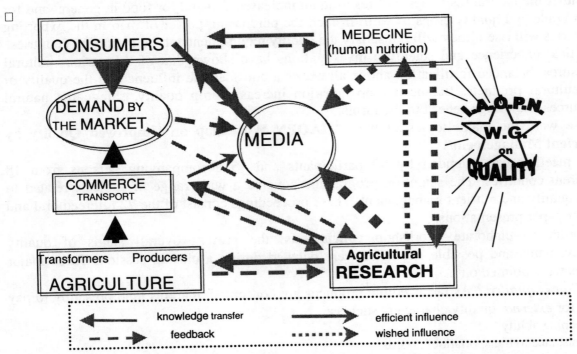

Figure 1 - Actual and wished relationships for achieving quality of edible produce.

☐ on the other hand, to pursue researches in view of achieving *similar or higher standards of 'true' quality* by means more compatible with the quantitative food demand by the world population in the coming decades.

3. *How to improve quality at the markets scale* whereas the quantitative increase of production is still required to continue ?

Of course a first step is to *intensify the research* on the qualitative effects of all agricultural practices, especially nutritional management, with more focus on the most crucial issues, and more reciprocal cooperation between the concerned teams. This is why IAOPN proposes to set up a permanent **Working Group** aiming to prolong and extend the links created during this Workshop. Any interested scientist is heartily invited to apply.

However, even excellent research results are inefficient if they are not put into application. Figure 1 was proposed by the second author to help defining an adequate strategy. All participants agreed on the scheme. As information exchanges between agricultural and medical scientists appeared insufficient, invitation of some of the latter to the next fertilization/quality workshops was considered. The extreme importance of the mass media (mostly press and TV) was stressed, whilst they tend to be much more influenced by fashions or pressure groups than by sound scientific evidences ; thus scientists were highly recommended to communicate more with the 'great press' and the TV.

"Improved Crop Quality by Nutrient Management" Workshop

STEERING COMMITTEE

Dilek Anaç   Convener
Nevin Eryüce
Bülent Okur   Secretariat

SCIENTIFIC COMMITTEE

Dilek Anaç   Ege University (Izmir-TURKEY)
Nevin Eryüce   Ege University (Izmir-TURKEY)
Bülent Okur   Ege University (Izmir-TURKEY)
Pierre Martin-Prevel   President, I.A.O.P.N Teyran-FRANCE

WORKSHOP ASSISTANTS

B. Eryüce	C.C. Kılıç
Ö. Gürbüz	E. Irget
H. Çakıcı	S. Delibacak
L. Tuna	B. Çokuysal
G. Yolcu	G. Yönter
B. Yağmur	

Ege University Faculty of Agriculture
Department of Soil Sciences 35100 Bornova-İzmir-Turkey

International Association for the Optimization of Plant Nutrition as at 1 July 1997

BOARD

P. Martin-Prevel, President
M.M. El-Fouly, Vice Presented
M. Braud, Secretary and Treasurer
A. Boudonnou , Assistant Secretary
G. de Monpezat, Assistant Treasurer

STANDING COMMITTEE

D. Anaç, Turkey
J. Baier, Czech Republic
B. Braud, France
A. Boudonno, Italy
G. de Monpezat, France
M.M. El-Fouly, Egypt
M.U. Gaviton, Spain
D.K.L. MacKerron, United Kingdom
P. Martin-Prevel, France
L. Mansson, Sweden
E. Portela, Portugal
N. Rossi, Italy
V. Römheld, Germany
G.W. Welkie, United States of America

1. Crop quality – nutrient management by conventional fertilization

Chapter 1

Factual background for improving quality through nutritional management (short version)

Keynote paper

Pierre Martin-Prével

President IAOPN – 3 rue St.Hubert, 34820 Teyran, France - Phone/fax : +33 4 6770 2295 - E-mail : pfmp@cirad.fr

Key words :   produce quality, plant nutrition, plant nutrients, fertilization, crop species

Abstract :   From the CABI data-bases 257 abstracts of papers issued in 1997 and early 1998 were summarized into 6 huge synoptic tables. Trends are examined by commodity groups and species for all crop types. Perspectives for the future are stressed from the frequencies of beneficial, neutral and detrimental effects of essential and non-essential elements with specific attention to N, K and organic matter. *Nota :* tables and references are presented only in an extended version containing more comments and available on order.

1. Purpose and Method

As first bibliographical comparison basis to help discussions and future research, a punctual sampling among the current research tendencies on all commodities was reviewed.

From the last CABI's CD-roms (1996, 1997, first half of 1998) bulked for on-line consulting in the CIRAD Documentation Centre, about 400 entries of papers *published in 1998 or 1997* were extracted by crossing all developments of 'fertilizers' and 'plant nutrition' with all developments of 'quality' in the descriptors. **257 references** actually covered original results about nutrition/fertilization effects on quality ; their *abstracts as compiled by CABI* and references were transferred into a WP data-base with additional fields for *commodity groups,* specific *crops, nutrients* involved, *senses* of effects, quality *criteria* with reported changes, *other useful informations* in abstract. Sorting and shaping of the whole resulted in 6 huge synoptic tables, included with the references and wider comments in a complete version of this article. However appreciation on the seriousness of each paper is left to the reader after securing the original full publication.

2. Results

Cereals

Criteria were fairly variable for cereals but **protein** content was a major one, followed by factors of **bread-making quality** (S favourable on wheat) or **eating quality** (minutiously determined in rice). Variable effects were registered on non-dietary criteria when grains were produced as seed material.

Nitrogen was by far the major nutritional concern. Conflictory results in fact mostly reflected differences in *levels as compared to the need* for maximum yield (N scarcely detrimental if no over-fertilization), *timings,* relations with *other factors* such as water availability, grain *destinations* (high N content good for food but not for malting). Advantages and disadvantages of **organic manuring** were not obvious from the few alluding references.

Fruits

This group counted with the greatest number of references and diversity of studies.

Criteria.- *Keeping ability* was greatly considered, either as a whole (shelf life under various conditions, spoliage) or assessed by its components. *Visual* and *organoleptic parameters* (colour, neatness, firmness, skin thickness, TSS, acidity, individual metabolites, panel notation) also received major attention. Resistances to *physiological and pathological disorders* (bitter pit, scald, skin cracks, fungi...) and even to cold damage to trees in field were also recorded, whilst concern towards the *consumer's health* was scarce (just a few mentioned components such as ascorbic acid).

Nutrients.- *Nitrogen* was again the most studied, with frequent negative effects when applied at rate super-optimal for crude yield, sometimes even at lower rate. *Potassium*, and *phosphorus* to some extent, were more often quoted for favourable than unfavourable effects, even on apples (pigments) in spite of its known antagonistic effect on *calcium* and therefore on bitter-pit incidence. Ca was fairly used for improving shelf life (peach, apricot) or other criteria, as pre-harvest spray on leaves or fruit or as post-harvest dip. Unexpectedly, *magnesium* was scarcely mentioned. *Sulfur* appeared once as such (mango) but was also involved as form of applied K or N, and quoted favourable in each instance. *Silicium* was mentioned once (apple), but from a composite fertilizer of which the action of each part was not established. All *essential trace elements* except Mo were extensively tested, either individually or as mixtures, most frequently as leaf sprays ; they showed always beneficial for various quality parameters. Even *rare earth* were successfully tested on apples. Trace elements could possibly explain a part of the favourable effects of *organic matter* and *microbial preparations or extracts* quoted by some references.

Vegetables

As expected great similarities appeared with fruits, especially for **fruity** vegetables. But *organic matter* (in 'organic' or conventional farming) was more frequently cited (pepper, cucumber, tomato), as *phosphorus* and *sulfur* on tomato and legumes (faba bean, clusterbean, chickpea). Special attention was paid to *nitrates*

in **leafy** vegetables, reflecting a widespread current concern and leading to avoid excessive N fertilization. However conflicts between quality and yield did not take place at moderate N rates and most authors defined a « reasonable *yield + quality optimum* ».

In the **onion** group, special attention was paid (but in a not fully clear abstract) to effects on *flavour and pungency components* by *sulphur* and also *selenium* ; the latter is not recognized as essential for plants but is important for animals and humans. **Potato** endured conflictory effects of given elements according to the considered quality parameters ; its fertilization may require different *NPK balances and forms* (KCl effects on *fungal rot*) according to the local major quality problems. *Trace elements* were specially tested in **cabbage** species and varieties : boron, zinc and even *iodine*, but unfortunately the sense of its effect on nitrate and ascorbic acid was not reported in the CABI abstract whereas a yield increase was stipulated.

Fodder Crops

The group includes composite pastures and herbages, mono- or oligo-specific grassy turfs (both as lawns or for cattle feeding), legumes, grass-legume mixtures, fodder trees or shrubs.

Nitrogen was confirmed to favorize grasses when competing with legumes. Mineral fertilizing, especially with N, was quoted detrimental to **biodiversity** in composite stands whereas species mixing was considered a health-insurance for cattle. About the 'greenhouse effect', a study on alpine *Carex* grassland concluded that adequate NPK fertilization should prevent litter deterioration.

Digestibility was appreciated globally or through analysis by neutral and acid detergents, in vitro digestibility, etc., with an interesting *cultivar-photoperiod-fertilizer (N or K) interaction* in *Cynodon*. Effects on **crude protein** varied with nutrients, species and conditions : *N* showed always very efficient, *P, Mg and trace elements* favourable on *Leucaena* and *Na* detrimental on *Lolium*. **Risks for cattle** were also evidenced : *tetany* by cation imbalance, *Na* deficiency in blood if not liming a vetch stand, *Se* deficiency induced by *sulfur*. For **turfs**, *N* was also the major quality factor through soil coverage and herb colour, but with interference by *mowing rhythm or height*.

Ornamentals

Nitrogen appeared once more as a major quality factor up to relatively high rates but reference cross-checking was hindered by non-comparable units (e.g. ppm without solution volume) whilst many authors attributed to « N » the effects of the whole fertilizing. Beside *visual appeal* or its components, *keeping ability* of potted plants or cut flowers and *sanitary* issues were studied. Excess of «N» could show detrimental in these respects, but *P, S, Ca*, metallic *minor elements* and probably *K* were beneficial in all tested instances. As a whole, good « *market quality* » resulted from a nutrition fairly high but not excessive and well balanced in macro- and micro-nutrients. Opposite effects of Ca and NH_4 on *Botrytis* in potted roses were particularly noteworthy.

Industrial Crops

Fiber crops.- On *fiber yield* and *fiber gross properties*, again *nitrogen* showed generally favourable up to a level ensuring a sub-maximum crude yield. N was easily detrimental but P, Ca, Zn favourable to physico-chemical indexes. N internal metabolites also modified boll opening and *bollworms* occurrence.

Wood or biomass for industry.- Internal *N leaf level* was associated with lower wood density in pines and K fertilization depressed biomass combustibility in another species.

Oleaginous and proteaginous crops.- *N, P and B* exerted *slightly conflicting effects* on similar criteria. A case on severely pruned olive trees provided another evidence of quality improved by N (here associated with K) as far as it is *required for normal growth*. A case on sunflower recalled that N may be detrimental to quality if not *balanced* with other nutrients but beneficial if balanced. As expected, *sulfur* was beneficial for oil and protein in sunflower.

Sugar crops.- Most of the available references only considered *sucrose %* in the crude harvest or extract. Factors influencing juice extraction (milling quality for sugarcane) or purification (ash, « Na- or K-impurities », N-derivatives) were sometimes added. *Nitrogen* was frequently found detrimental in these respects, requiring a compromise between crude yield and sugar content plus extraction-purification hindrances. By contrast, all five *other macro-nutrients* showed beneficial, or at least neutral, on sugar % when tested. *Zinc* was quoted beneficial in one instance, four metallic *micro-nutrients* as neutral in another one.

Smoking materials.- The known adverse effects of Cl on tobacco quality were not mentioned. A better value but a lower nicotin content were quoted for the flue-cured tobacco from fertilized plots : conclusion is left to the consumer's personal wish toward nicotin !

Aromatic and stimulating crops.- In basil, lemongrass, mint, vetiver, turmeric, essential oils and/or some *characteristic compounds* were quantified but few fertilizer influences mentioned in abstracts. In coffee and tea, caffein or similar compounds and a fair number of flavour components were studied. K *chloride* was found detrimental to coffee quality when compared to sulphate. For tea, *N* fertilization resulted as requiring extreme precautions whereas *internal levels* showed correlations (positive for *P, S, K, Mg, Mn, Cu*, eventually stimulated by mycorrhizal inoculation, negative for N and *B*) with a whole set of criteria.

3. Conclusion

The much greater number of 'positive' than 'negative' tendencies in table 1 reveals **great existing possibilities** for increasing quality by fertilization. Every commodity specialist may now start from the extended article (available on order) and complete his own list before choosing the original papers to be secured.

Table 1 - Summary of reported effects of individual nutrients on quality (numbers of citations)

	-	(-)	=	(+)	+	?	Total
N	23	11	24	17	37	33	145
P		1	8	5	24	9	47
S				2	12	1	15
Cl	1						1
Si					1		1
K	4	2	9	5	18	6	44
Mg	1		2		6	1	10
Ca			1	1	11		13
Na	1	1					2
Fe			2	1	5	1	9
Mn			1		3		4
Zn	1		1	1	13	1	17
Cu	1		1		2		4
Mo							0
B	2		2	1	6		11
I						2	2
Se						1	1
RE					1		1
OM	1	2	8		11	4	26
MO			1		2		3
Total	35	17	60	33	152	59	355

- (-)	: detrimental effect ; detrimental tendency
+ (+)	: favourable effect ; favourable under conditions
= ?	: neutral or doubtful ; not described in abstract
RE	: rare earths
OM MO	: organic matter ; micro-organisms

The same holds true for nutrients. The numerical importance of **nitrogen** in the review is thought to result from current pollution concerns ; repeated remarks in text will serve as conclusions for this element. On *organic matter*, further research is definitely required as all reported comparisons with mineral fertilizers failed to consider its thorough mineral content (only N or NPK mentioned).

References on *potassium* were insufficient for a specific conclusion but many of them fit into a former synthesis (Martin-Prével, 1989, no figure included), here diagrammatically reported as figure 1. Absissae are supposed to be adapted to each crop, so that K effects on its crude yield coincide with the continuous curve. When quality is paid by the market (dotted curves), K can induce increased or decreased profits, according to its relative level and to other factors such as Mg, Ca or N nutrition.

This example also recalls that at present time markets tend to react only to rough quality factors such as sugar yield and shelf decay. It can be hoped that the 'true' quality, i.e. nutritional value and palatability in #1 and #2 ranks respectively, will some day become still more influent. But « this is another story... » !

Quantitative yield (continuous line)
Profitability (dotted lines)

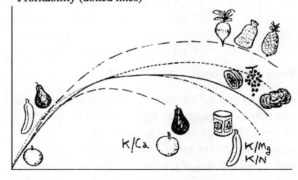

K nutritional status

Figure 1 - Scheme of economic results of K-induced effects on post-harvest quality.

4. References

Martin-Prével P (1989). Physiological processes related to handling and storage quality of crops. Methods of K Research in Plants ; 21st Colloquium International Potash Institute, Bern ; 1989, 255-283 ; 149 ref.

Chapter 2

Organic manures and mineral fertilizers

A.E. Johnston

Soil Science Department, IACR-Rothamsted, Harpenden, Herts AL5 2JQ

Key words: Organic manures, mineral fertilizer

Abstract: Until the mid 19th century organic manures were the only way of returning plant nutrients to soil and in some countries organic manures are still the principle source of nutrients. The increasing use of mineral fertilizers during the last 150 years has made a major contribution to increasing food production worldwide. Mineral fertilizers are easy to handle and can be applied with precision with reasonable certainty about their effectiveness. Organic manures are bulky, expensive to transport and contain small amounts of nutrients per unit volume. In addition, the amount and rate of release of nutrients from organic manures is often difficult to predict with accuracy, especially for nitrogen. Some of the benefits and concerns about the use of organic manures are discussed in this paper.

1. Introduction

By the 1830s it was generally accepted that plants acquired their carbon, hydrogen and oxygen from carbon dioxide in the atmosphere and water in the soil. The "mineral" elements, like phosphorus (P), potassium (K), sulphur (S) and chlorine (Cl), found in plant ash, were taken up by the roots from the soil solution. There was, however, great uncertainty about the source of nitrogen (N) found in living plants and farmyard manure (FYM) but not in plant ash.

In the early years of the 19th century besides FYM, produced on the farm, other organic manures were available, these included guano and cotton and rapeseed cake. Ammonium sulphate, a by-product of coal gassification to produce town gas, and Chilean saltpetre, from South America, were two sources of N whilst K, as the sulphate and chloride, were produced by leaching wood ashes. In England at that time bones and bone products were known to be useful soil amendments on some but not all soils. They were not effective on the soils of Rothamsted Farm, a small agricultural estate near Harpenden, some 40 km north of London. J.B. Lawes, owner of the estate, like others, treated bones with sulphuric acid to make superphosphate, a highly effective source of P on many soils. He patented his process in 1842 and by June 1843 was producing superphosphate commercially at a factory in London; the start of the fertilizer industry (Johnston, 1994). By the middle of the 19th century farmers had available to them a range of fertilizers as well as the traditional organic manures.

After many centuries in which farmers had equated the productivity of a soil to its organic matter content, many of those farming in the mid-19th century had difficulty in understanding how a few kilogrammes of an inorganic salt could increase yield as much as many tonnes of an organic manure. However, in many temperate regions yields did not decline with the continued use of fertilizers.

In more recent times something of an issue has developed around the use of organic manures and fertilizers. (As used in this paper organic manures are not synonomous with organic farming.). In 1949 E.M. Crowther, then Head of the Chemistry Department at Rothamsted, discussed the difficulties of working with the complex biological, chemical and physical properties of soil. He noted that, "it is easy to rationalise personal preferences by formulating elastic hypotheses in modern polysyllabic jargon. A liking for muck (FYM) or (organic) compost sounds more scientific when expressed in terms of feeding or stimulating the microorganisms which produce mycorrhiza, vitamins, antibiotics or polyuronide gums, but, important as these and many other effects of soil organic matter may ultimately prove to be, there is at present little opportunity for testing their significance".

Fifty years later, much of what Crowther (1949) said about personal preferences and beliefs often still dominates the discussion about the merits of organic manures and fertilizers. This paper briefly discusses some past and ongoing research at Rothamsted and Woburn which compares organic manures and fertilizers.

Early comparison of fertilizers and organic manures

Besides becoming a fertilizer manufacturer, Lawes was also a farmer keen to understand more about both crop and animal nutrition. In June 1843, Lawes appointed a chemist, J.H. Gilbert, to come and help organise, establish and supervise the experiments he wanted to start on Rothamsted Farm (Johnston, 1994). All the experiments on crops compared FYM with various combinations of plant nutrients supplied by fertilizers. The results in the first few years quickly established that it was necessary first to overcome soil P deficiency. Then it was necessary to supply a readily available source of N, as nitrate or ammonia. This finding led to a bitter controversy with Justus von Liebig at Giessen in Germany (Johnston, 1991a). For many years Liebig maintained that plants acquired their N from the air.

These early experiments also showed that the same yields could be obtained with fertilizers as with FYM (Table 1). Lawes and Gilbert appreciated that the greater availability of fertilizers would help supply plant nutrients to grow crops to feed the rapidly increasing urban population.

Recent experiments with organic manures

Some of the experiments started at Rothamsted between 1843 and 1856 still continue. Yields in the 1960s and 1970s on fertilizer- and FYM-treated soils were still comparable (Table 2). This was so even though the FYM-treated soils, receiving 35 t ha^{-1} FYM each year, by then contained about 3 times as much soil organic matter (SOM) as did the fertilizer-treated soils. The presence of the extra SOM did not appear to be important. Summarizing data from many experiments, including those at Rothamsted, Cooke (1967) came to the tentative conclusion that, "... in modern farming systems in temperate countries, where adequate amounts of fertilizers are properly used ... economic yields may be obtained in permanent cropping systems without any special action to add organic matter ...". Cooke restricted this comment to temperate regions; it would not apply to tropical soils where the importance of SOM has long been recognised as has the difficulty of increasing SOM in these soils.

Effects of organic matter in soil

Crowther (1949) noted that FYM can have many effects. One of the problems is to separate the effects of the nutrients in organic manures from any effects of the organic matter itself.

More recently experiments on the silty clay loam soil at Rothamsted and the sandy loam at Woburn are now showing benefits where soil contains extra SOM (Johnston, 1986, 1991b).

At Woburn, yields of potatoes and spring barley, both testing four amounts of N and grown on soils with two levels of SOM, were largest on the soil with more SOM and given most N. At the lower level of SOM it was not justified to apply more than 200 kg N ha^{-1} to potatoes (Table 3).

In these long-term experiments benefits from having extra SOM are not because SOM levels have changed recently, they haven't. Rather it is because recently introduced cultivars have a larger yield potential than older cultivars. To achieve this potential requires optimum soil conditions. One of these conditions is good soil physical conditions and SOM has a major effect on soil structure.

The extent to which commercially viable (and available) supplies of organic manures can effect SOM is probably small however. In temperate climates SOM changes slowly with time and large inputs are required to make appreciable changes. Only after more than 150 years of applying 35 t ha^{-1} FYM each year is SOM in Rothamsted soil approaching a new equilibrium value appreciably above that in the 1840s (Jenkinson *et al.*, 1994).

Effects of nutrients in organic manures

More important than the supply of organic matter is the supply of nutrients in organic manures.

Potassium In many organic manures nearly all the K is water soluble and available to crops. The extra SOM supplied by the manure will often provide additional cation exchange sites to retain K.

Phosphorus In animal manures, like FYM and slurry, as much as 60-80% of the total P content is inorganic phosphate. In soil this behaves exactly as the P from mineral fertilizers. Organic P in organic manures has to be mineralized to break down the organic molecules and release inorganic phosphate ions which react as do those in mineral fertilizers.

Although the data for Olsen P show no effect of FYM compared to superphosphate there is one important difference between the differently treated soils. Table 4 shows the total-, Olsen- and $CaCl_2$-P in FYM- and PK-treated soils in two long-term experiments at Rothamsted. (Other similar data are in Johnston and Poulton, 1992.) The increase in total and Olsen P in the two soils was very similar (Table 4). However the P extracted by 0.01 M $CaCl_2$ was much larger in FYM- than in PK-treated soils. Also one treatment in the Barnfield experiment had FYM and superphosphate applied together. For this combined treatment the increase in both total- and Olsen-P were additive when compared to where the two P sources were added separately but the increase in $CaCl_2$-P was much larger than the sum. The larger concentration of $CaCl_2$-P in FYM-treated soil could be due to the extra organic matter in these soils providing more low energy bonding sites for P. From this it could be inferred that the P will more easily replenish that in the soil solution and hence be more available to crops.

Nitrogen Without doubt the increased use of N since the 1950s has played a major role in increasing food production worldwide. The important issue, however, is to match demand and supply. Any unused N input which remains as nitrate in autumn is at risk to loss either by leaching or denitrification when rainfall exceeds evapotranspiration. This is a financial loss to the farmer and can lead to the possibility of environmental problems.

Matching supply and demand is not easy with mineral fertilizers, it is very much more difficult with organic manures in which most of the N is in organic combinations. Because of this it is very difficult to accurately quantify the amount of readily mineralizable organic N and even more difficult to predict when it will be mineralized. Although organic N is mineralized in spring there is often an appreciable increase in mineralization in autumn when there is little crop demand and a considerable risk of loss of nitrate. This is because soil moisture and temperature favour increased microbial activity once the microbes have a readily available energy source from incorporated crop residues.

Separating nutrient and organic matter effects

Yields of winter wheat grain from FYM- and PK-treated plots in the Broadbalk experiment in 1979-84 have been used to try to separate the N effects of SOM and other effects (Johnston, 1987). Rates of N were tested on both soils and an exponential plus linear response curve of the same form was fitted to both sets of yield data for each year. The six FYM and PK curves could be brought into coincidence by horizontal and vertical shifts. Then the two combined curves were brought into coincidence by superimposing the maximum yields. Similar curve shifting to estimate the N and SOM effects of ploughed in grass clover leys has been discussed by Johnston *et al.* (1994).

2. The Future

As noted previously, the use of organic manures in food production should not be confused with organic farming. The latter has very strict protocols and only manures produced in a totally organic system can be used. But large quantities of manure are produced each year in other systems. These should be used effectively and efficiently because of the valuable nutrients they contain and their organic matter content.

Unless large quantities of manure are applied regularly the effect on SOM in the short-term is likely to be small. But if the additions slow or halt the decline in SOM in mainly arable cropping systems then this is important. The nutrient elements are a valuable resource. Within the concept and practice of integrated plant nutrient management, farmers must be encouraged to use organic manures effectively. The P and K will add to soil reserves. The use of fertilizers to supplement the reserves on the basis of soil critical values for each nutrient has been discussed elsewhere (Johnston, 1999). The major problem with integrating organic manures into a nutrient policy for any farm is to make an appropriate allowance for the amount and availability of the N they contain. Much further research on this topic would be justified.

3. References

Cooke, G.W. (1967) *The Control of Soil Fertility*. Crosby Lockwood & Son Ltd, London, 526 pp.

Crowther, E.M. (1949) The effects of plants and animals on soil fertility: A review of Rothamsted work. Specialist Conference in Agriculture, Australia 1949, Session D, Soil Fertility, pp. 1-8.

Jenkinson, D.S., Bradbury, N.J. and Coleman, K. (1994) How the Rothamsted Classical Experiments have been used to develop and test models for the turnover of carbon and nitrogen in soil. In: R.A. Leigh and A.E. Johnston (eds) *Long-term Experiments in Agricultural and Ecological Science*, CAB International, Wallingford, UK, pp. 117-138.

Johnston, A.E. (1986) Soil organic matter, effects on soils and crops. *Soil Use and Management* **2**, 97–105.

Johnston, A.E. (1987) Effects of soil organic matter on yields of crops in long–term experiments at Rothamsted and Woburn. INTECOL Bulletin **15**, 9–16.

Johnston, A.E. (1991a) Liebig and the Rothamsted experiments. In: G.K. Judel & M. Winnewisser (eds) *Symposium "150 Jahre Agrikulturchemie"*. Justus Liebig-Gesellschaft zu Giessen, Giessen, pp.37-64.

Johnston, A.E. (1991b) Soil fertility and soil organic matter. In: W.S. Wilson (ed) *Advances in soil organic matter research: the impact on agriculture and the environment*. Royal Society of Chemistry, Cambridge, pp. 299-313.

Johnston, A.E. (1994) The Rothamsted Classical Experiments. In: R.A. Leigh & A.E. Johnston (eds) *Long-term experiments in agricultural and ecological sciences*. CAB International, Wallingford, pp. 9-37.

Johnston, A.E. (1999) Efficient use of nutrients in agricultural production systems. *6th International Symposium on Soil and Plant Analysis*, Brisbane, Australia (to be published)

Johnston, A.E. and Poulton, P.R. (1992) The role of phosphorus in crop production and soil fertility: 150 years of field experiments at Rothamsted, United Kingdom. In: J.J. Schultz (ed) *Phosphate Fertilizers and the Environment*. International Fertilizer Development Centre, Muscle Shoals, USA, pp. 45-64.

Johnston, A.E., McEwen, J., Lane, P.W., Hewitt, M.V., Poulton, P.R. and Yeoman, D.P. (1994) Effects of one to six year old ryegrass-clover leys on soil nitrogen and on the subsequent yields and fertilizer nitrogen requirements of the arable sequence winter wheat, potatoes, winter wheat, winter beans (*Vicia faba*) grown on a sandy loam soil. *Journal of Agricultural Science, Cambridge* **122**, 73-89.

Table 1. Yields, t ha[-1], of winter wheat, spring barley, mangolds and beans at Rothamsted

Field name	Crop	Years	Treatment			
			None	PK	N	NPK
Broadbalk	Winter wheat grain	1852-61	1.12	1.29	1.63	2.52
Hoosfield	Spring barley grain	1872-81	0.85	1.10	1.66	2.62
Barnfield	Mangolds roots	1876-84	9.6	11.3	25.6	46.0
Geescroft	Field beans grain	1847-62	1.11	1.50	-	1.64

Table 2. Yields, t ha[-1], of winter wheat and spring barley, grain at 85% dry matter, and roots of mangolds and sugar beet at Rothamsted

Field name	Crop	Years	Yield with	
			FYM[a]	NPK fertilizers[b]
Broadbalk	Winter wheat	1852-56	2.41	2.52
		1902-11	2.62	2.76
		1955-64	2.97	2.85
		1970-75	5.80	5.47
Hoosfield	Spring barley	1856-61	2.85	2.91
		1902-11	2.96	2.52
		1952-61	3.51	2.50
		1964-67(PA)[c]	4.60	3.36
		1964-67(MB)[c]	5.00	5.00
Barnfield	Mangolds[d]	1876-94	42.2	46.0
		1941-59	22.3	36.2
	Sugar beet	1946-59	15.6	20.1

[a] FYM 35 t ha[-1] containing on average 225 kg N   [b] N fertilizer per ha; barley, 48 kg; wheat, 144 kg; root crops, 96 kg
[c] PA Plumage Archer given 48 kg N ha[-1], MB Maris Badger given 96 kg N ha[-1]   [d] Mangolds *Beta vulgaris*

Table 3. Yields of potatoes and spring barley on a sandy loam soil at Woburn

% C in soil	Fertilizer N applied			
	N0	N1	N2	N3
	Potatoes, tubers, t ha[-1], 1973 and 75			
0.76	25.7	35.6	41.7	43.2
2.03	27.1	40.6	50.7	59.0
	Spring barley, grain, t ha[-1], 1979			
0.76	2.19	5.00	6.73	7.05
1.95	2.58	5.12	6.85	7.81

N0 N1 N2 N3: 0, 100, 200, 300 kg N ha[-1] for potatoes; 0, 50, 100, 150 kg N ha[-1] for barley

Table 4. Total, Olsen P and $CaCl_2$-P in surface soils (0-23 cm) from two long-term experiments at Rothamsted

Experiment and year started	Year soil sampled	Treatment	Total P (mg kg[-1])		Olsen P (mg kg[-1])		$CaCl_2$-P (µg l[-1])	
Barnfield, 1843	1958	Control	670		18		0.5	
		Superphosphate (P)	1215	(545)[b]	69	(51)	3.0	(2.5)
		FYM[a]	1265	(595)	86	(68)	12.8	(12.3)
		FYM plus P	1875	(1205)	145	(127)	22.3	(21.8)
Broadbalk, 1843	1966	Control	580[c]		8		0.2	
		Superphosphate	1080	(500)	81	(73)	6.6	(6.4)
		FYM	1215	(635)	97	(89)	19.5	(19.3)

[a] FYM = farmyard manure   [b] Figures in parenthesis are increases over the control   [c] 1944 samples for total P

Chapter 3

Use of potash fertilizers through different application methods for high yield and quality crops

M. Marchand[1] and B. Bourrié[2]

[1]IPI WANA office - SCPA 2 place du Général de Gaulle, BP 1170 68053 Mulhouse Cedex, France
[2]SCPA Research Centre in Aspach le Bas-France

Key words: potassium, vegetables, quality

Abstract: Potassium and nitrogen are two major nutrients in crop production, since a deficiency in one or both nutrients causes yield losses. In vegetable cropping, both nutrients play also a key role on quality parameters. Application methods can enhance the uptake of some elements or can correct deficiencies in particular situations. As for other field crops, a balanced N x K fertilization enhances growth and improves the uptake of both nutrients, which in turn reduces nitrate losses during and after the cropping season. The quality of the yield is also depending on the N/K ratio and the fertilizer grades.

Field experiments were carried out in 1996 and 1997 on potato in order to study the influence of various potash grades (potassium sulfate, potassium chloride, potassium nitrate) applied through drip irrigation. In other experiments, the effect of potassium applied as foliar spray was studied on vegetables and tobacco. The following measurements were recorded: yield parameters, K contents, and main quality parameters according to the crop.

The response of the three grades of potash have shown a positive effect of potassium on quality and yield on potato. Potassium sulfate appears as the most suitable grade of potash as far as keeping quality and tuber size are concerned. Foliar applications of potassium can also improve yield and quality, especially in heavy clay soils where potassium is not readily available for the plants, or in sandy soils.

1. Introduction

How to use fertilizers is an important aspect in plant nutrition. Agronomic recommendations usually take into account the general requirements of the crop, a yield target and local adaptation to climatic conditions. Then, farmers have to meet economical and practical conditions to apply fertilizers.

With the development of techniques like drip irrigation in water management and spraying for plant protection, different ways to apply fertilizers have been used to complete or to replace soil application. The main positive effect claimed is a better efficiency of major and micro elements through a better placement according to the uptake of the plant (Mengel and Kirkby, 1987).

Fertigation is the most important way used in plant nutrition after soil application. The main reason is the development of irrigation all over the world. In the beginning of this century, 41 million of hectares were under irrigation. In 1950, irrigation concerned 105 million of hectares in the world and nowadays, 222 million of hectares are irrigated, i.e. 15% of the world cropping area (Anonymous, 1998). With the lack of water available for agriculture and the competition with towns and industrial uses, the development of irrigation in the future will concern more and more the drip irrigation techniques, which allow also fertigation.

13

14

Foliar application is one way to apply nutrients during the cropping season. It allows to feed the plants with small quantities but could not be the only way to meet all the requirements of a crop. However, it is a convenient mode to correct deficiency or to enhance crop quality.

In order to evaluate the possibilities of using potash fertilizers with fertigation and foliar spray, series of field experiments have been carried out on vegetable and tobacco during the last three years in SCPA Research Center, in eastern France.

2. Experiments with fertigation

2.1. Potato, field experiment 96:

2.1.1. Materials and Methods:

Field experiment was carried out on potato cropped under drip irrigation in SCPA Research Center located in Alsace, France, in 1996. The target of this experiment was to study the effect of different grades of potassium applied through drip irrigation on potato yield and quality.

Soil is a loamy soil with a satisfactory K_2O content. Each plot received 100 kg/ha K_2O, 70 kg/ha N and 100 kg/ha P_2O_5 at the end of March, four days before planting. 11 applications through drip irrigation system with the same amount of nutrients allowed to supply 396 kg/ha K_2O, and 120 kg/ha N as SOP + urea, NOP or MOP + urea. Each treatment had 6 replications. Potato experiment took place for 129 days (Bintje variety). Total rainfall during this period was 336 mm and irrigation brought an additional 136 mm of water.

2.1.2. Results of the 96 field experiment :

The following figures show the main results of this experiment. The three grades of potash give similar results in terms of total yield with a small advantage to SOP. But in terms of losses, we have found a significant difference between NOP on one side and SOP or MOP on the other

side. Consequently, SOP treatment gives the best marketable yield, followed by MOP treatment.

Dry matter content required by processing industry for Bintje is between 19 and 22 %. No difference was found between the three grades. Dry matter was between 19.25 and 20 % in the experiment.

K content in tubers is high, between 2.35 and 2.45 % without significant difference. Fertigation enhances K availability for the plant.

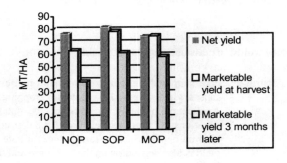

Figure 1.Effect of different grades of K on potato yield and quality (Aspach Research Center-France 96)

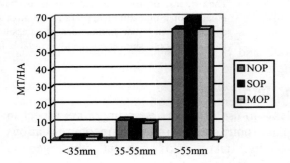

Figure 2. Effect of different grades of K on potato size (Aspach Research Centre-France 96)

Table 1: Soil analysis for the 1996 potato experiment (P_2O_5 Joret Hebert)

Clay (%)	Silt (%)	Sand (%)	pH	O.M (%)	CEC meq/1000g	P_2O_5 mg kg⁻¹	K_2O mg kg⁻¹	MgO mg kg⁻¹	CaO mg kg⁻¹
16.6	65.1	17.1	6	2.02	96	177	183	181	1920

Table 2: Soil analysis for the 1997 potato experiment (P_2O_5 Joret Hebert)

Clay (%)	Silt (%)	Sand (%)	pH	O.M. (%)	CEC meq/1000g	P_2O_5 mg kg⁻¹	K_2O mg kg⁻¹	MgO mg kg⁻¹	CaO mg kg⁻¹
24.4	66.8	4.5	5.9	2.53	141	235	205	251	2864

Fig.1 illustrates the high yield obtained with drip irrigation and fertigation, in line with previous results (El Khadi, 1997). Fertigation allows to enhance K effect using a better water management.

The keeping quality of the potatoes is also illustrated in the above figure. After 3-months storage, marketable yields are significantly different : SOP and MOP keep a real advantage when losses in NOP treatment reach 50 % due to sanitary problems with late N application.

Fertigation with SOP allows to produce a larger proportion of big tubers as shown in Fig. 2.

2.2. Potato, field experiment 97

2.2.1. Materials and Methods

A second field experiment was carried out in 1997, in the same location. The target of this experiment was to study the effect of different NOP and SOP combinations applied as basal dressing and through fertigation.

Soil is still a loamy soil with a satisfactory K_2O content. The soil analysis show small differences.

In 1997, each plot received 250 kg/ha K_2O, 70 kg/ha N and 100 kg/ha P_2O_5 at the end of March, four days before planting.

Six applications through drip irrigation system with the same amount of nutrients allowed to supply 280 kg/ha K_2O, and 80 kg/ha N as SOP + urea or NOP. Each treatment had 6 replications. The following treatments was studied :

T1 : NOP as source of K for basal dressing and fertigation,

T2 : SOP for basal dressing and NOP with fertigation,

T3 : SOP as source of K for basal dressing and fertigation.

Potato experiment took place for 131 days (Bintje variety). Total rainfall during this period was 270 mm and irrigation brought an additional 210 mm of water.

2.2.2. Results of the 97 field experiment :

Figure 3 illustrates the results in terms of production and size. The same difference as in 1996 is observed between NOP and SOP with an advantage to SOP (Marchand and Bourrie,1997).

As already mentioned, dry matter content required by processing industry for Bintje is between 19 and 22 %. No significant difference was found between the three treatments. Dry matter was between 20.12 and 21.38 % in the experiment.

K content in tubers is high, between 2.05 for SOP and 1.83 % for T2, without significant difference.

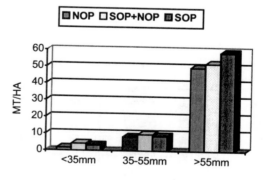

Figure 3. Effect of combinations of different grades of K on potato production and size (Aspach Research Centre – France 97)

3. Experiments with foliar spray

Foliar spray of soluble potassium sulfate have been evaluated in various cropping conditions on vegetables and tobacco. The main results are presented hereafter.

3.1. Spinach

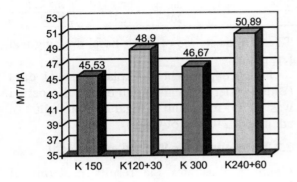

Figure 4. Effect of foliar application of soluble SOP on spinach (GAAS-China 98)

This experiment compares two doses of SOP applied as basal dressing and basal plus foliar application for the same amount of K_2O : 150 and 300 kg/ha. The results are widely in favor of split application with foliar spray.

The main quality parameters are also concerned : foliar spray increases the percentage of dry matter, decreases the nitrate content in the leaves and increases the K content.

3.2. Pea

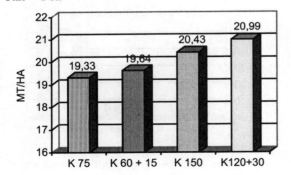

Figure 5. Effect of foliar application of soluble SOP on pea (GAAS-China 98)

This experiment also compares two doses of SOP applied as basal dressing and basal plus

foliar application for the same amount of K_2O : 75 and 150 kg/ha. The same effect of foliar application on yield is observed.

Regarding the quality parameters, foliar application increases the percentage of soluble sugar and the K content.

3.3. Tomato

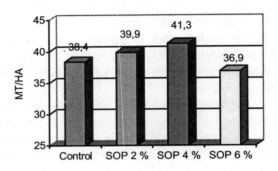

Figure 6. Effect of foliar application of solubke SOP on tomato (South China Ag. University-China 98)

In this experiment the same basal dressing is applied on each plot. Three foliar sprays are applied every 10 days, starting 20 days after transplanting. A depressive effect is observed with the highest concentration due to leaf damages.

Regarding the quality parameters, foliar application increases the K content and the proteins in the fruits. No significant change is observed for the other quality parameters.

3.4. Tobacco

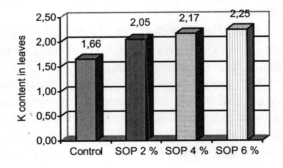

Figure 7. Effect of foliar application of soluble SOP on tobacco (Hong He Research Station – China 97)

In this experiment the same basal dressing is applied on each plot. Three foliar sprays are applied every 10 days, starting 40 days after transplanting. A good response is observed as a K content of 2 % is considered as the minimum for a fair combustibility of tobacco leaves (Jones and Rasnake,1985). In the same time, nitrate content is decreased that is considered as quality criteria for tobacco maturation.

4. Conclusion

The results of these experiments confirm the interest of application of nutrients through drip irrigation and foliar spray. From all the experiments carried out in various conditions we can draw the following conclusions of K use in vegetable production:

1) K increases yield, quality and profit.

2) K uptake is enhanced when using fertigation as a consequence of an application adapted to the instant plant requirements.

3) Foliar applications of potassium increase the K content in the leaves and allow a better quality of the crop, especially in high fixing heavy soils.

5. References

Anonymous. Encarta. Irrigation statistics, Microsoft 1998.

El Khadi, M Fertigation experiments, DRC 1997.

Jones, JL. & Monroe Rasnake. Effects of KCl vs K2SO4 on the yield and quality of Virginia sun-cured tobacco, Tob. Sci. 29:12-13, 1985.

Marchand, M & B. Bourrié. Use of potash fertilizers for high yield and quality production of potatoes. IPI workshop, Bornova - Turkey, 1997.

Mengel, K & E.A. Kirkby. Principle of plant nutrition, IPI 1987.

Chapter 4

Citrus and tomato quality is improved by optimized K nutrition

O. Achilea

Haifa Chemicals Ltd. P.O.Box 10809, Haifa Bay, Israel, 26120. (oded@haifachem.co.il)

Key words: Citrus, tomato, potassium, potassium nitrate, foliar feeding, fertigation, fruit size, sugar content, acid content, dry matter, shelf life, tangerine, tangelo, creasing, splitting,

Abstract: Internal and external qualities of produce are becoming increasingly important market factors. One of the major elements in determining crop quality is potassium availability to the plant. The most important function of potassium in plant metabolism is enzyme activation. Accumulation of carbohydrates, organic acids and other crop-specific ingredients is highly dependent on optimal photosynthesis, the intensity of which is related to K status in the plant.

Fertigation and foliar feeding with potassium nitrate have proven to be highly efficient means of fulfilling the potassium requirements for many crops. The combination of potassium and nitrate in this fertilizer has been found to be highly beneficial in improving fruit size, dry matter, colour, taste and integrity and resistance to biotic and abiotic stresses, for citrus and tomato fruit. Moreover, the integration of potassium nitrate in routine management or in specific growth stages results in remarkably positive Benefit/Cost ratio.

1. Introduction

The gradual improvement in the standard of life in both developing and developed countries has created a clear demand for quality goods. In this framework, quality of agricultural produce is becoming an increasingly important factor in determining crop market values. People are willing to pay higher prices for foods, that are richer in nutritional ingredients, larger and more colourful, better tasting, aesthetically appealing, marketed throughout the year, etc.

Potassium is ubiquitous in the soil/plant system and is unique among the basic plant nutrients in its multi-functional contribution to plant metabolic processes. The interaction of K with plant systems and constituents is varied and extensive (Bailey, Grant and Choudary, 1993). While K does not become a part of the chemical structure of plants, it plays many important regulatory roles, some of which are specified as follows:

- Potassium is required to activate at least 60 different enzymes involved in plant growth and metabolism. (Anon., 1987)
- K regulates water balance of the plant through root osmotic gradient, and the functioning of stomata guard cells.
- The role of K in photosynthesis involves the activation of specific enzymes, and the control of ATP formation.
- Adequate supply of K is a prerequisite for regular transport of sugars from the leaves through the phloem to sugar sinks, to be consumed or stored.
- Translocation of nutrients through the xylem is dependent upon K availability.
- K is required for every major step of protein synthesis.

Multi-K (potassium nitrate, KNO_3) is a binary, high-quality speciality fertilizer. It is fully soluble and chloride-free, consisting of 13% nitrate-nitrogen and 46% potassium as K_2O (or 38% K). Potassium nitrate has become increasingly important in the recent decade, and world demand is now estimated at 1.1 million MT/Y, representing 3% of total K_2O consumption. The physical and chemical properties of Multi-K make it a highly versatile fertilizer for advanced agriculture and horticulture. The comprehensive line of Multi-K products includes various forms and formulae, matching specific needs of many crops and growing conditions and providing K-optimised plant nutrition.

Citrus is a subtropical-tropical crop grown on app. 6.9 million ha worldwide, and on 77,600 ha in Turkey, (FAO), making it an important crop. Extensive experimentation activity carried out by Haifa Chemicals LTD over several years has elucidated many crop quality improvements achieved by application of Multi-K.

Tomatoes are probably one of the most extensively grown vegetable crop, grown on app. 3.2 million ha worldwide, and on 158,000 ha in Turkey, (FAO).

Tomato fruit quality is positively correlated with potassic fertilization of the plant. It has been shown (Adams and Grimmett, 1986) that the higher the K content in the nutrient-solution, the higher is the dry matter, titratable acidity and potassium content in the fruit, while hollow fruit disorder is inversely correlated with K content in the nutrient-solution.

2. Results and Discussion

Citrus

"Mineola" tangelo (a tangerine x pomelo cross) is a relatively new variety with appealing taste, and attractive shape and colour. The taste can be markedly enhanced by foliar application of Multi-K. Foliar feeding with Multi-K significantly increased acid (and thereby-vitamin C) and sugar contents of the fruit. Moreover, fruit size was significantly increased

as well. These results are shown in Table 1 (Fuente and Ramirez, 1993).

Navel and Valencia oranges and Nova and Murcot tangerines are subject to rind irregularity problems, making them less attractive to the consumer, and severely shortening their shelf life. The main symptoms are creasing and splitting. Multi-K foliar sprays, as seen in table 2, can easily correct this irregularity (Lavon et al., 1992 and Bar-Akiva, 1975).

Small size fruit is a serious problem, drastically reducing grower's return for oranges, grapefruits, and tangerines (3.6, 0.24, and 1.56 million ha, respectively, world-wide; and 37,200, 2,100 and 22,000 ha respectively, in Turkey, FAO). Citrus fruit size can be significantly increased by spraying the trees with Multi-K in the spring-summer season, at concentrations of 3-6%. The beneficial results of Multi-K treatments on tangelos, oranges, grapefruits and tangerines can be found in tables 1 & 3. Sakovich, 1995, reported similar results for lemons (data are not shown).

Citrus groves are often infected by various pests that markedly reduce the aesthetic look of the fruit, and consequently diminish the grower's income. Florida Wax Scale *(Ceroplastes floridensis, comstock)* is a pest, that covers the fruit and the leaves of the infected tree with extensive sooty mold. This irremovable cover seriously reduces marketability of the fruit. Spraying with Multi-K at 4% in the spring, when first signs of the pest infestation are seen, prevents diffusion of the pest throughout the grove. The mechanism is most probably based on creation of unfavourable conditions for the crawlers of the pest. Repeated treatment can replace the application of Organo-Phosphates, thus enabling the establishment of integrated pest management (IPM) in the grove, and avoiding deleterious environmental implications (Yardeni and Shapira, 1994), see Table 4. Similar suppression effects were obtained while spraying Multi-K against the citrus leaf miner *(Phylocnistis citrella)* and other pests (data are not shown).

Tomatoes

An experiment was carried out in order to optimize fertigation management of processing tomatoes. Potassium chloride, potassium sulphate and Multi-K (potassium nitrate) were compared regarding the effect on major parameters of crop quality. The parameters were dry matter yield, waste rate, and health, color uniformity and mean weight of the berries. The results depicted in Table 5 show that: Multi-K was consistently superior to the other two potassic sources, regarding two of the parameters checked. The effect on waste rate was not consistent Considerably higher beneficial effect was found when application rate was increased by 50%.

3. References

Adams, P. and Grimmett, M.M., 1986. Some responses of tomatoes to the concentration of potassium in recirculating nutrient solutions. Acta Horticulturae 178: 29-36.

Anon.1987. Functions of potassium in plants. In: Potassium for Agriculture, Armstrong D.L. (Ed.).

Bailey LD, Grant CA and Choudary M (1993) The Environmental Enhancing Nutrient. Fertilizer International 327: 25-30.

Bar-Akiva A (1975) Effect of foliar applications of nutrients on creasing of "Valencia" oranges. HortScience 10 (1): 69-70.

Boman BJ (1995) Effects of fertigation and potash source on grapefruit size and yield. In: Dahlia Greidinger International symposium on fertigation, Technion, Haifa, Israel. 55-66.

FAO, FAOSTAT Database Gateway on the Internet.

Fuente Orozco H and Ramirez A (1993) Nitrato de potassio (KNO$_3$) foliar para mejorar la calidad en citricos. Faculdad de agronomia, Universidad de Caldas, Colombia.

Kanonitz S, Lindenboum H and Ziv J (1995) Increasing Shamouti fruit size with 2.4.D and NAA. Alon Hanotea 49: 410-413. (Hebrew).

Lavon R, Shapchiski S, Muhel E and Zur N (1992) Nutritional and hormonal sprays decreased fruit splitting and fruit creasing of "Nova". Hassade 72: 1252-1257. (Hebrew).

Rabber D, Soffer Y and Livne M (1997) The effect of spraying with potassium nitrate on Nova fruit size. Alon Hanotea. 51: 382-386. (Hebrew).

Sakovich N (1995) Don't overlook potassium deficiencies. California Grower, May 1995: 23-28.

Yardeni A and Shapira E (1994) Thinning the population of the Florida wax scale *Ceroplastes floridensis comstock*, as an option for IPM procedure in citrus groves. Alon Hanotea 48: 194-200. (Hebrew).

Note for all tables: Figures in the same column followed by different letters are significantly different (P<0.05).

Table 1. Effect of Multi-K on *Mineola*'s acid and sugar contents, and fruit size.

Treatment	Acid (%)	Sugars (Brix°)	Mean Fruit diameter (cm)
Control	0.55 a	8 a	8.2 a
4 x foliar spray: Multi-K 6%	0.8 b	11.2 b	9.0 b

Table 2 .Effect of Multi-K on citrus fruit rind disorders

	Nova tangerine	*Valencia* orange
Treatment	Fruit splitting (%)	Fruit creasing (%)
Control	62.2 a	42.6 a
2 x foliar sprays Multi-K 5% *	19.6 b	27.6 b

  *Multi-K was applied in June and on the first half of August (the latter with 20ppm 2.4.D).

Table 3. Effect of Multi-K on citrus total yield and fruit size

Treatment	*Shamouti* orange* Yield parameters		*Ruby* grapefruit** Yield parameters		*Nova* tangerine*** Yield parameters	
	Total Kg/tree	Medium – jumbo size share (%)	Total Box/tree	Mean wt. of a fruit unit (g)	Total Kg/tree	% fruit >62mm in diameter
26.6 Control	102.8 a	36 a	7.9 a	356 a	55.9 a	26.6 a
Multi-K	125.3 b	61 b	8.8 ab	374 ab	71.3 b	51.7 b

*One foliar spray, Multi-K 4% + 2.4.D 20ppm. (Kanonitz et al., 1995)
** Sixteen fertigation applications with total of 1060 Kg/ha Multi-K (Boman, 1995)
***One foliar spray. Multi-K suspension 15%. (Rabber et al., 1997)

Table 4. Combined use of Multi-K & summer oil in pest management in a citrus grove.

Years	Pest management	Number and character of sprays required for clean fruit	
		Organo-Phosphates (No. of sprays/year)	Multi-K + summer oil (No. of sprays/year)
1983-1986	Non selective	2 – 4	0
1987	IPM	0	0
1988- 1993	IPM	0	1 - 2

1987: pest management accomplished exclusively by Multi-K 4%+summer oil 1-2% sprays.

Table 5. The effects of several potassic fertilizers on tomato fruit quality components.

K treatment	Mean berry weight (g)	Waste (% of total yield)	Dry matter yield (Ton/ha)
Control: 0 K	124.2	15.3	2.3
KCl : 92 Kg/ha K_2O	124.4	12.5	2.6
K_2SO_4 : 92 Kg/ha K_2O	126.8	9.6	2.7
Multi-K : 92 Kg/ha K_2O	129.0	12.0	2.9
Multi-K : 138 Kg/ha K_2O	120	10.1	4.5

Chapter 5

Evaluation of different fertilization strategies on orange

F. Intrigliolo and G. Roccuzzo[1]

Istituto Sperimentale per l'Agrumicoltura - Corso Savoia 190 - 95024 Acireale, Italy.

[1] *Present address: Istituto Sperimentale per la Nutrizione delle Piante, Via della Navicella 2-4 - 00184 Roma, Italy*

Key words: plant nutrition, fertigation, fruit quality

Abstract: The effects of timing and modes of nutrient application on yield, vegetative and nutritional status of 30-year old 'Valencia late' sweet orange trees [*Citrus sinensis* (L.) Osbeck] grafted on sour orange (*C. aurantium* L.) were evaluated. The treatments compared were: A - single nutrient fertilizers (N, P, K) topsoil dressed in March; B - single nutrient fertilizers topsoil dressed in July; C - fertigation from the 2nd decade of March to the 1st decade of September. A simple randomized scheme was adopted. Trees in all treatments were spray-irrigated and received in each of the trial years the following nutrient supplies: 450, 400 and 600 g per plant of N, P_2O_5 and K_2O, respectively. No significant differences between the various treatments were observed regarding yield, fruit regreening, fruit drop, some fruit quality parameters and trunk circumference. Earlier ripening was observed in treatment C, due to significant differences in acidity and TSS/Acidity ratio. Trees in treatment A showed a lower fruit weight, probably due to the significantly lower K level in leaves. Higher leaf N levels in treatment A were noted, whereas trees in treatment C were found to have lower levels of Fe. No differences were observed in the leaf contents of the other nutrients.

1. Introduction

The rationalization of cultural practices, and mainly of fertilization, is one of the main tools used to improve yield and fruit quality and to increase net profit. Moreover, the best use of fertilizers may reduce environmental pollution.

Among fertilization techniques, fertigation enables growers to reduce fertilizer distribution costs. According to Bester et al. (1977) and Koo (1980) it can also increase nutrient use efficiency by reducing losses due to leaching or immobilization, since this technique allows nutrient supply to be split. Many studies have been carried out to compare soil dressing to fertigation (Koo, 1980; Koo and Smajstrla,1984; Legaz et al., 1981), to split distribution or to slow release fertilizer application (Legaz et al., 1981; Willis et al., 1991). The results of these papers indicate that there are no differences as far as yield and fruit quality are concerned among different timings and modes of nutrient application, even though for fertigation proper covering of the portion of soil explored by roots seems to be essential (Bielorai et al., 1984; Koo, 1984; Scuderi and Raciti, 1980).

The period of nutrient distribution can influence the availability of nutrients to plants and interfere with yield and fruit quality. In particular, when fertilization is carried out before the harvest negative interferences such as fruit drop increase and worsening of fruit quality might occur (Scuderi et al., 1973).

The aim of this study was to evaluate the effects of timing and modes of nutrient application on yield, vegetative and nutritional status of 'Valencia late' sweet orange trees [*Citrus sinensis* (L.) Osbeck] in the warm-arid conditions of Sicily.

2. Materials and Methods

The study was carried out in the eastern part of Sicily between 1990 and 1995 on 30-year old

Table 1. Main chemical and physical characteristics of the soil

Sand %	Silt %	Clay %	pH (water 1:2.5)	Total lime %	EC mS cm^{-1}	Organic matter %	Total N %	available P$_2$O$_5$ mg kg^{-1}	exchang. K$_2$O mg kg^{-1}
68.24	15.91	15.85	7.53	0.26	0.84	1.69	1.25	44	275

'Valencia late' sweet orange trees grafted on sour orange (*C. aurantium* L.), with a planting distance of 4 m on the row and 6 m between the rows. The medium sandy soil had a fair content of organic matter (according to the Italian situation), total N and available P$_2$O$_5$, and a very high content of exchangeable K$_2$O (Table 1). Soil analysis were performed according to the SISS methods (1985).

Three treatments were compared: single nutrient fertilizers (N, P and K) topsoil dressed in March; single nutrient fertilizers topsoil dressed in July; fertigation from the 2nd decade of March to the 1st decade of September.

Trees in all treatments received annually the following nutrient supplies: 450, 400 and 600 g per plant of N, P$_2$O$_5$ and K$_2$O, respectively, and were spray-irrigated from April to September every 21 days. Irrigation volumes were calculated utilizing evaporation data detected by means of a Class A pan evaporimeter.

A simple randomized scheme was adopted. Data were collected from 12 plants per treatment.

The data collected yearly were related to: yield, number of fruits, average fruit weight and the amount of fruit regreening as estimated at harvest (between May and June during the trial); fruit quality parameters, sampling 20 fruits per plant in the first decade of May; plant growth, measuring the trunk circumference 20 cm under the scion grafting in the month of November; plant nutrition, sampling in the month of October 20 terminal, spring-cycle leaves per plant from nonfruiting and nonflushing shoots (Embleton et al., 1973). Leaf analysis was carried out on the homogenized sample oven dried at 65°C to constant weight. Total N was determined using the micro Kjeldahl method. K, Ca, Mg, Fe, Zn and Mn were determined by atomic absorption spectrophotometry (Perkin Elmer Mod. 2380)

after ashing at 550 °C and extraction in 10% hydrochloric acid. P was determined by vanadate-molybdate yellow colorimetric method.

3. Results and Discussion

Data for yield, fruit quality parameters and increase of scion trunk circumference (ISTC) are provided in Table 2.

Table 2. Effect of treatments on yield and fruit quality (mean 1990-1995)

Parameter	A	B	C
Yield (kg/tree)	126.0	130.1	127.7
Fruit drop (kg/tree)	3.5	3.1	2.8
Fruit weight (g)	199 B b^z	216 A a	207 AB a
Fruit regreening (kg/tree)	5.68	4.98	6.09
Juice (%)	49.0	49.2	50.1
Rind thickness (mm)	5.00	5.24	5.29
Central axis (mm)	11.0 B c	11.6 B b	12.6 A a
TSS (%)	10.2	10.0	10.1
Acidity (%)	1.43 a	1.40 a	1.35 b
TSS/Acidity ratio	7.31 b	7.34 b	7.48 a
ISTC (cm)	3.40	3.95	3.33

^{z} Mean separation in each row by Tukey's multiple range test. Small letters, capital letters significant at P $\leq$ 0.05, 0.01, respectively. Absence of letters indicates that the data are not significantly different.

No significant differences between the treatments were observed either in the mean or in the single years (data not shown) as far as yield and fruit regreening were concerned, even though some seasonal fluctuation was noted. Fruit drop was minimal and without significant differences between treatments. No significant differences between treatments were found regarding ISTC.

The plants in treatment A showed a lower value of fruit weight (covariance total fruit number) compared to those in treatment B (p $\leq$ 0.01) and C (p $\leq$ 0.05). These results were confirmed in nearly all the trial years (data not shown).

Similar values of juice percentage and rind thickness were recorded for the treatments in the final mean, even though some differences were recorded in only a few years (data not shown). Central axis diameter in treatment C trees was higher (p ≤ 0.01), and trees given treatment B showed a higher value when compared to those given treatment A (p ≤ 0.05). As far as juice quality parameters are concerned, no difference was recorded for total soluble solids (TSS), whereas fruits from treatment C trees were found to have the lowest value of acidity and the highest TSS/acidity ratio. No difference regarding juice quality parameters was found between treatments A and B.

The analysis of data reveals that plants topsoil dressed in July (B) and fertigated (C) gave better productive results. In both cases, a similar yield level was combined with the larger fruit size. Moreover, the best ripening ratio was recorded for those trees given treatment C due to the reduction of acidity, based on Koo's (1980) fertigation studies.

According to leaf analysis guide by Embleton et al. (1973) nutrient levels were optimal, with the exception of N and K (table 3).

Highly significant differences among treatments were found for K levels, which were the lowest in treatment A plants. The application of the element in a dry period without irrigation probably prevented its migration in the soil, thus reducing its availability in the soil layers explored by roots. Carrier effect of water and related enhancement of the uptake was evident both in treatment B plants, in which the element was distributed during the irrigation season, and in fertigated plants. Low K levels in treatment A plants did not seem to be due to either the interference of N, whose levels were not higher as to affect its metabolism (Intrigliolo et al., 1990; Intrigliolo et al., 1993; Intrigliolo and Intelisano, 1997),

nor to the high yield and the related depletion of plant supplies (Rapisarda et al., 1995).

Table 3. Effect of treatments on mineral composition of leaves (dry weight; mean 1990-1995)

Parameter	A	B	C
N (%)	2.38 a [z]	2.32 b	2.33 b
P (%)	0.121	0.126	0.124
K (%)	0.49 B	0.69 A	0.68 A
Ca (%)	5.01	4.95	5.12
Mg (%)	0.46	0.49	0.48
Fe (mg kg^{-1})	141 A	135 A	122 B
Zn (mg kg^{-1})	25	26	27
Mn (mg kg^{-1})	27	28	29

[z] Mean separation in each row by Tukey's multiple range test. Small letters, capital letters significant at P ≤ 0.05, 0.01, respectively. Absence of letters indicates the data are not significantly different.

N content in all treatments was at the upper limit of the low class, and treatment A showed a significantly higher value. The slightly lower values of nitrogen in the leaves of treatments B and C plants might be due to the conditions opposite to those reported for potassium, that is the higher level of leaching by irrigation water, favored by the high irrigation volumes in a soil with high sand content.

As far as micronutrients are concerned the only difference noted was the statistically lower value of Fe in treatment C plants. In all treatments, however, the levels of the element were high and the interferences on plant metabolism probably insignificant.

4. Conclusions

The modes and timing of nutrient distribution adopted in the trial did not significantly affect many nutritional or yield parameters. However, the slight differences noted in yield and plant development are important. Moreover, it is evident that nitrogen distribution before the harvest did not affect fruit regreening or fruit drop, as reported for other orange cultivars (Scuderi et al., 1973).

It seems that no any element, except K, played a role in the differences in fruit quality recorded among treatments. The lower fruit weight values of the plants fertilized in March might be due to the poorer utilization of K.

These results indicate that, in conditions of high K availability in soil, not only the amounts to distribute but distribution modes capable of increasing plant assimilation should be considered.

It should also be noted that only when foliar levels of a nutrient are greatly different, as for K levels in our trial, can real changes in yield and fruit quality parameters be observed. These changes are generally due to the direct effect of the element and to the interference on other nutrients' metabolism, although there was no evidence of this last effect in our trial. On the contrary, the recorded differences regarding nitrogen, even though statistically significant, were not wide enough to interfere with plant physiology and therefore to influence yield or fruit quality.

Finally, as shown in other studies (Koo, 1980), it seems evident that fertigation improved parameters which describe fruit maturation, such as central axis diameter, acidity and ripening ratio.

5. References

Bester DH, Fouchè PS and Veldman GH 1977 Fertilizing through drip irrigation systems on orange trees. Proc.Int.Soc.Citric.1, 46-49.

Bielorai H, Dasberg S, Erner Y and Brum M 1984 The effect of fertigation and partial wetting of the root zone on production of Shamouti oranges. Proc.Int.Soc.Citric. 1, 118-121.

Embleton T W, Jones WW, Labanauskas CK and Reuther W 1973 Leaf analysis as a diagnostic tool and guide to fertilization. The Citrus Industry 3, 183-210.

Intrigliolo F, Tropea M, Sambuco G and Giuffrida A 1990 Nutrizione minerale dell'arancio - 4° contributo: influenza delle concimazioni sulla nutrizione e sulla produzione del "Tarocco" nucellare e vecchia linea. Annali Ist.Sper.Agrum. 23, 69-110.

Intrigliolo F, Fisichella G, Tropea M, Sambuco G and Giuffrida A 1993 Influence of nitrogen nutrition on nutritional status and yield of 'Navelina' orange. In: (M.A.C. Fragoso and M.L. van Beusichem eds.) Optimization of Plant Nutrition: 439-444.

Intrigliolo F and Intelisano S 1997 Effects of differential nitrogen application on nutrition, growth, yield and fruit quality in young lemon trees. Acta Hort. 448, 499-512.

Koo RCJ 1980 Results of Citrus fertigation studies. Proc.Fla.State Hort.Soc. 93, 33-36.

Koo RCJ and Smajstrla AG 1984 Effects of trickle irrigation and fertigation on fruit production and juice quality of 'Valencia' orange. Proc.Fla.State Hort.Soc. 97, 8-10.

Legaz F, Ibañez R, de Barreda DG and Primo Millo E 1981 Influence of irrigation and fertilization on productivity of the 'Navelate' sweet orange. Proc.Int.Soc.Citric.2, 591-595.

Rapisarda P, Intrigliolo F and Intelisano S 1995 Fruit mineral analysis of two "Tarocco" clones of sweet orange to estimate fruit mineral removals. Acta Hort. 383, 125-133.

Scuderi A, Raciti G and Licciardello G 1973 Indagini sugli effetti della concimazione azotata dell'arancio nel periodo inverno-primaverile. Annali Ist.Sper.Agrum. 6, 36-48.

Scuderi A and Raciti G 1980 Prove comparative quadriennali di concimazione dell'arancio con formulati liquidi, idrosolubili e tradizionali. Tecnica agricola 32, 69-87.

SISS 1985 Metodi normalizzati di analisi del suolo. Edagricole, Bologna, Italy.

Willis LH, Frederick SD and Graetz DA 1991 Fertigation and growth of young 'Hamlin' orange trees in Florida. Hortsci. 26, 106-109.

Chapter 6

Effect of KNO₃ applications on fruit yield and N, P, K content of leaves in *Vitis Vinifera* grapes

Ş. Ceylan[1] and İ.Z. Atalay[2]

[1]*Ege University Vocational High School Ödemiş-Izmir-Türkiye*
[2]*Ege University Faculty of Agriculture Department of Soil Sciences İzmir-Türkiye*

Key words: grape, KNO₃, foliar application

Abstract: The aim of this study was to determine the effect of different K fertilizers and different K doses on yield and N, P, K content of leaves in grapes. The treatments were the control;1 %, 2% and 3% KNO₃;.and 1% KH₂PO₄ + NH₄H₂PO₄ foliar applications which were realized weekly at full blossom and ripening. Yield of the grape was regularly recorded for two years. The findings of the investigation are as follows:1. The effect of KNO₃ applications on leaves were found to be significant on fruit yield. The highest yield was obtained at 3 % doses. In 1996-1997 years, KNO₃ applications increased fruit yield compared to control plot 35 % and 17 % as a respectively.2. KH₂ PO₄ and NH₄H₂PO₄ applications increased the yield according to control plot 34 % in 1996 and 5 % in 1997 year. 3. It was found that the effect of KNO₃ applications was significant on K content of petiol and P, K content of blade.

1. Introduction

Turkiye with all its suitable ecological conditions, is one of the largest grape production centers of the world. 30 % of this production is realized in Ege region where majority of seedless grapes are grown. İzmir in this region plays an important role by its 27 825 ha of vineyard and 248 820 ton of production (Anonymous, 1988).

At present for maximum production, quality and quantity, KNO₃ foliar applications have gained importance for its fast efficacy.

It is reported that KNO₃ affect the pigment density, blooming, fruit formation, sugar/acidity ratio both internally and externally while affecting the product directly when sprayed (Haifa Chemicals Ltd.).

Many studies were conducted on this subject (Lafon and Cauillaud, 1953; Rose, 1980; Kahraman, 1992). Çokuysal (1992) specially reports berry enlarging affect of KNO₃ equivalent to giberallic acid.

In İzmir, where this research was carried on 64 % of the vineyard plantations are stated to be potassium deficient (Kovancı et al., 1984).

This study has been performed to determine the effects of KNO₃ applications on productivity and on leaf N, P, K contents in Sultani seedless grapes.

2. Materials and Methods

The study was carried on in Ödemiş district of İzmir for two years. The physical and chemical characteristics of the soil is shown in Table 1.

The experiment was conducted in completely randomized block design with 4 replications and with 10 vines in each parcel. 60 kg da⁻¹ of 15:15:15 complete fertilizer were incorporated to each experimental plot as basal dressing and 20 kg da⁻¹ of NH₄NO₃ as a supplement.

The first group of treatments were control, 1%, 2 % and 3 % of KNO₃ foliar applications which were realized every week in the month of

June, in 1996 and 1997. The second group of treatments were again realized in the same years in June and were as 1 % KH_2PO_4 and $NH_4H_2PO_4$ successive foliar sprays every week. In the present study, leaves opposite the 1st fruit cluster of the shoots were sampled (Kovancı and Atalay, 1975).Blades and petioles were separately analyzed for their N by distillation method (Kacar, 1962), P by colorimeter (Lott et al., 1956) and K by flamephotometer (Kacar, 1962).

Yield was evaluated as kg da^{-1} and brix was measured in the fruits by a refractometer.
The results of the analysis has been evaluated statistically.

Table 1. Physical and chemical properties the experimental soil.

Soil Properties	Depth (cm.)	
	0-30	30-60
Sand (%)	77.3	77.3
Silt (%)	16.0	16.0
Clay (%)	6.7	6.7
Texture	Loam.Sand	Loam.Sand
PH	6.2	6.3
$CaCO_3$ (%)	1.3	1.1
Salt (%)	<0.03	<0.03
O.M (%)	1.74	1.40
Total N (%)	0.090	0.084
Ext P mgkg^{-1}	16.6	13.1
Ext P mgkg^{-1}	280	180

3. Results and Discussion

KNO_3 foliar applications in 1996 and 1997 have significantly increased (1 %) the yield to considerable levels (Table 2). In the study the highest yield was obtanied in 3 % KNO_3 application treatment. By this application, in 1996 35 % and in 1997 17 % increases were observed compared to that of controls. In 1996, the yield of 3 % KNO_3 application treatment and the yield of KH_2PO_4-$NH_4H_2PO_4$ applications were quite similar. When the latter application results are compared with the control, 34 % increase, is seen in the yield. These results are in agreement with data presented by Lafon and Cauillaud (1953) who have stated 28 % increase, by Rose (1980) 15 % increase and by Kahraman (1992) 18 % increase in the yield by KNO_3 applications.

Results revealed that the applications have affected the leaf blade P, K and petiol K contents (Table 3).
Brix, the blade N and the change in petiol N, P contents related to these applications are not found statistically significant.

Table 2. Effects of K on productivity and brix

Treatments	Yield kgda^{-1}	1996	
		Relative yield	Brix
	Aver.	%	%
Control	1562 C	--	21.3
1% KNO_3	1713 BC	10	19.8
2% KNO_3	1780 B	14	20.7
3% KNO_3	2109 A	35	19.7
1%KH_2PO_4-$NH_4H_2PO_4$	2093 A	34	19.8

Treatments	Yield kgd^{-1}	1997	
		Relative yield	Brix
	Aver.	%	%
Control	1692		22.0
1% KNO_3	1796	6	21.2
2% KNO_3	1821	8	21.5
3% KNO_3	1978	17	20.5
1%KH_2PO_4-$NH_4H_2PO_4$	1771	5	21.7

Figures having common letter (s) in a column and row do not differ significantly at 5 % level of significance.

Table 3. Effects of K on blade and petiol N, P, K contents

Treatments	1996 Year					
	Blade			Petiol		
	N %	P %	K %	N %	P %	K %
Control	2.60	0.37	0.78	1.27	0.69	2.24 A
1%KNO_3	2.52	0.35	0.76	1.34	0.64	2.08 AB
2%KNO_3	2.51	0.37	0.81	1.32	0.65	2.20 A
3%KNO_3	2.60	0.35	0.86	1.29	0.65	2.03 AB
1%KH_2PO_4-$NH_4H_2PO_4$	2.49	0.42	0.78	1.32	0.68	1.51 B

Table 3 continued

Treatment	1997 Year					
	Blade			Petiol		
	N%	P%	K%	N%	P%	K%
Control	2.29	0.40 B	0.64 D	0.99	0.71	1.57
1%KNO₃	2.48	0.38 B	0.80 C	1.02	0.71	1.57
2%KNO₃	2.45	0.44 AB	0.90 B	0.93	0.72	1.57
3%KNO₃	2.44	0.39 B	1.07 A	0.96	0.68	1.63
1%KH₂PO₄ -NH₄H₂PO₄	2.31	0.55 A	0.68 D	1.00	0.76	1.58

Figures having common letter (s) in a column and row do not differ significantly at 5 % level of significance.

4. Conclusion

In conclusion, results showed that KNO_3 foliar applications lead to considerable increases in yield of Sultani seedless grape. Above findings put forth that KNO_3 foliar applications as a supplement is a necessary agricultural practice for the seedless Sultani grapes of İzmir region.

5. References

Anonymous., 1998. Tarımsal Yapı ve Üretim Dev. İst. Ens. Yayınları. Ankara.

Atalay, İ.Z., 1977. İzmir ve Manisa Bölgesi Çekirdeksiz Üzüm Bağlarında Bitki Besini Olarak N, P, K, Ca ve Mg'un Toprak Bitki İlişkilerine Dair Bir Araştırma. E.Ü. Ziraat Fak. Yayın No: 345 (Doktora).

Çokuysal, B., 1990. Sultani Çekirdeksiz Üzümde Güberelik Asit ve Yaprak Gübrelemesinin Yaprak Besin Elementleri ve Ürün Kalitesi Üzerine Etkileri . (Doktora)

Haifa Chemicals Ltd. Foliar Nutrition with Potassium Nitrate.

Kacar, B., 1962. Plant and Soil Analysis Univ. Of Nebraska, Col of Agr. Lincoln, Nebraska, U.S.A.

Kacar, B., 1972. Bitki ve Toprağın Kimyasal Analizleri. II. Bitki Analizleri. A.Ü. Ziraat Fak. Yayın.453. Uygulama Kılavuzu No: 155. A.Ü. Basımevi, Ankara.

Kahraman, A., 1992. Sultani Çekirdeksiz Üzüm Bağlarında Yapraktan KNO3 Uygulamalarının Verim ve Bazı Kalite Özelliklerine Etkileri. (Doktora)

Kovancı, İ., Atalay, İ.Z., 1975. Manisa Bölgesi Sultani Çekirdeksiz Üzüm Bağlarında Bitki Besin Elementlerinden N, P ve K'nın Mevsimsel ve Pozisyonal Değişiminin İncelenmesi. Bitki, Cilt 2. Sayı: 4. 453-493.

Kovancı, İ., Atalay, İ.Z., Anaç, D., 1984. Ege Bölgesi Bağlarının Beslenme Durumunun Toprak ve Bitki Analizleri ile İncelenmesi. Bilgehan Basımevi, Bornova.

Lafon, J., Cauillaud, P., 1955. Extrait de la Revve. La Potasse p. 265.

Lott, W.L., Nery, J.P., Gallo, J.R., Medcaff, J.C., 1956. Leaf Analysis Technique in Coffee Research IBEC, Resc. Inst. II. 9, 21-24.

Rose, J., 1980. Effect of Supplemental Foliar and Drip Irrigation Applications of KNO3 on Grapes. M. Sc. Thesis California State University.

Chapter 7

The effect of NK fertilization on growth patterns and leaf nutrient concentration of carob-trees (*Ceratonia siliqua* L.)

P.J. Correia[1]; I. Anastácio[2]; M.A. Martins-Loução[3]

[1]*U.C.T.A.- Universidade do Algarve, Campus de Gambelas, 8000 Faro Portugal.*
[2]*AIDA- Apartado 51, 8100 Loulé Portugal.*
[3]*D. B.V.– Faculdade de Ciências de Lisboa, Bloco C2, piso 4, 1700 Lisboa Portugal.*

Key words: Carob-tree; nitrogen; potassium; fertilization; nutrients; vegetative growth; inflorescences; LAI

Abstract: The purpose of this experiment was to analyse soil nutrient availability as factors controlling vegetative and reproductive growth in carob (*Ceratonia siliqua*) trees. The orchard with 10 year-old trees, was established on a calcareous soil (total calcium carbonate: 65.2 %; active lime: 17.5 %). Four fertilization levels were tested: no fertilizer (C); 0.8 kg N.tree^{-1} (N treatment); 1 kg K$_2$O.tree^{-1} (K treatment) and 0.8 kg N.tree^{-1} plus 1 kg K$_2$O.tree^{-1} (NK treatment). No irrigation was applied during the experimental period. Branch length increment, inflorescence number, leaf area index, fruit and leaf mineral content were registered. Correlations between growth patterns and climatic variables were evaluated in order to discriminate between fertilization effects and abiotic stress, typical of Mediterranean climate, such as drought. The preliminary results of a NK fertilization trial are here presented and tree responses to N and K application were discussed in terms of source-sink effects, namely vegetative versus reproductive growth. The knowledge of these growth patterns could be important for making decisions related to fertilization. Thus, modifying orchard fertilization regimes may be a helpful strategy to improve yield on these particular droughty sites.

1. Introduction

Due to current environmental concerns, fertilizers supply should meet the potential growth of the crop. In mature carob-trees, N application increased fruit production and vegetative growth in short and long-term experiments (Lloveras and Tous, 1992; Correia and Martins-Loução, 1995). This response was strongly affected by late Spring precipitation. However, yield irregularity still occurred in a four-year period, independently of N fertigation applied during Summer. This was mainly due to an absence of inflorescences (Correia and Martins-Loução, 1996) which led to a decrease in fruit number.

It was observed that foliar K application as KNO$_3$ increased the number of inflorescences (Ataíde and S.José, 1997) and K addition had a positive effect on plant size and canopy cover (Davis, 1994).

This study aimed to evaluate the contribution of K and N fertilization in vegetative and reproductive growth of mature carob-trees. The nutritional status of these trees was also studied.

2. Materials and Methods

The orchard presented 10 years-old carob trees (cv. "Mulata") and was located in the south region of Portugal (37° 03' N; 7° 35' O). The soil, at 30-40 cm depth, was constituted as: 65.2 % of total CaCO$_3$, 17.5% of active CaCO$_3$; pH (H$_2$O), 7.5; 25 ppm of P$_2$O$_5$, 324 ppm of K$_2$O and 1.6 % of organic matter. Climatic parameters were measured *in situ* using an automatic micrometeorological station. Reference evapotranspiration (ETo) was calculated for each month, through the modified Penman method using the computer program CROPWAT (FAO).

The fertilizer was applied once per year, in April 1997 and April 1998. N-fertilizer had 26 % of nitrogen with equal parts of nitrate and ammonium and 10 % of CaO. K-fertilizer had 50 % of K_2O. No irrigation was applied in the orchard during the experimental period. Four treatments were tested: No fertilizer (C); 0.8 kg N tree^{-1} (N treatment); 1 kg K_2O tree^{-1} (K treatment); 0.8 kg N tree^{-1} plus 1 kg K_2O. tree^{-1} (NK treatment). Each treatment consisted of 3 replicates of 4 trees in a randomized complete block design.

For each treatment, 6 trees were selected for phenological measurements. On these trees, 8 branches with a 1 m branch length were marked on the canopy external site. Branch length increment (BLI) was recorded at monthly intervals, from May (time 0) until September 1997. The number of inflorescences at the beginning of the flowering period (October) was also registered.

Leaf mineral concentration was determined on three samples per treatment. For each sample 30-40 mature leaves were randomly collected. Fruit mineral concentration was evaluated using the same procedure. Leaves were collected in May 97 and April 98 and fruits in June 97 and April 98. N was determined by the Kjeldhal method, K by flame photometry and P colorimetrically. Concentrations were expressed as dry matter percentage. Leaf area index (LAI) was determined at sun-set using LAI2000 (Li-COR, USA). Measurements were performed in trees with similar canopy size. Relationships between evapotranspiration values and branch length increment were examined by linear regression analysis. Multifactor and one-way ANOVA were used to analyse data and means were compared by Duncan Multiple Range Test (DMRT). All data analysis was made by the Statgraphics 5.0 STSC program.

3. Results

Table 1 shows leaf and fruits mineral concentration determined in two sampling periods.

Table1. Leaves and fruits N, P and K concentrations (as % dry weight) for all treatments. In each month, means followed by the same letter are not significantly different at p<0.05 (DMRT).

	N (% dw)		P (% dw)		K (% dw)	
Leaves	May97	Apr 98	May97	Apr 98	May97	Apr 98
Control	1.50 ab	1.85 a	0.19 a	0.21 a	0.49 a	0.73 a
N	1.69 b	2.30 b	0.18 a	0.17 a	0.55 a	0.71 a
K	1.46 a	1.85 a	0.15 a	0.16 a	0.45 a	0.69 a
NK	1.55 ab	1.91 a	0.17 a	0.16 a	0.47 a	0.67 a
Fruits	June97	Apr98	June97	Apr98	June97	Apr98
Control	1.55 a	1.45 a	0.16 a	0.15 a	1.23 a	1.40 a
N	1.63 a	1.72 a	0.15 a	0.16 a	1.25 a	1.37 a
K	1.54 a	1.69 a	0.15 a	0.16 a	1.31 a	1.39 a
NK	1.58 a	1.55 a	0.15 a	0.16 a	1.20 a	1.27 a

Trees fertilized only with nitrogen (N treatment), showed the highest leaf N concentration levels compared to the other treatments. Leaf P and K concentrations were similar for all treatments (Table 1). Apparently, there was an increase in the studied leaf macronutrient concentrations from 1997 to 1998 in all treatments.

In 1998, fruit sampling was anticipated, since fruit maturation started earlier. No significant differences were observed among treatments, for fruit macronutrient concentrations.

Branch length increments registered from June until September 97, for all treatments, were closely related to climatic conditions in the same period (Figure 1).

Trees fertilized with N and K showed the highest vegetative growth in 97 (results not shown).

Figure 2 shows the number of inflorescences in October 97. The highest value was recorded in K treatment, although no significant differences from N treatment were observed.

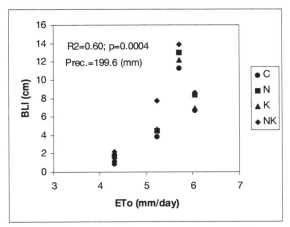

Figure 1. Linear relation between branch length increment registered in 1997 (BLI) and reference evapotranspiration.

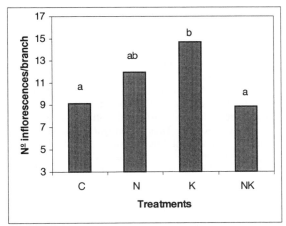

Figure 2. Number of inflorescences per branch for all treatments, recorded in October 97. Columns followed by the same letter are not significantly different at p<0.05 (DMRT).

Leaf area index registered in April 98 and May 98 is presented in Figure 3. Since no vegetative growth was observed during Winter, LAI values recorded in April 98, are expected to be mainly influenced by leaf shedding.

Trees fertilized with N and K showed the lowest LAI value. In the following month trees fertilized with NK increased LAI values, while in the other treatments there was a decrease. Compared to the control treatments, K and NK fertilized trees present significantly higher LAI values.

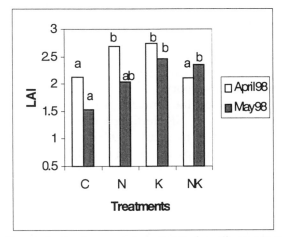

Figure 3. Leaf area index (LAI) for all treatments determined in April and May 1998, in the beginning of the new growing season. Columns followed by the same letter are not significantly different at p<0.05 (DMRT).

4. Discussion

In Mediterranean conditions, Spring-time is a crucial period for carob-tree, since shoot growth occurs simultaneously with flowering induction. At the same time, mature trees also support developing fruits from the previous season. Climate variables should be considered in order to improve yield and vegetative growth by means of minimum fertilizers inputs. In our conditions, they account for 60% of branch length increments during Spring and Summer, providing that considerable rainfall occurs during Winter and Spring (Figure 1).

Fertilizer application may enhance growth response, particularly when using N-fertilizers (Correia and Martins-Loução, 1995). This can be attributed to an increment on leaf N concentration which was observed for N and NK fertilized trees. This fact normally happens in several species (e.g. Nilsson and Wiklund, 1994). Since only a small vegetative growth increment was registered in 1998 for all treatments (data not shown), it may be expected a rather low N translocation from leaves to the shoots. This could explain the increase in leaf N content from 97 to 98 and the lack of significant differences among treatments. A similar explanation can be suggested for leaf K. The absence of differences for leaf P between treatments and years is in accordance with its low translocation.

Fruit nutritional status is not dependent on fertilization treatments (Table 1) but it is rather time-dependent. However more data is needed to support this theory.

In spite of the branch growth increment during 1997, in NK fertilized trees, leaf area index recorded in April 1998 did not reflect increase in canopy cover (Figure 3). This may be due to an higher leaf shedding and less fruits, since those trees present significant low number of inflorescence recorded in October 97 (Figure 2). However, this trend was not maintained in May 98. A close analysis of climatic conditions during April/May 98 may help to explain LAI variations since an high air evaporative demand was observed to increase leaf shedding in carob-tree (Correia and Martins-Loução, 1997).

Apparently, the application of N and K simultaneously, had no effect in the number of inflorescence, which may suggest that the addition of both ions induces a negative effect on flowering induction. K availability is normally associated with fruit quality improvement (e.g. Marschner, 1997) but foliar applications of KNO_3 may also led to an increase in inflorescence number (Ataíde and São José, 1997). Apparently, N application improves flowering induction, as occurs in apple trees, and K also have positive effects on flower formation (Marschner, 1997). In our conditions, K application led to a significant increase in inflorescence number and higher yields are expected in 1998. Similar values were obtained with N application but not with NK association. More data is need to confirm these trends and to understand the role of K and N on flowering induction of carob-tree.

These preliminary results emphasise the importance of ions interaction on yield improvement.

Acknowledgements

We are thankful to Eng. Maria da Fé Candeias (DRAAG-Tavira) for chemical analysis. This investigation was financially supported by PRAXIS and INTERREG projects n° 3/3.2/HORT/2168/95 and 20/REGII/6/96, respectively.

5. References

Ataíde, EM (1997) Influência do nitrato de potássio no percentual de floração de mangueira cv "Tommy Atkins". Actas de Hortic. 18: 114-117.

Correia, PJ and Martins-Loução MA (1995) Seasonal variations of leaf water potential and growth in fertigated carob-tree (*Ceratonia siliqua* L.). Plant Soil 172: 199-206.

Correia, PJ and Martins-Loução MA (1996) Effect of N fertilization and irrigation on the development of carob-tree (*Ceratonia siliqua* L.) In: 3th International Carob Symposium. Tavira, Portugal. (in press).

Correia, PJ and Martins-Loução MA (1997) Leaf nutrient variation in mature carob (*Ceratonia siliqua* L.) tree in response to irrigation and fertilization. Tree Physiol 17: 813-819.

Davis, JG (1994) Managing plant nutrients for optimum water use efficiency and water conservation Advances in Agronomy. 53: 85-119.

Lloveras, J and Tous, J (1992) Response of carob-trees to nitrogen fertilization. Hortscience 27: 849.

Marschner, H (1997) Mineral nutrition of higher plants. Academic Press. London.

Nilsson, LO and Wiklund, K (1994) Nitrogen uptake in a Norway spruce stand following ammonium sulphate application, fertigation, irrigation, drought and nitrogen-free-fertilization. Plant Soil 164: 221-228.

Chapter 8

Effect of mineral fertilizers on leaf nutrients, yield, and quality properties of F_1 hybrid tomatoes grown in plastic house

İ. Doran[1] , S. Kesici[2] , A. Aydın[2] , A. Nizamoğlu[3]
[1] *Agricultural Faculty of Dicle University, Diyarbakir/Turkey*
[2] *Alata, Horticultural Research Institute, Erdemli-İçel/Turkey*
[3] *Province Food Control Laboratory, Yenimahalle-Ankara*

Key words: tomato, greenhouse, fertilization, yield and quality properties, nutrients

Abstract: Balanced fertilisation has been an important issue for high productive hybrid varieties in recent years in İçel Province of Turkey, the second significant production area of vegetables grown under greenhouses. The research has been carried out to determine the effects of N, P_2O_5, K_2O rates on yield, quality, nutrient levels of leaves of tomato plants (single crop Noria F_1 hybrid) grown under plastic cover. This research was carried out in the plastic houses of Alata Horticultural Research Institute. 48 combinations of N (0-200-400-600 kg ha^{-1}), P_2O_5 (0-150-300 kg ha^{-1}), K_2O (0-200-400-600 kg ha^{-1}) fertilisers were applied in addition to 60 tons ha^{-1} farmyard manure given as top dressing. According to the statistical analysis, N, P_2O_5, K_2O rates were significantly effective on yield and on all quality properties of fruit, and on N, K levels of leaves. Significantly positive relations have been determined between yield and N, K as the similar relations between TSS, vitamin-C and Zn, Mn interactions, while significantly negative relations were between TSS and N, Mg interactions in the leaves. It has been concluded that combination of $N_2P_1K_2$ (400 kg N + 150 kg P_2O_5 + 400 kg K_2O ha^{-1}) in addition to 60 tons ha^{-1} farmyard manure, had the best effect on yield, quality properties and nutrient levels of leaves.

1. Introduction

After 1970's, Turkey has realised fast development in using hybrid vegetable seeds and growing vegetables under greenhouses due to the advantages of the Mediterranean climate.

By 1996, total greenhouse area increased to 15.1 ha and the production in general to 1.680.000 tons. Ninety five percent of the greenhouse area is used for vegetables and 55% of this amount is used for tomato production (Anonymous,1997).

Fertilisation has been an important issue due to growing high yielding hybrid varieties in greenhouse vegetable production. Hybrid varieties deplete more nutrients from soil because of fast growing (Sevgican, 1989).

This study has been carried out to determine the effect of mineral fertilisers on yield, quality and nutrient levels of leaves of Noria F_1, an important hybrid tomato variety for İçel province.

2. Materials and Methods

The study was carried out in the plastic houses of Alata Horticultural Research Institute, and Noria F_1 hybrid tomato variety was used.

The experiment was performed as randomised block design with three replications and lasted in two years. In this study, 48 combinations of N (0-200-400-600 kg ha^{-1}), P_2O_5 (0-150-300 kg ha^{-1}), K_2O (0-200-400-600 kg ha^{-1}) were applied, in addition to 60 tons ha^{-1} farmyard manure and 75 kg ha^{-1} magnesium sulphate as top dressing. Furrow irrigation was used and the plastic house was heated in the cold days.

1/4 of ammonium nitrate was applied at the first fruit cluster and rest at 15 day intervals for three times. All of the triple super phosphate and magnesium sulphate were applied at pre-planting and before last tillage, respectively. 1/4 of potassium sulphate was applied pre-planting and the other 1/4 at the first fruit cluster. Rest of the fertiliser was applied at 15 day intervals for two times.

Before the basal dressing, soil samples were taken from 0-30 cm depth to determine some of the properties of the experimental soil (texture, lime, pH, soluble total salinity, organic matter, available P, available K, available Ca and available Mg) (Chapman et al., 1961).

Samples were taken from young mature leaves of the plants in April and were analysed for N, P, K, Ca, Mg, Fe, Zn, Mn, and Cu. Total N was determined by the Kjeldahl method and P was determined colorimetrically. Other nutrients were assessed by using Atomic Absorption Spectrophotometer (Chapman et al., 1961; Geraldson, 1973; Roarda and Smilde, 1981).

In May, fruit samples were taken from each replication and analysed for soluble solids, titratable acidity, pH, vitamin-C and invert sugar (Anonymous, 1973).

Results were analysed with factorial analysis of variance and LSD tests (Düzgüneş, 1987).

3. Result and Discussion

Experimental soil was sandy loam in texture; alkaline in reaction; rich in lime, P, K, Ca, and Mg; poor in organic matter and had no salinity problem. The best growing media for tomato plants is a loamy textured soil with a total soluble salt concentration no more than 0.19 %. The soil should have an organic matter of 6-10%, of lime 1-5%, of pH 5.5-7.5 and should be rich in nutrients (Macit and Agme, 1980). According to physical and chemical properties, the experimental soil is suitable for tomato growing.

Effects of mineral fertiliser rates on nutrient contents of the leaves were examined and average values of nutrients were compared with the reference values given for tomato plants (Roarda and Smilde, 1981). Results are given in Table 1.

According to Table 1, effect of ammonium nitrate rates on N content of leaves and effect of potassium sulphate rates on K content of leaves were found significant, whereas effect of triple super phosphate rates on P content and the effects of those three fertilisers on leaf Ca, Mg, Fe, Zn, Mn and Cu contents were found insignificant.

Nutrient levels of tomato leaves were generally close to the reference values given for tomato leaves and no deficiency symptoms were observed but increase in N content of leaves at N_2 rate and K content of leaves at K_2 rate and fluctuation in P content at different P rates may be attributed to the application of $MgSO_4$, farmyard manure, and nutrients existed in the soil before applications.

Some significant positive and negative correlations were found between leaf nutrients. For example, leaf N was positively related to leaf K and Ca. On the other hand, leaf N was negatively correlated with Mn. Other positive correlations were likewise between leaf K and leaf N, P, Ca and Zn. Whereas, leaf K was negatively related to leaf Mg and Mn. Similarly, leaf P had positive relations with leaf K, Ca and Zn contents. On the other hand , negatively related to Mg, Fe and Cu.

The effects of N, P_2O_5, K_2O rates on yield and all quality properties of fruit were found significant. It has been concluded that 400 kg N ha^{-1}, 150 kg P_2O_5 ha^{-1} and 400 kg K_2O ha^{-1} in addition to 60 tons ha^{-1} farmyard manure had the best effect on yield, quality properties and nutrient levels of leaves. This combination was the most profitable application for tomato plants grown under plastic cover.

In view of fertiliser combinations, yield ranged between 83 tons ha^{-1} ($N_0P_0K_0$) and 169 tons ha^{-1} ($N_3P_1K_2$). The most economical yield was 163 tons ha^{-1} ($N_2P_1K_2$); ranking the third after 169 tons ha^{-1} ($N_3P_1K_2$) and 165 tons ha^{-1} ($N_3P_2K_2$).

In tomato growing under greenhouse conditions, it is necessary to apply, 80-100 kg super phosphate, 40-50 kg potassium sulphate and 8-10 tons farmyard manure as a basic fertiliser. By one ton yield 2.18 kg N, 0.60 kg P_2O_5 ,4.03 kg K_2O and 0.46 kg MgO is taken up (Çolakoğlu and Pekcan, 1990).

Güzel (1985) claims that 45 kg N, 8 kg P_2O_5, 90kg K_2O, 55 kg CaO and 12 kg MgO is removed by 13 tons of yield. He also states that 68 kg N, 17 kg P_2O_5, 140 kg K_2O, 90 kg CaO and 19 kg MgO is removed by 20 tons.

With the previous findings, research results are in agreement. It can be concluded that sufficient and balanced fertiliser program increases tomato yield significantly.

The effects of fertiliser rates on total soluble solids, Vitamin-C, total acidity, and pH of tomato fruits were significant. Regarding years, the effects of fertiliser rates on quality properties except Vitamin-C was found significant.

The effects of N, P_2O_5 and K_2O fertiliser rates (and interactions) on invert sugar, vitamin-C, total acidity and total soluble solids (TSS) of tomato fruits were found significant. It is observed that quality properties of tomato fruits are improved with increasing rates of potassium sulphate and triple super phosphate, and decreasing rates of NH_4NO_3.

While there was significant positive relation between the N content of tomato leaves, yield and pH, there was significant negative relation between the N content of tomato leaves and Vitamin-C and total soluble solids. Increasing N content of leaves increased the fruit yield but decreased the quality properties.

Table 1. Effects of mineral fertiliser rates on the leaf nutrients

Types	Doses	N	P	K	Ca	Mg	Fe	Zn	Mn	Cu
	0	3.490 b	0.337	3.493	3.244	0.534	90.61	92.33	55.39	18.95
Ammonium	I	3.635 a	0.344	3.468	3.386	0.523	90.81	89.17	57.69	18.18
Nitrate	II	3.712 a	0.362	3.511	3.422	0.560	101.86	84.56	56.25	19.11
	III	3.683 a	0.351	3.379	3.431	0.599	89.11	87.22	57.39	17.50
(LSD)		0.144 *	n.s	n.s	n.s	n.s	n.s	n.s	n.s	n.s
	0	3.646	0.345	3.273 b	3.489	0.575	101.06	89.33	56.81	19.35
Potassium	I	3.594	0.341	3.396 ab	3.350	0.569	87.36	88.67	55.81	18.92
Sulphate	II	3.609	0.357	3.547 a	3.339	0.551	87.97	85.94	56.86	18.28
	III	3.671	0.351	3.635 a	3.306	0.522	96.00	89.33	57.25	17.18
(LSD)		n.s	n.s	0.258 *	n.s	n.s	n.s	n.s	n.s	n.s
	0	3.766	0.378	3.759	3.365	0.553	88.83	86.35	58.06	19.53
Triple Super	I	3.865	0.374	3.709	3.413	0.554	95.29	89.33	55.50	17.52
Phosphate	II	3.883	0.384	3.783	3.335	0.556	95.17	89.27	56.48	18.25
Optimum Levels		2.80	0.40	2.70	2.40	0.30	100	22.0	55	10.0
(Roarda and		-	-	-	-	-	-	-	-	-
Smilde, 1981)		4.90	0.60	5.80	7.20	0.80	391	85.0	385	16.0

n. s, * : insignificant or significant at P= 0.05 , respectively.

a, b : Mean separation significant within column by LSD test

It can be concluded that N doses to a certain level increase the yield and then increase in N rates cause decreases in quality properties of tomato fruits.

According to correlation analysis, there were not significant relations between yield and quality properties. There were significant positive relations between vitamin-C and TSS ratio and invert sugar. Likewise, there were significant positive correlations between TSS ratio and total acidity and invert sugar. Also, there was a significant negative relation between total acidity and pH, while there were significant positive relations between invert sugar and total acidity and pH.

Regarding the correlation analysis, while there were significant positive correlations between the N content of leaves and yield and pH, there were significant negative relations between N and vitamin-C and TSS ratio. Increasing N content of leaves increased fruit yield but decreased quality properties.

While a significant negative relation was found between the P content of leaves and TSS ratio, no statistically significant high level positive correlation was determined between fruit yield and P content of leaves.

There were significant negative relations between the K content of leaves and vitamin-C and pH, while there were significant positive relations between the K content and yield and total acidity.

Another significant negative correlation was found between the Mg content of leaves and TSS ratio. There was a positive relation between Zn and Mn content of leaves and Vitamin-C and TSS ratio of tomato fruits.

Other nutrients were at optimum levels in tomato plants. There were significant positive relations between yield and N and P and K contents of leaves due to sufficient and balanced fertilisation of plants.

4. References

Anonymous (1973).FAO technical report Olive Research Inst. İzmir-TURKEY.

Anonymous (1997).Annual report. (in Turkish) Citrus and Greenhouse Research Inst. Antalya.

Cao W and Tıbbıts TW (1993). Study of various NH4/NO3 mixtures for enhancing growth of potatoes. Journal of Plant Nutrition, 16(9), 1691-1704.

Chapman HD Pratt PF Ratt and. Parker F (1961). Methods of analysis for soils, plants and waters. U. of California. Div. of Agric. Sci. 309 p.

Çolakoğlu H and Pekcan T (1990) The effect of drip irrigation on the development and nutrient uptake of tomato plant in greenhouse condition. (in Turkish). 5th Greenhouse Symposium. 131-136. İzmir-TURKEY

Düzgüneş O Kesici T Kavuncu O and Gürbüz F (1987). Research and trial methods (Statistical Methods-II) (in Turkish) AUAF 1021 Ankara-TURKEY.

Geraldson CM (1973). Soil testing and plant analysis, Soil Sci. Soc.of America Inc. Madison Wisconsin, U.S.A.

Güzel N (1985). Private interview CUAF Soil Sci Adana-TURKEY.

Kacar B (1983).General Plant Physiology (in Turkish) AUAF 246 Ankara

Macit F and Agme Y (1980). The fertilisation of vegetables (in Turkish) Bilgehan Publication İzmir-TURKEY.

Mooray J and Graves CG (1980). The effects of root and air temperature on the growth of tomatoes. Acta Hort.98: 29-43.

Özbek H Kaya Z and Tamcı M (1984).The nutrition and metabolism of plants (in Turkish) ÇUAF 162 Adana-TURKEY.

Roarda Von E and Smilde E (1981).Nutritional disorders in glasshouse tomatoes, cucumbers and lettuce Centre for Agri. Wageningen.

Sevgican A (1989). The vegetable crops in greenhouse (in Turkish) TRF 19 Yalova.

Chapter 9

Effect of combined N and K nutrition on yield and quality of spinach

D.Anac, N.Eryüce, Ö.Gürbüz, B.Eryüce, C.Kılıç, M.Tutam
Ege University, Faculty of Agriculture Soil Dept. 35100 Bornova - Izmir /Turkey

Key words: spinach, NO_3, N fertilization

Abstract: Vegetables require a balanced fertilization in order to achieve higher yields and good quality as for taste, vitamin, nitrate and mineral contents are concerned. These issues are very important for human health and conservation during storage and transportation. Spinach is one of the highly consumed leafy vegetables in Turkey. Considering that potash fertilizers could increase the quality of vegetables through a better balanced fertilization, a field trial was performed to evaluate the use of different doses of N and K on yield and some quality criteria of spinach. Two rates of N on K_2O were applied, 80 and 160 $kgha^{-1}$ of each respectively. Highest yield was obtained at highest N-K treatments. Vitamin C contents were the highest in parcels that received less N but more K. Potassium applications significantly decreased NO_3 concentrations in leaves. Leaf total N and K concentrations increased with respect to treatments. No steady changes were measured in the case of leaf P, Cl, Fe, Cu, Mn, Zn, contents. The effect of K nutrition on leaf Mg was negative, contradicting to leaf SO_4 content which increased by K levels.

1. Introduction

Vegetables require a balanced fertilization in order to achieve high yields and good quality as taste, vitamin, nitrate and dry matter contents are concerned. These issues are very important for human health and the situation in closely related to the N/K fertilizer ratio. It is very well know that nitrates and some heavy metals often accumulate in the leaves due to mismanagement in fertilisation (Venter, 1980 and Futz, 1983).
Spinach which is also a leafy vegetable is highly consumed in Turkey.
Objectives
Considering that potash fertilizers could increase the quality of vegetables through a better balanced fertilisation, the purpose of this study is to evaluate the use of different doses of N and K on yield and quality of spinach.

2. Materials and Methods

A field experiment was performed with 2 rates of N (80 and 160 kg ha^{-1}) and 2 rates of K_2O (80 and 160 kg ha^{-1}) applied in combination. Split plot was the layout of the experiment with 4 replications. Recommended amounts (100 kg ha^{-1}) of P_2O_5 was also incorporated. Potassium, P_2O_5 and 2/3 of N was given as basal dressing and 1/3 of N as side dressing 36 days after sawing. Potassium was given in the form of K_2SO_4 and N in the forms of $(NH_4)_2SO_4$ and NH_4NO_3. Experimental plots were 35m^2 and the field soil was slightly alkaline in reaction medium in N,P,K contents, loamy in texture, high in $CaCO_3$ content and have a high CEC.

At the end of the experiment two harvest were realised. Total yield was recorded, ionic composition of leaves were measured and Vitamin C contents were determined.
Variance of analysis was used to evaluate the results statistically. Correlations were also determined between some variants .

3. Results and Discussion

Yield increases were determined with respect to increasing N and K rates. Highest yield was achieved at highest N and K treatments ,,as

22250 kg ha^{-1} (Table-1). Regional spinach yield under farmers condition is 20000 kg ha^{-1} which is lower than the yield obtained in the study (unpublished and compiled data). Also, a significant positive correlation (r=0.849**) was found between N treatments and the yield.

Table 1: Yield (kg ha^{-1}) with respect to N and K rates

N rates (kg ha^{-1})	K$_2$0 rates (kg ha^{-1})	
	80	160
80	13060	15100
160	18600	22250
F values	N** K*	NxK^{ns}

Vitamin C content of leaves was measured highest in parcels that received less N but more K$_2$O. This results was realised for both of the harvests. Enhancements in K$_2$O rates increased the Vitamin C content of leaves and the effect was found significant for the first harvest (r=0.661**). On the other hand, N decreased this attribute (Table 2).

Table 2: Vitamin C concentrations (mg 100g^{-1}) with respect to N and K$_2$O rates.

21.10.1997		
N (kg ha^{-1})	K$_2$O (kg ha^{-1})	
	80	160
80	3.40	4.14
160	3.35	3.92
F values	N^{ns} K**	NxK^{ns}
25.11.1997		
N (kg ha^{-1})	K$_2$O (kg ha^{-1})	
	80	160
80	6.11	6.42
160	5.99	6.36
F values	N^{ns} K^{ns}	NxK^{ns}

As for the leaf NO$_3$ concentrations, increasing N rates significantly increased this parameter. Highest measurement was realized at high N and low K$_2$O rate (Table 3). The correlation coefficient between the applied N and NO$_3$ content of leaves were r= 0.560* and r= 0.451^{ns} for the 1st and 2nd harvests. On the other hand, K$_2$O treatments significantly (1%) decreased this parameter (r= -0.774**). It is reported that NO$_3$ accumulation in leaves depend upon many factors one of which is the N fertilizer (Marschner, 1984, Cantliffe, 1973). Minotti (1978) states that 1050 mg kg^{-1} is the limit value for spinach.

Table 3: Leaf NO$_3$ concentrations (mg kg^{-1}) with respect to N and K$_2$O rates.

21.10.1997		
N (kg ha^{-1})	K$_2$O (kg ha^{-1})	
	80	160
80	1588	920
160	1813	1763
F values	N* K**	NxK**

28.10.1997		
N (kg ha^{-1})	K$_2$O (kg ha^{-1})	
	80	160
80	868	128
160	1005	637
F values	N* K**	NxK^{ns}

Results showed that leaf total N contents increased with N treatments parallel to the NO$_3$ concentrations, however, data were not statistically significant. Similar to above results, K treatments significantly (1%) increased the leaf K contents (r= 0.619*). On the other hand, K$_2$O rates had a negative effect on leaf Mg concentrations which was significant at 5% level. In the case of leaf P, no steady trend with respect to treatments were determined (Table 4). Results obtained in this study are in agreement with the previous findings of Bergman (1986) and Scaife-Turaer (1983) for leaf N, P, K, Ca and Mg.

Table 4: Leaf N, P, K, Ca, Mg contents (%) with respect to N and K$_2$O rates

N (kg ha^{-1})	N		P	
	K$_2$O (kg ha^{-1})		K$_2$O (kg ha^{-1})	
	80	160	80	160
80	4.42	4.33	0.36	0.34
160	4.52	4.56	0.35	0.39
F values	N^{ns}	K^{ns}	NxK^{ns}	
	N^{ns}	K^{ns}	NxK^{ns}	

N (kg ha^{-1})	K		Mg	
	K$_2$O (kg ha^{-1})		K$_2$O (kg ha^{-1})	
	80	160	80	160
80	7.26	7.58	0.74	0.57
160	7.28	7.70	0.63	0.70
F values	N^{ns}	K**	NxK^{ns}	
	N^{ns}	K*	NxK**	

Leaf SO$_4$ concentrations increased significantly (5%) with respect to increasing K$_2$O treatments. The increase is most probably related to SO$_4$ radical of the applied K$_2$SO$_4$ fertilizer.

Leaf Cl and microelements (Fe, Cu, Mn and Zn) showed no consistent relations with the treatments.

Table 5. Leaf SO_4 and Cl concentrations (%) with respect to N and K_2O

N (kg ha^{-1})	SO_4		Cl	
	K_2O (kg ha^{-1})		K_2O (kg ha^{-1})	
	80	160	80	160
80	0.33	0.37	1.20	1.61
160	0.28	0.33	1.33	1.24

F values N^{ns} K^{*} NxK^{ns}
              N^{ns} K^{ns} NxK^{**}

Table 6. Leaf Fe, Cu, Mn and Zn concentrations (mg kg^{-1}) with respect to N and K_2O.

N (kg ha^{-1})	Fe		Cu	
	K_2O (kg ha^{-1})		K_2O (kg ha^{-1})	
	80	160	80	160
80	342	349	12.3	12.5
160	338	359	12.0	13.3

F values N^{ns} K^{ns} NxK^{ns}
              N^{ns} K^{ns} NxK^{ns}

N (kg ha^{-1})	Mn		Zn	
	K_2O (kg ha^{-1})		K_2O (kg ha^{-1})	
	80	160	80	160
80	62	63	121	123
160	64	61	126	126

F values N^{ns} K^{ns} NxK^{ns}
              N^{ns} K^{ns} NxK^{ns}

4. Conclusion

Hundred sixty kg of N and 160 kg of K_2O can be recommended to the regional spinach growers who generally apply only N-P complete fertilizers to spinach.

Acknowledgement

This research work is partially supported by International Potash Institute.

5. References

Bergmann, W. 1986. Farbatlos. Emahnugs strorungen bei kultupflanzen V&B Guntav Ficher Verlay. Jana.

Cantliffe, D.J.B 1972. Nitrate accumulation in table beets and spinach as affected by N, P, and K nutrition and light intensity. Agronomy Journal, Vol. 65. 563-565.

Fritz, D. 1983. Nitrat in Gemuse und Grundwasser. In: Nitrat in Gemuse und Grundwasser. Vortragstagung Bonn, 1-7. Universitats druck erei. Bonn.

Marschner, H. 1984. Mineral Nutrition of Higher Plants. Academic Press, London.

Minotti, P. L. 1978. Critique of "Potential nitrate levels in edible plants parts.

Scaife, A. and Turner, M., 1983. Diagnosis of Mineral Disorder in Plants. Volume. 2 Vegetables.

Venter, F. 1978. Untersuchungen überden nitratgehalt in Gemüse. Der Stickstoff 12. 13-38.

Chapter 10

The effect of different amounts of N-fertilizers on the nitrate accumulation in the edible parts of vegetables

M. J. Malakouti, M. Navabzadeh and S. H. R. Hashemi
Soil and Water Reesearch Institute, North Kargar Ave., Tehran, Islamic Republic of Iran.

Key words: N fertilization NO$_3$ accumulation, vegetables

Abstract: Eight important vegetables i.e. spinach, lettuce, celery, cabbage, parsley, leek, radish and cress were grown at low to high rates of nitrogen fertilization. Seven N-fertilizer treatments consist of 0, 100, 200, 300, 400, 500, and 600 N kg/ha conducted in Varamin in 1995 for spinach, lettuce, celery cabbage and on the other four vegetables the five treatment contained, i.e. 0, 100, 200, 300, and 400 N kg/ha carried out in Karaj region in 1996. At harvest time, two leaf samples were taken: one in the early morning and the other in the evening and nitrate content was measured by sulfosalicylic method. The results revealed that there was a positive relationship between the amount of N-fertilizer and nitrate accumulation. The maximum nitrate accumulation occured with 400 kg N/ha treatment. In these treatments, leaf samples taken in the evening, had less nitrate in comparison with the morning harvest. The results also revealed that the yields were increased up to 400 kg N/ha treatment significantly but additional N-fertilizers did not increased their yield may be due to the ammonium toxicity.

1. Introduction

Efficient use of N fertilizer in vegetable production is an important consideration. Excessive use of N increase vegetable production costs and low recovery by the vegetables can contribute to nitrate contamination of vegetables. Much has been reported concerning the amount of N required different vegetables and the influence of N on vegetable quality. Nitrogen fertilization had considerable influence on the degree of nitrate accumulation. At levels of N where growth was restricted, accumulation of nitrate was very low. With increasing N fertilization, the rate of accumulation increased. In an experiment with table beets, fertilization with 560 kg/ha N as compared to none, increased nitrate accumulation in the petioles 70 times, in the blades 30 times, and in the roots 20 times (Lorenz, 1978).

The effects of different levels of N fertilizer on the nitrate accumulation in lettuce and spinach has been tested in Tehran Province in 1994. The results revealed that various N-levels had a significant effect on their yield and nitrate accumulation. Harvested samples showed that all morning harvested samples had more nitrate content than afternoon and the differences were significant. Comparing between spinach edible parts at a constant nitrogen consuming level, identified that petioles had the highest nitrate content (0.50% NO$_3$-N). In lettuce stems nitrate content was 0.44% (NO$_3$-N), and lower nitrate content was in interal lettuce leaves (Zarei and Malakouti 1996).

Accumulation of NO$_3$-N in edible part of cabbage and celery, seven different levels were tested in Karaj area in 1994. The results showed that with increasing the rate of N fertilizer in cabbage and celery, the production of protein and accumulation of NO$_3$-N were increased.

Considering the accumulation of NO$_3$-N, there were remarkable differences between morning samples and evening samples in both vegetables. The percentage of NO_3-N in morning samples was 0.57% for cabbage and 0.67% for celery, while in the afternoon samples were 0.33 and 0.63 percent respectively. In order to have an acceptable yield in cabbage and celery and also to have a low concentration of NO_3-N, it is prescribed to use 100 N kg/ha and harvest them in the late afternoon to minimize the accumulation of NO_3-N (Behtash and Malakouti, 1996). By increasing the urea application up to 600 kg/ha, the yield as well as NO_3-N contents in potato tubers were increased, but with 800 kg/ha urea, NO_3-N contents were decreased. This may be due to toxicity of ammonium accumulation. There had been considerable significant differences in nitrate accumulation potential between cultivars. Therefore, 400 kg/ha urea was recommended for higher potato yield. Amount of nitrate content at 600 kg/ha urea was higher than standard limit (300 mg/kg FW). Therefore, in calcareous soils with lower organic matter, application of more than 400 kg/ha urea should be seriously prevented (Tabatabaei and Malakouti, 1998).

The use of 180 kg N/ha accompanied by iron and manganese foliar spray, resulted in a 240 mg/kg of dry weight for nitrate concentration with 76 tons/ha onion yield. The application of this amount of N with iron and zinc brought about a 327 mg/kg dry weight of nitrate concentration and 78 tons/ha of onion yield. Maximum nitrate concentration (1462 mg/kg dry weight) was observed with 500 N kg/ha without consuming any of the micronutrients (Bybordi and Malakouti, 1998).

2. Materials and Methods

Eight important vegetables, i.e. spinach, lettuce, celery, cabbage, parsley, leek (Lipidium sativum), radish, and cress (Allium sp.) were grown at low to high rates of nitrogen fertilization. Seven N-fertilizer treatments were 0, 100, 200, 300, 400, 500, and 600 N kg/ha in Varamin and five N-fertilizer treatments for the last four vegetables were 0, 100, 200, 300, and 400 N kg/ha in Karaj region in the form of completely randomized block design with four replications. These experiments were carried out in two consecutive years (1995-1996). Soil samples were taken before plantation and all their physico-chemical characteristics were determined according to the conventional methods. At harvesting time, two samples were taken once in the early morning and the other in the evening. Leaf samples were dried at 65°C, ground, and analyzed for total N, and nitrate was measured by sulfosalicylic method.

3. Results and Discussions

Physico-chemical characteristics of the studied soils were shown in Table 1. As it can be seen from tables, all studied soils were calcareous, with lower organic matter, higher $CaCO_3$, and various amounts of available phosphorous and potassium.

Table 1. Physicochemical characteristics of the studied soils.

Place	PH	Texture	Clay (%)
Karaj	7.8	CL	30
Varamin	7.5	SCL	25
	N(%)	P(mg kg^{-1})	K (mg kg^{-1})
Karaj	0.05	38	450
Varamin	0.06	5	145

The results of studied plots in spinach and lettuce showed that there were significant relationship between the nitrate accumulation and amount of N fertilizer applications. In spinach, the highest nitrate accumulation (0.50% D.W.) was obtained from the morning samples while the level of nitrate in the same plot was in standard limit (0.20% D.W). In lettuce, the highest amount of nitrate accumulation was 0.58% D.W. in the morning samples and it did not change very much in the late afternoon samples (0.45% D. W.). Time of harvesting had not effect on the nitrate reduction in the edible parts of the lettuce.

Application of various amounts of N fertilizers on the yield and NO_3-N concentration were tested in cabbage, celery leek, radish, cress, and parsley and the results revealed that there was a significant and positive relationship between N-fertilizer yield, and nitrate accumulation.

It can be seen from all these data that there was a positive relationship between amount of N-fertilizers and vegetable yield and it increased up to 400 N kg/ha but beyond that there was a reverse effect due to ammonium toxicity. This is corresponded with the various workers especially.

The results of all these experiments showed that highest percentage of nitrate were tested in spinable and parsley (Barker et al. 1961; Carter and Bosma 1974; Lorenz 1978; Locascio et al. 1984, and Tabatabaei and Malakouti 1998).

So the results revealed that there was a positive relationship between the amount of N-fertilizer and amount of nitrate accumulation. Their nitrate concentration was increased up to 0.65 in lettuce and celery, 0.57 percent in spinach and cabbage, and 0.30 percent in parsley on dry weight basis. Since the normal nitrate level in the edible parts of vegetables is about 0.20 percent, the maximum nitrate accumulation occured with 400 and 200 kg N/ha treatments. In these treatments, leaf samples taken in the evening, had less nitrate in comparison with the morning harvest. The results also revealed that the vegetable yield increased up to 400 kg N/ha treatment significantly but additional N-fertilizers did not increase their yield.

4. References

Barker, A. V., Peck, N. H. and Mac Donald, G. E. (1961) Nitrate accumulation in vegetables, Agron. J., 63:126-129.

Behtash, B., and Malakouti, M. J. (1996) Effect of different amounts of N-fertilizers on the yield and nitrate accumulation in the edible part of cabbage and celeny. 5th Iranian Soil Science Congress Karaj, Iran.

Bybordi, A., and Malakouti, M. J. (1988) Study on the effect of N, Fe, Mn, and Zn on the yield and quality of the onion. J. of Soil & Water, Vol. 12, No. 4, SWRI, Tehran, Iran.

Carter, J. N. and Bosma, M. (1974) Effect of fertilizer and irrigation of nitrate-nitrogen and total nitrogen in potato tubers, Agron. J. 66:263-266.

Lorenz, O. A. (1978) Potential nitrate level in edible plants part. pp. 201-220, In: D. R. Nielsen et al. (eds.). Nitrogen in the environment, Vol.2, Soil- Plant - Nitrogen Relationship, Academic Press, New York.

Maynard, D. N. and Barker, A. V. (1979) Regulation of nitrate accumulation in vegetables. Acta Horticulture, 93:123-159.

Tabatabaei, S. J., and Malakouti, M. J. (1998) Studies on the effect of N, P, and K-fertilizers on the potato yield and nitrate accumulation in potato tuber. J. of Soil & Water, Vol. 11, No. 1, Tehran, Iran.

Zarei, H., and Malakouti, M. J. (1996) Effects of different amount of N-fertilizers on the yield and nitrate accumulation in the edible parts of spinach and lettuce. 5th Iranian Soil Science Congress Karaj, Iran.

Soil and Water Research Institute, North Kargar Ave., Tehran, Islamic Republic of Iran.

Chapter 11

Effects of different forms and rates of nitrogen on yield and quality of Colona spp.

T.Demirer, N.M.Müftüoğlu, C.Öztokat

18ᵗʰ March University, Faculty of Agriculture, Çanakkale,Turkey

Key words: nitrate, nitrogen fertilisers, crisphead lettuce

Abstract: Ammonium nitrate, ammonium sulphate, calcium ammonium nitrate and urea were applied (as fertilisers) at the levels of 50-100-150 and 200 kg N ha⁻¹. The highest yield and the lowest nitrate content were obtained from Colona spp. fertilised by ammonium sulphate.

1. Introduction

Vegetables like lettuce and head salad are generally consumed in every season and are produced as the main or secondary crops. All lettuces (green and fresh) are nutritive, good sources of vitamins and minerals with appetite. A hundred gram of lettuce contains 95 % water, 1-2 % carbohydrate, 1-2 % protein and 0.25 % fat (Becker, 1956).

In addition to other nutrients, crisphead lettuce is sensitive particularly to nitrogen since it grows very fast. Optimum nitrogen dose for lettuce is recommended to be 50 kg N da⁻¹ (Karaçal and Türetken, 1992b). Whermann (1963) reported that N should be applied to lettuce in small quantities. Often, dry matter, sugar and Vitamin C contents of head lettuce decrease with increasing doses of N (50-200 kg N ha⁻¹) while nitrate content increase (Poulsen et al., 1995).

Karaçal and Türetken (1992b) investigated the effects of N doses on yield and nutrient elements of lettuce and found the optimum N dose as 50 kg N da⁻¹.

Fertilising plants above the metabolic needs causes nitrate accumulation at the roots and vegetative parts (Maynard et al., 1976). Although nitrate accumulation is not toxic, it causes serious health problems in humans. Also high N containing feeds cause abortus in cattles (Wright and Davidson, 1964).

Nitrate accumulation in plants depend on many environmental and genetic factors in addition to the rate and form of N source (Maynard et al., 1976). But most important of all is the rate and form of nitrogen.

The objective of this study was to examine the effects of different rates and forms of N fertilizers on yield and quality of crisphead.

2. Material and Methods

A pot experiment was carried out in Çanakkale 18ᵗʰ March University, Faculty of Agriculture. The test plant was Colona spp.

Texture of the soil used in this research was sandy-clay. The soil contained 0.4% organic matter, 5.2 % $CaCO_3$, 17.1 me 100 g⁻¹ Ca+Mg, 28 kg ha⁻¹ P_2O_5, 1.8 mg kg⁻¹ Fe, 2.4 mg kg⁻¹ Mn, 1.0 mg kg⁻¹ Zn, 1.2 mg kg⁻¹ Cu and it's pH was 8.1.

Ammonium nitrate (A.N. 33%), ammonium sulphate (A.S. 21%), calcium ammonium nitrate (CAN 26 %) and urea (46 %) were used as N sources at the rates of 50-100-150-200 kg N ha⁻¹.

The layout of the experiment was completely randomised block design with 5 replications. The total pot number was 85. Thirty ton ha^{-1} composted manure, 100 kg P$_2$O$_5$ ha^{-1} (TSP), and 120 kg K$_2$O ha^{-1} K$_2$SO$_4$ were given to each pot preplanting.

Manure was completely mixed to the soil in the pot while phosphorus was given locally and potassium to the effective root depth. Twenty five percent of nitrogen was given to 2-3 cm depth before planting, and the rest at the time of head binding.

Data were statistically analysed using "Tarist" programme.

3. Results and Discussion

Yield and quality parameters are presented in Table 1. Fertilizer types and rates significantly affected the yield (P<0.01), and the highest yield was observed in ammonium sulphate at the level of 150 kg N ha^{-1}.

diameter was obtained in CAN (26 %) treatment.

Fertiliser rates were effective on dry matter contents (P<0.05) which were decreased with increasing rates.

Nitrate content of head lettuce was significantly affected by the types and rates of fertilizers (P<0.01). Fertilisers with nitrogen in nitrate form caused higher nitrate contents in head lettuce compare to the fertilisers with nitrogen in ammonium form.

Scientists report that (Ikeda and Osawa, 1984; Van Der Boon and Fri., 1990; Tepecioğlu and Yalçın, 1997) high NH$_4$ in the soil solution result in less plant nitrates.

All the other parameters were found unimportant and related results has been given in Table 1. There was no bitterness problem in any application.

Table 1. Yield and quality properties in relation to source and rate of N

Fertilizers and rates	Yield kg ha^{-1}	Head Weight g	Head Diameter cm	Leaf Number	Dry Matter %	Nitrate Mg kg^{-1}	Sugar %	Vitamin mg kg^{-1}
1)A.N.	33757 (2)	219.2 (3)	11.96 (3)	24.8 (4)	4.9 (5)	1517.048 (1)	1.80 (2)	58.153 (3)
2)A.S.	33252 (1)	216.5 (2)	11.76 (1)	24.5 (2)	4.7 (3)	1392.652 (3)	1.78 (5)	46.615 (1)
3)CAN	32904 (4)	212.6 (1)	11.68 (2)	24.3 (3)	4.7 (4)	931.001 (4)	1.71 (3)	41.423 (4)
4)ÜREA	32828 (3)	211.1 (4)	11.25 (4)	23.9 (1)	4.6 (1)	717.912 (2)	1.53 (4)	31.897 (5)
5)Control	26542 (5)	189.7 (5)	9.87 (5)	22.2 (5)	4.3 (2)	247.488 (5)	1.50 (1)	31.516 (2)
	**	NS	**	NS	NS	**	NS	NS
1)50 kg N ha^{-1}	34095 (3)	234.5 (2)	12.25 (4)	26.0 (3)	4.9 (1)	1314.928 (4)	1.79 (4)	57.444 (1)
2)100 kg N ha^{-1}	33440 (2)	224.3 (3)	11.74 (3)	24.8 (4)	4.8 (2)	1126.012 (3)	1.77 (2)	41.171 (2)
3)150 kg N ha^{-1}	30610 (4)	197.1 (4)	11.60 (2)	23.9 (2)	4.6 (3)	839.664 (2)	1.51 (1)	40.068 (4)
4)200 kg N ha^{-1}	29282 (1)	183.8 (1)	11.05 (1)	22.9 (1)	4.4 (4)	564.277 (1)	1.48 (3)	39.044 (3)
	**	**	NS	NS	*	**	NS	NS

*: p< 0.05; **: p< 0.01, NS: nonsignificant

Similar results were reported by Karaçal and Türetken (1992a and b), Bayraktar (1970), Sorensen and Fri., (1994), Kaptan and Almaca (1997).

The effects of fertiliser types on head weight were unimportant (P>0.05). However, fertilizer rates were significantly effective on head weight (P<0.01), and the highest value was measured in 100 kg N ha^{-1} treatment.

While fertilizer rates did not cause any significant changes in head diameter (P>0.05), fertilizer types resulted in significantly different head diameters (P<0.01). The highest head

4. Conclusion

The highest yield and the lowest nitrate accumulation were obtained from the crisphead lettuces fertilized by ammonium sulphate (21 %). Results suggested that 150 kg N ha^{-1} rate given in small portions was enough to generate optimum yield.

5. References

Bayraktar K. (1970) Sebze Yetiştiriciliği. E.Ü. Ziraat Fak. Yay II. Cilt: 436-437. İzmir.

Becker J. (1956) Gemüsebau. Paul Parey Verlop. Wolf Garten.

Ikeda H. and Osawa T. (1984) Lettuce growth as influenced by N source and temperature of the nutrient solution. Proc. Of the 6th Int. Conf. Of Soilles Culture 273-284.

Karaçal I. And Türetken I. (1992b) Effect of application of increasing rates of ammonium sulphate fertiliseron yield and nutrient uptake in the lettuce plant. Yüzüncü Yıl Üniv. Agr. Fac. Journal 2:2, 95-106.

Kaptan H. and Almaca A. (1997) The determination need of the nitrogenous fertiliser of lettuce and head lettuce. Journal of the Faculty of Agriculture (1) 3.

Maynard D.N., Barker A.V., Minotti P.L. and Peck N.H. (1976) Nitrate accumulation in vegetables. Advances in Agronomy 28: 71-118.

Paulsen N: Johhansen A.S. and Sorensen J.N. (1995) Influences of growth conditions on the value of crisphead lettuce 4 quality changes during storage. Plant Foods for Human Nutrition 47:2, 157-162.

Sorensen J.N. Johhansen A.S. and Poulsen N. (1994) Marketable and nutritional quality as effected by nitrogen supply culivare and plant age. Plant Foods for Human Nutrition 46:1, 1-11.

Topçuoğlu B. and Yalçın S.R. (1997) Effects of different nitrogenous fertiliser applications on the yield and quality of greenhouse grown crisp lettuce. Journal of Faculty of Agriculture Akdeniz Üniv. (10):1 211-222.

Van Der Boon J., Steenhuizen J.W. and Steingröver E. (1990) Growth and nitrate concentration of lettuce as effected by total nitrogen and chloride concentration NH_4/NO_3 ratio and temperature of recirculating nutrient solution. Journal of Hort. Sci. 65 (3): 309-321.

Wright M.G. and Davidson K.C. (1964) Nitrate accumulation in crops and nitrate poisoning of animals. Adv. In Agronomy 16: 197-247.

Wehrmann J. (1963) Landw. Forsch. 16:30.

Chapter 12

Influence of the supply and uptake of nitrogen on the quality of potato tubers

D K L MacKerron and M W Young
Scottish Crop Research Institute, Invergowrie, Dundee, DD2 5DA, UK

Key words: Nitrogen

Abstract: Potato processors impose strict quality criteria in assessing the suitability of tubers for processing. One of the more critical criteria is the dry matter concentration, [DM], in the tubers at harvest time when, generally, [DM] should exceed 20%. The [DM] of developing tubers increases with time and has been shown (Jefferies et al., 1989) to be related to the passage of thermal time qualified by soil moisture deficit at harvest. There are also reports that potato crops grown with higher N supply produce tubers with lower [DM]. The work to be reported here was conducted in two series of experiments, over the four years 1984, 1985, 1990, 1991 and using two varieties, Maris Piper and Cara. The treatments were level of nitrogen supply ranging from 0 to 240 kg N / ha and two of the experiments included split applications of (80+80) kg N / ha, applied at planting and tuber initiation. There were four replications in all experiments and all were irrigated. Harvests were taken at 10, 8, 5, and 5 intervals during the growing seasons of the four years, respectively, when full sets of growth analysis measurements were made, including [DM], [N], and the partitioning of N within the plants between tuber, leaf, and stems. The results show that rate of N-supply modifies the development of tuber [DM] by delaying the maturity of the crop. The use of thermal time removed the most of the non-linearity observed in the course of development referred to normal time. The differences in [DM] between crops given differing rates of N supply were statistically significant. Relating tuber [DM] to N-uptake in the tuber provided a useful means of combining data from the several harvests and nitrogen treatments. In contrast, there was no useful relation between [DM] and the total N-uptake. Further analysis of the results is designed to quantify the delay in maturity by reference to N-uptake rather than N-supply and to provide generality across years.

1. Introduction

The concentration of dry matter, [DM] tends to increase progressively during the growth of the crop (MacKerron and Davies, 1986; Jefferies and MacKerron, 1987, 1989) but the pattern of increase differs between crops and seasons and varieties differ in the values of [DM] that they attain. In potato crops intended for processing, 20% is generally considered the minimum acceptable and, so, the varieties grown for processing are those that can regularly attain that level. Even so, attaining 20% [DM] can be problematic in some seasons. Jefferies et al. (1989) drew on data from sixteen crops of three varieties to show that variation in [DM] was best accounted for by a regression model that was a function of thermal time above a base of 0°C accumulated from plant emergence, and soil moisture deficit at harvest. When they validated their regression against independent data, there was a linear relation between observed and estimated [DM] that accounted for 79% of the variance.

The supply of N influences maturity in the growing potato crop (MacKerron and Davies, 1986) including the level of tuber [DM], although Burton (1989) only cited two studies that quantified the effect. Although Jenkins and Nelson (1992) reported significant effects of N-application on tuber [DM] that differed in two years, they did not analyse the relation between [DM] and the uptake of nitrogen.

The work reported here is designed to build on the earlier work of Jefferies et al. (1989) to include the effects of nitrogen supply on the progress on tuber [DM].

2. Materials and Methods

Data were available from four field experiments on potato, conducted at the Scottish Crop Research Institute over four years, in which nitrogen fertilizer was applied at differing rates. In 1984 and 1990, the rates of application were 0, 40, 80, 160, (80+80), and 240 kg N / ha. The (80+80) kg N / ha treatment involved a split application of half the N at planting and half after tuber initiation. In 1985 and 1991, the treatments were the same but the split application was omitted. The variety Maris Piper was grown in all four experiments. Those in 1990 and 1991 also included the variety Cara. All experiment had four replications and all had irrigation provided. There were four replications in all experiments and all were irrigated. Harvests were taken at 10, 8, 5, and 5 intervals during the growing seasons of the four years, respectively, when full sets of growth analysis measurements were made, including leaf, stem, and tuber dry weights, and tuber [DM]. Nitrogen concentration, [N], was analysed in the three tissue types from each harvest from which total nitrogen uptake, $N_{upTotal}$, and nitrogen uptake by the tubers, $N_{upTuber}$, were calculated. In 1984 and 1985, analysis of [N] was done by the Kjeldahl method. In 1990 and 1991 analysis was by Dumas combustion (Young et al., 1993). Soil moisture was measured by neutron probe in 1984 and 1985 from which soil moisture deficits, SMD, were calculated. In 1990 and 1991 SMD were calculated by the Penman-Monteith method. Elapsed time, t, and thermal time, Tt, were both reckoned from the date of 50% emergence, E50. Tt was calculated above 0°C with no allowance for optimum or maximum (units - Kelvin days, K days).
The data were examined by regression analysis using the statistical package GENSTAT.

3. Results and Discussion

The data from all four years were combined for analysis and the variates, t, Tt, SMD, $N_{upTotal}$, and $N_{upTuber}$ were regressed on [DM] (Table 1). The level of applied N fertilizer, $N_{applied}$, was included as a factor in the analysis. Time and thermal time were almost equivalent in their effect but thermal time accounted for a larger part of the variance and was used thereafter in the analyses.

Table 1 Percentage variance accounted for by regression models and level of statistical significance (P) of the model (first term) or of the change in the model.

Model	% Variance	P
SMD	5.1	-
$N_{upTotal}$	2.9	-
$N_{upTuber}$	28.4	<0.001
$N_{upTuber}$ + SMD	39.5	<0.001
t	76.5	<0.001
$t + t^2$	82.6	<0.001
Tt	79.2	<0.001
$Tt + Tt^2$	86.1	<0.001
$Tt + Tt^2$ + SMD	86.1	-
$Tt + Tt^2 + N_{upTotal}$	88.9	<0.001
$Tt + Tt^2 + N_{upTuber}$	87.4	<0.001
$Tt + Tt^2 + N_{applied}$	89.2	<0.001
Logistic Tt	86.5	<0.001
Logistic Tt + $N_{applied}$ (separate a)	90.5	<0.001
Logistic Tt + $N_{applied}$ (separate a, c)	90.8	<0.001
Logistic Tt + $N_{applied}$ (separate a, c, b, m)	91.3	<0.001

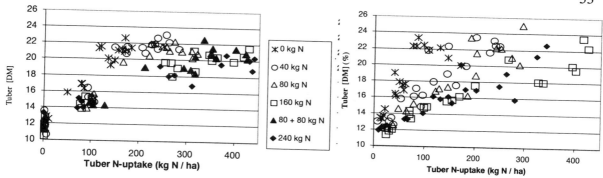

Figure 1. Relation between Tt and [DM] in variety Maris Piper in the two years, 1990 (left) and 1991 (right)

The relation between Tt and [DM] is not linear (Fig. 1) and so both quadratic and logistic relations were investigated. For clarity the data shown are treatment means and from two years separately.

The analyses of the effects of t and Tt on [DM] agree with the findings of Jefferies et al. (1989). However, SMD had little effect in the data examined here.

The specimen data in Fig. 1 shows [DM] being slightly higher at most harvest times, with lower levels of N-application while indicating the strong influence of thermal time. This effect was supported by the good relation between [DM] and $(Tt + Tt^2 + N_{applied})$ (Table 1) but that cannot be a useful relation for general application since it ignores the contribution of nitrogen from the organic matter in the soil.

Instead, the influence of N applied as fertilizer, on the performance of the crop should be considered through its effect on nitrogen taken up by the crop. On its own, $N_{upTuber}$ accounted for a small but statistically significant proportion of the variance, whereas $N_{upTotal}$ did not (Table 1). The specimen data in Fig. 2 illustrates the variation in the relation between $N_{upTuber}$ and [DM]. However, the two variates, $N_{upTuber}$ and $N_{upTotal}$, served almost equally well in improving the fit of the quadratic model of Tt on [DM] (Table 1). The quadratic equation has the form:

$$Y = aX^2 + bX + c$$

And the effect of adding a second, linear variate gives it the form:

$$Y = a + bX1 + c\,X1^2 + dX2$$

The estimated values for the fitted regression are given in Table 2.

This relation cannot be displayed directly, but Fig. 3 shows the distribution of the residuals with the fitted values for the polynomial regression of [DM] on $(Tt, Tt^2$ and $N_{upTuber})$.

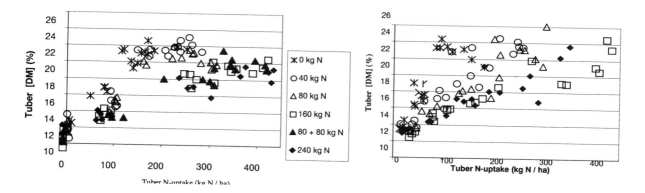

Figure 2. Relation between $N_{upTuber}$ and [DM] in variety Maris Piper in the two years, 1990 (left) and 1991 (right)

54

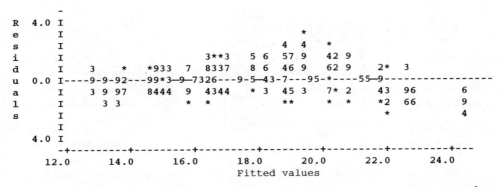

```
R   4.0 I        -
e       I
s       I                                                    *
i       I                                          4   4     *
d       I                          3**3     5 6   57 9     42 9
u   0.0 I    3      *    *933   7   8337   8 6   46 9     62 9       2*   3
a       I----9-9-92---99*3-9-7326---9-5-43-7---95-*----55-9----------------
l       I    3 9 97    8444  9  4344   * 3   45 3    7* 2     43  96      6
s       I    3 3           *   *          **    *  *    *2  66      9
        I                                        *                4
   4.0  I
        -+---------+---------+---------+---------+---------+---------+---
       12.0      14.0      16.0      18.0      20.0      22.0      24.0
                                   Fitted values
```

Figure 3. Distribution of residuals with fitted values from the polynomial of [DM] on Tt, Tt^2, and $N_{upTuber}$

The logistic equation, which provides a slightly better fit to all the data (Table 1), has the form:

$$Y = a + c / (1 + \exp(-b*(X =- m)))$$

The second logistic relation uses a separate constant term, a, for each level of fertilizer application. The final one provides completely separate curves for each level of application. However, as stated earlier, it is unsatisfactory to use fertilizer levels as an explanatory variable and the logistic curve suffers from the disadvantage that it is a function of only one variate and, at most, one factor. It cannot be used, therefore, to describe the relation with both thermal time and nitrogen uptake.

Table 2. Constants and regression coefficients (± Standard error) for the polynomial and logistic models]

[DM] = a + bTt + cTt^2 + $dN_{upTuber}$

Parameter	a	b	c	D
Value	7.0	0.017	-4.6×10^{-6}	-0.0045
SE	0.25	0.0005	2.3×10^{-7}	0.00037

[DM] = a + c / (1 + exp(-b * (X – m)))

Parameter	a	c	B	M
Value	10.4	11.6	0.0038	757.5
SE	0.64	0.77	0.00036	32.4

It appears that for predictive purposes, it will be necessary, at present, to use a polynomial relation such as that described (Table 2). Although the use of $N_{upTotal}$ in the regression accounts for slightly more of the variance than does $N_{upTuber}$, it may be preferable to use the latter because of the ease or accuracy of estimating it. There is a reasonable and progressive relation between thermal time and $N_{upTuber}$ but not with does $N_{upTotal}$. The reason for this is that the potato crop, given an ample supply of N, will take up luxury levels of N into the vacuoles of the leaves for later re-distribution within the plant (Millard and MacKerron, 1986).

This model has not yet been tested on independent data and it may be that, in order to provide general applicability, a completely different approach will need to be taken in the analysis of such experimental data.

Acknowledgements

The work reported here was funded by the Scottish Office Agriculture, Environment and Fisheries Department, and by the British Potato Council.

4. References

Burton WG (1989) The Potato. Longman Group (UK) Ltd., London, pp. 1-742.

Jefferies RA and MacKerron DKL (1987) Observations on the incidence of tuber growth cracking in relation to weather patterns. Potato Research 30: 613-623.

Jefferies RA and MacKerron DKL (1989) Radiation interception and growth of irrigated and droughted potato (*Solanum tuberosum*). Field Crops Research 22: 101-112.

Jefferies RA, Heilbronn TD and MacKerron DKL (1989) Estimating tuber dry matter concentration from accumulated thermal time and soil moisture. Potato Research 32: 411-417.

Jenkins PD and Nelson DG (1992) Aspects of nitrogen fertilizer rate on tuber dry-matter content of potato cv. Record. Potato Research 35: 127-132.

MacKerron DKL and Davies HV (1986) Markers for maturity and senescence in the potato crop. Potato Research 29: 427-436.

Millard P and MacKerron DKL (1986) The effects of nitrogen application on growth and nitrogen distribution within the potato canopy. Annals of Applied Biology 109: 427-437.

Chapter 13

Effect of nitrogen and iron sources applied to calcareous soil on the growth and elemental content of wheat

M.S.A. Dahdoh; B.I.M Moussa; A.H.El-Kadi & A.M.M. Abd El-Kariem

Desert Research Center, Mataria, Cairo, Egypt

Key words: Wheat, N source, Fe source, Calcareous soil

Abstract: A field experiment was carried out to estimate the effect of different sources of both N and Fe fertilizers and their interactions on the yield and elemental composition of wheat plants grown on calcareous soil. The N sources are NO_3^-, NH_4^+ and urea which were added to the soil using rates of 30, 60 and 120 kg N/acre. The Fe sources are $FeSO_4$ and FeEDDHA which were added at the rates of 0, 10 and 20 kg Fe/acre. Results show a significant increase of wheat yield when using the high rate of both N and Fe sources. The nitrate and EDDHA forms for N and Fe respectively show the superiority effect on grain yield compared to the other sources.

1. Introduction

Utilization of the newly reclaimed lands becomes a necessity for increasing agricultural production to meet the demands of the increasing population in Egypt. These reclaimed soils are generally of marginal potentiality poor in fertility, organic matter and nutrient supply thus their fertilization is a very necessary practice.

Nitrogen is a vital and essential element for all growing plants, where plants differ in their nitrogen requirements from the different sources.

The objective of this study is to discern the mutual effects between iron and nitrogen sources as soil application on wheat production and chemical constituents of wheat plants grown on calcareous soil.

2. Materials and Methods

A field experiment was conducted on wheat plants (Hordeum sativa, c.v. Sakha 8) grown in calcareous soil of South Sinai (Wadi Sudr) irrigated with saline water (7000 ppm) to evaluate the interaction effect between N and Fe sources. Some of the relevant physical and chemical characteristics of the investigated soil are presented in Table (1).

The experiment included 45 treatments with four replicates, all treatments were arranged in factorial complete block design. The area of each plot was (3 x 3.5m). Three N sources namely; ammonium nitrate, ammonium sulphate and urea with three rates for each source 30, 60 and 120 kg N/acre were applied to the soil in two equal doses; before cultivation and after 45 days of cultivation. Two Fe sources, $FeSO_4$ and FeEDDHA, with three rates for each source 0, 10 and 20 kg Fe/acre were added to the soil in one dose together with the second dose of N. Phosphorus and K fertilizers were added at the rates of 30 kg P_2O_5/acre and 50 kg K_2O/acre before cultivation and after 45 days of cultivation for P and K, respectively.

Table (1) Particle size distribution and some chemical properties of the investigated soil.

Particle size distribution (%)				Chemical properties												
Sand %	Silt %	Clay %	Texture	PH	EC	O.M.	CEC	CaCO₃	Available elements (mg kg⁻¹)							
					dS/m	(%)	meq/100 g soil	(%)	N	P	K	Ca	Mg	Fe	Mn	Zn	Cu
41	37.8	22.2	S.C.L	8.1	3.18	0.78	13	52	20	12	43	16	13	3.12	1.69	0.43	0.34

At the end of the experiment the yield of wheat grains was recorded and statistically evaluated. Samples of wheat grains were ground and wet digested then subjected to thedetermination of N, P, K, Ca and Mg according to the methods described by Van Schouwenburg (1968). Also, micro-nutrients Fe, Mn, Zn and Cu were determined using Atomic Absorption Spectrophotometer.

3. Results and Discussion

Data presented in Table (2), show that application of N from the different sources increased significantly the yield of wheat grains. The more effective source was ammonium nitrate, while the least was ammonium sulphate and urea lies between those extremes. Also, increasing N rates led to proportional increases in yield of grains for all nitrogen sources. The least effect associated with ammonium sulphate addition could be related to its possible volatilization if compared with the other N sources. Undoubtedly, supplementary application of all N sources that enhanced the production of leaves may enhanced the activity of photosynthesis and, in turn, increases the production of wheat grains (Anderson and Parkpian ,1988). The yield of wheat grains under addition of Fe EDDHA is generally higher than FeSO₄ under any N treatments, however, the superior effect corresponds to NO₃ addition. This may be attributed to its chelation that makes it available.EDDHA is one of the few chelates known to maintain iron available in very alkaline soil as occured in Egypt (Chen and Barack, 1982).

Effect of N rates and sources on plant elemental content:

Data in Table (2) show that plant N concentrations increased gradually and its uptake increased significantly with increasing N

addition. The increase was apparently clear with the addition of nitrate form followed by urea and sulphate forms. This may be attributed to the possible evolution of NH₃ form due to the the reaction of ammonium sulphate with CaCO₃ (Fenn and Miyamoto,1981).

Results also show that increasing N rate of all the studied N sources increased the plant P, K, Ca, Fe, Mn, Zn and Cu concentration, while Mg increased only under the addition of nitrate form but decreased under addition of the other two sources.

The uptake of all nutrients increased significantly with increasing N rate for all N sources. The increase of these nutrients in plant may be attributed to the competition between NO₃ and other anions on anion exchange sites of plant roots (Farid and Stroehlein 1990). The increase with N rate even under all N sources may be attributed to N sources induced acidity on the availability of native elements especially in root zone which increases most of the studied elements (all the micro-nutrients, P and Mg) as described by Schwamberger and Sims (1991). Also, the increase of K may be attributed to substitution of K by NH₄⁺ or due to the higher growth of wheat by increasing addition of N to soil for all the studied sources (Chen and Mackenzie, 1992).

Effect of Fe on plant elements content:

The application of Fe at with any of the studied N and Fe sources increased the concentration and significantly the uptake of N over the control (Dahdoh et al., 1994). Inorganic iron supplied as ferrous sulphate is absorbed several times greater as iron supplied in chelate form by the root of iron deficient plant, but in fact less than 10% of the absorbed quantity is taken by the root cells (Foder and Cseh, 1993).

Generally, the more effective source for Fe and N contents is the chelate form (Fe EDDHA). In contrast, addition of both Fe sources decreased the concentration of P, Ca, Mg, Mn, Zn and Cu while increased significantly their uptake due to the increase of the grains yield. The chelate form show the superior effect in decreasing the concentration of the previous nutrients, except for Ca and Mg where the sulphate form show the least obtained values, however, the decrease of Ca is very slight between the two sources and control (Moussa et al.1996). The decrease of P may be due to the precipitation of P in Fe-phosphate form, while increasing N and K could be attributed to the enhancement effect of Fe on plant growth through its effect on chlorophyll and being as carrier for elements through the uptake process. The decrease of Mn, Zn and Cu is in harmony with many investigators who found an antagonistic relationship between Fe & Mn (Agarwala et al. 1988), Fe & Zn (Anderson & Parkpian, 1988) and Fe & Cu (Anderson and Parkpian, 1988) .

Also, decreasing Ca and Mg by Fe may be due to the replacement of Ca and Mg by Fe on the soil particles and the reaction of these two elements with the soil components which can reduce their availability to the grown plants by precipitation or chelation. As it is known that absorption of ions from solution may be affected by other ions these decreases or enhance access to the sites of absorption, to competition at the site of absorption and interaction with metabolic controls of absorption. Each of these three types of interactions can occur within the roots.

4. References

Agarwalla SC; Catterjee C; Cupta S & Nautiyal M (1988) Iron-Manganese interaction in Cauliflo was J. Plant Nutr., 11 : 1005-1014.

Anderson WB & Parkpian P (1988) Effect of soil applied iron by-product on micronutrient concentrations in sorghum cultivars. J plant Nutr. 11, 6-11, 1333-1345.

Chen JS & Mackenize AF (1992) Fixed ammonium and potassium as affected by added nitrogen and potassium in three Quebec soils. Commun. Soil Sci. Plant. Anal, 23 (11&12), 1145-1159.

Chen Y & Barack P (1982) Iron nutrition of plants in calcareous soils. Adv. in Agron, 35 : 217-240.

Dahdoh, MSA; El-Dosouky M and El-Mashhady HH (1994) Effect of phosphorus-iron interaction under variable moisture levels on barley grown in calcareous soil. Egypt J Appl, 9, 6: 328-344.

Farid AA & Strochlein JL (1990) Response of tomato to rates and methods of placement of urea and ammonium sulphate as affected by Dwell, A nitrification inhibitor. Commun Soil Sci. Plant Anal, 21, (17 & 18), 2163-2171

Fenn LB & Miyamoto S(1981) Ammonia loss and associated reactions of urea in calcareous soils. Soil Sci Soc Amer J 45 : 537-540.

Foder F & Cseh E (1993) Effect of different nitrogen forms and iron chelates on the development of stuging nettle. J. Plant Mutr. 16, 11: 2239-2253.

Moussa BIM, Dahdoh MSA & Shehata HM (1996) Interaction effect of some micronurtients on yield, elemental composition and oil content of peanut. Commun Soil Sci, Plant Anal. 27 (5-8), 1995-2004.

Schwamberger E.C. & Sims, J.I. (1991). Effects of soil, pH, nitrogen source, phosphorus and molybdenum on early growth and mineral nutrition of burley tobacco. Commun. Soil. Sci. Plant Anal. 22 (7&8), 641-657.

Van Shouwenburg J Ch (1968) International Report of Soil and Plant Analysis Lab. of Soil and Fertilizers Agric., Univ. Wageningen, Netherlands.

Table 2. Effect of N and Fe sources applied to calcareous soil on the yield and nutrients uptake of wheat

Fe treatment		N treatment		Yield of grains ton/acre	Macronutrients concentration kg / acre						Micronutrients concentration g/acre		
Source	Rate kg/acre	source	Rate kg/acre		N	P	K	Ca	Mg	Fe	Mn	Zn	Cu
Control	0	NO_3^-	30	0.90	8.80	2.70	9.90	3.51	2.43	313	267	223	10.80
			60	1.30	16.2	5.33	17.29	5.59	5.98	455	468	411	22.10
			120	1.80	35.64	8.28	27.00	10.08	9.90	684	691	601	38.00
		NH_4^+	30	0.83	7.06	2.41	10.13	2.74	2.82	184	237	203	9.96
			60	1.22	14.64	3.54	14.60	4.39	3.54	311	390	329	15.86
			120	1.65	27.72	7.43	20.63	6.77	2.81	465	566	454	26.40
		Urea	30	0.95	9.60	3.23	11.40	2.66	3.04	219	282	257	11.40
			60	1.43	17.59	5.29	18.88	4.58	3.58	356	473	396	18.59
			120	1.70	19.88	8.16	23.29	6.46	3.06	490	581	520	25.50
Sulphate form	10	NO_3^-	30	0.95	9.80	2.28	11.70	2.76	2.38	504	227	208	10.45
			60	1.41	19.18	4.23	21.71	5.08	5.50	770	388	357	18.33
			120	1.98	39.80	6.53	36.23	8.91	9.90	1146	590	572	31.68
		NH_4^+	30	0.90	8.82	1.80	11.80	2.70	2.34	374	283	198	8.10
			60	1.40	17.92	3.64	19.60	4.76	3.36	630	466	318	15.40
			120	1.80	30.60	6.30	25.70	7.02	2.70	846	628	430	21.60
		Urea	30	0.85	9.27	1.96	11.20	2.55	2.55	365	264	179	8.50
			60	1.36	17.68	3.94	22.17	4.76	2.86	608	456	311	14.96
			120	1.86	28.83	6.14	36.10	5.39	3.35	885	640	435	24.18
	20	NO_3^-	30	0.96	10.75	1.44	9.89	2.69	1.73	581	215	183	8.64
			60	1.62	27.52	3.73	21.71	5.02	5.35	1004	403	363	17.82
			120	2.11	44.52	7.60	34.60	7.39	9.50	1393	608	547	29.54
		NH_4^+	30	0.86	8.6	1.72	9.46	2.32	1.98	363	198	172	6.88
			60	1.50	21.15	3.30	18.30	4.65	3.15	788	363	318	15.00
			120	1.72	32.34	4.31	22.53	5.85	2.41	963	497	375	18.92
		Urea	30	1.01	12.10	2.22	11.01	3.54	2.73	486	293	190	9.09
			60	1.54	23.25	3.70	19.25	6.01	3.85	755	477	342	16.94
			120	1.92	36.29	4.99	25.34	8.64	3.26	1068	618	445	23.04
EDDHA form	10	NO_3^-	30	1.11	14.32	2.55	12.43	3.44	3.44	588	266	243	7.77
			60	1.78	26.90	4.98	24.92	7.48	6.76	995	479	418	17.80
			120	2.18	49.49	6.54	36.19	10.46	9.59	1275	621	567	23.98
		NH_4^+	30	0.98	11.66	2.16	10.88	2.74	3.14	478	217	196	4.90
			60	1.52	25.08	4.10	19.46	5.62	4.86	763	395	318	12.16
			120	1.98	40.39	5.94	25.94	8.51	3.96	1024	556	436	17.82
		Urea	30	1.01	12.52	2.12	12.22	2.83	2.93	477	220	202	6.06
			60	1.67	24.55	4.34	21.71	6.01	4.51	852	406	351	13.36
			120	2.01	42.81	5.63	26.53	8.84	3.62	1043	551	428	20.10
	20	NO_3^-	30	1.21	17.10	2.42	7.62	3.39	3.39	845	266	218	7.26
			60	1.94	30.30	5.04	23.67	7.57	7.57	1387	481	431	13.58
			120	2.38	57.40	6.66	35.46	10.95	8.09	1785	619	569	21.42
		NH_4^+	30	1.09	14.06	1.96	11.01	3.05	2.73	655	220	205	5.45
			60	1.66	29.40	3.98	18.43	5.64	2.99	1054	398	329	13.28
			120	2.16	45.80	5.62	26.35	8.21	3.18	1469	551	451	17.28
		Urea	30	1.12	14.34	2.24	11.54	3.92	3.36	697	222	213	5.60
			60	1.82	26.94	4.55	21.29	7.46	4.55	1150	402	364	12.74
			120	2.19	47.74	5.69	27.16	10.51	3.29	1456	453	462	19.71

Chapter 14

Analysis of economic efficiency of top dressing by nitrogen of soil sown by wheat

B. Petrac[1], I. Juri[2], M. Crnjac[1]

[1] Faculty of Economics, University J. J. Strossmayer in Osijek, Gajev trg 7, 31000 Osijek, Croatia
[2] Faculty of Agriculture, University J. J. Strossmayer in Osijek, Trg Sv. Trojstva 3, 31000 Osijek, Croatia

Key words: economic efficiency, wheat, N fertilization

Abstract: Wheat is an arable crop that response to the application of nitrogen. It is necessary to analyse the economic efficiency of applied nitrogen related to rates, forms and application times Results showed that wheat yields increase with increased rate of soil applied nitrogen, over the range tested. Economic efficiency of nitrogen top dressing for all observed strategies including traditional and visual method decreases with increasing application of soil nitrogen.

1. Introduction

The traditional way of top dressing wheat with nitrogen during the growing season often leads to the wrong levels being given; too little or too much. Over fertilization may increase leaching of NO_3^- nitrogen, with undesirable costs and causing environment pollution. The problem of correct fertilization is important as nitrogen is the most mobile nutritive element. It occurs in several forms in soil with losses happening during the several transformations.

Wheat is an arable crop that responds to the application of nitrogen, which is a key element of growth and development, and so is an interesting crop for investigators and practitioners. In practical agriculture there is a continual quest for the correct rate, time, and form for the application of nitrogen. Several methods may be used to determine the best rate for top dressing with nitrogen. These include: visual impression of nutrition, leaf nitrogen concentration (foliar method), and, lately, a method based on the estimation of soil nitrogen concentration, the Nmin method. With this method, nitrogen top dressings are based on the mineral nitrogen concentrations in the soil at several depths. The range of depths depends on the development of the root system during the tillering and forking phases. The concentrations of NO_3^- and NH_4^+ are determined at 0-30cm and 30-60 cm depths during the tillering phase and also at 60-90 cm during the forking phase.

Because of the cost of nitrogen applied to wheat and to other arable crops in the Republic of Croatia, it is necessary to analyse the economic efficiency of the application of nitrogen at different rates and in different forms as well as their technological efficiency. Using such an analysis we established that fertilization with nitrogen is profitable.

2. Materials and Methods

The investigations were carried out in a field plot experiment on pseudogley dominant soil type of the Republic of Croatia according to split-plot design. The main factor (A) is a nitrogen rate: 30, 60,90 and 120 kg per hectar whereas factor (B) is a criterion in a wheat top dressing (strategy):

    1. visual impression of wheat nutrition

2. on the basis of a plant nitrogen foliar analysis

3. on the basis of soil concentration of NO_3

4. on the basis of soil concentration of $NO_3+30\%$ NH_4

5. on the basis of soil concentration of $NO_3+60\%$ NH_4

6. on the basis of soil concentration of $NO_3+100\%$ NH_4

The top dressing was carried out two times in vegetation period - during tillering and forking phase (ZADOKS GS 25 and GS 30) on the basis of the preceding soil mineral nitrogen concentration according to N-min method (Wermann and Scharpf, 1986) as well as plant nitrogen concentration (foliar analysis and impression of wheat nutrition by nitrogen).

Furthermore, top dressing was carried out with NH_4, NO_3 + $CaCO_3$ fertilizer whereas UREA $(CO(NH_2)_2)$ was used in a presowing period.

From the agronomic aspect the aim of this paper is to determine differences in methods (strategies) of wheat top dressing at different nitrogen rates till sowing. Economic anaylsis of efficiency of nitrogen top dressing of the soil sown by wheat was carried out from the economic reasons without details regarding agronomic moments of different strategies application.

The main investigation trend in this paper is economic efficiency of nitrogen application rates and ways which was analyzed from the two aspects:

Nitrogen top dressing efficiency means that wheat top dressing efficiency was calculated for each strategy regarding yields as a function of applied nitrogen rate per hectare, $(P=\eta N+a)$. Coefficient η, **a** of linear regression $(P=\eta N+a)$, will be computed after a method of the least squares by minimizing the function $F(a,\eta)=(\eta N_i+a-P_i)^2$, where N_i is the highest nitrogen level of I-te strategy, P_i return of I-te strategy. Parameter η represents top dressing efficiency.

$$\eta= \frac{\Delta P}{\Delta N}$$

P= yield increase

N= additional nitrogen rate

Economic efficiency of nitrogen top dressing measured by monetary value of 1 kg of obtained wheat compared to monetary value of 1 kg soil applied nitrogen can be computed as follows:

$$\eta= \frac{\text{income per ha (yield of t ha}^{-1} \text{ x 1 t price)}}{\text{costs of N top dressing per ha}}$$

Sampling was carried out twice: first one at full tillering (22nd March 1996) and second one at the beginning of wheat forking (29th April 1996). The samples were taken by a tube at 0-30 cm and 30-60 cm at wheat tillering and at 0-30 cm, 30-60 cm and 60-90 cm prior to wheat forking initiation.

3. Results and Discussion

On the basis of soil nitrogen content determined by the analysis at the first sampling (nitrate and ammonia) a wheat top dressing by nitrogen was carried out as it is presented in Table 1.

Table1. Wheat fertilization (first top dressing) according to soil nitrogen content

Soil N Appl. N kgha⁻¹	Strategy of Nitrogen kg ha⁻¹						
	1.	2	3.	4.	5.	6.	x
30	40.5	27.0	54.0	54.0	38.8	20.0	39.0
60	54.0	40.5	54.0	54.0	54.0	45.4	50.0
90	54.0	40.5	54.0	54.0	54.0	54.0	51.7
120	54.0	40.5	54.0	54.0	54.0	54.0	51.7
X	50.6	37.5	54.0	54.0	50.2	43.3	-

On the basis of soil nitrogen content at tillering phase (first top dressing) nitrogen rate for wheat top dressing amounted from 27.0 to 54.0 kg per hectare. The lowest rate was consumed on the basis of criterion "visual impression", 2nd strategy 37.5 kg per hectare. On the basis of soil nitrogen content at second sampling (nitrate and ammonia) the second wheat top dressing by nitrogen was carried out and presented in Table 2.

NO_3 content somewhat increased at depth of 0-30 cm whereas NH_4+ content decreased at depth of 30-60 cm on the average with all levels of anticipated-targeted nitrogen application in the second top dressing (forking phase). According to soil nitrogen content nitrogen rate consumed in the second top dressing amounted from 37.0 to 62.0 kg per hectare. The highest

Table 2. Wheat fertilization (second top dressing) according to soil nitrogen content

Soil N Appl. N kgha⁻¹	Strategy of Nitrogen kg ha⁻¹						
	1.	2.	3.	4.	5.	6.	x
30	40	53	26	25	23	21	31
60	56	68	45	42	38	33	42
90	54	68	43	40	30	38	15
120	54	68	60	59	58	56	59
x	51	62	46	44	37	37	-

consumed nitrogen rate was in 2nd strategy (visual impression) and the lowest one was in 4th and 5th strategy.

Table 3. Total nitrogen consumption in wheat top dressing (two top dressings) in kg ha⁻¹

Soil N Appl. N kgha⁻¹	Strategy of Nitrogen kg ha⁻¹						
	1.	2.	3.	4.	5.	6.	x
30	80	80	80	80	62	41	70
60	110	108	99	96	92	78	97
90	108	108	97	94	88	92	98
120	108	108	113	112	110	112	111
X	101	101	97	95	88	81	-

The highest nitrogen rate in total wheat top dressing was consumed at standart top dressing (1st strategy) and per visual impression (2nd strategy) 101 kg per hectare on the average whereas the lowest one was consumed at 6th strategy (according to y-total content of nitrogen forms NO_3^- and 100 % of NH_4^+) amounting 81 kg per hectare (Table 3).

The given schedule presents wheat top dressing according to N-min method with achieved high yields at fertilization treatments compared to yield achieved in the no-fertilization treatment (only 3.2 t ha⁻¹ -soil fertility potential). We assume that all yield above this mentioned one can be contributed to NPK fertilization efficiency. From the aspect of nitrogen influence on wheat yield it can be concluded that yield increases as total applied nitrogen increases. Thus average yield at 130 kg N ha⁻¹ in targeted step amounted 5.8 t ha⁻¹ whereas it was 6.62 t ha⁻¹ at 220 kg N ha⁻¹. Economic efficiency of nitrogen fertilization (strategy) was measured by monetary value of 1 kg of

achieved wheat compared to monetary value of 1 kg of soil applied nitrogen.

Table 4. Wheat yields top dressed according to N-min method in 1995/96 Feri~anci

Soil N Appl. N Kgha⁻¹	Strategy of Nitrogen kg ha⁻¹						
	1.	2.	3.	4.	5.	6.	x
No fert.							32.0
30	59.2	58.7	60.8	58.9	56.1	54.3	58.0
60	63.0	61.0	62.6	61.7	60.1	62.1	61.7
90	65.7	65.4	63.4	62.9	62.9	65.0	64.2
120	67.1	66.5	67.3	65.9	62.9	67.8	66.2
Effic. N top dressing	63.7	62.9	63.5	62.4	60.5	62.3	-

When comparing the lowest nitrogen rates it can be noticed that 3rd and 4th investigated strategies had the same efficiency level as traditional and visual one (about 15 USA $ of yield per 1 USA $ of investment in N). Contrary to this, economic efficiency of minimal 5th strategy is somewhat higher (17 USA $ of yield per 1 USA $ of investment in N) whereas the best economic efficiency was obtained by 6th strategy (approximately 20 USA $ of yield per 1 USA $ of investment in N).

According to the above mentioned, profitability of soil nitrogen enrichment investment is obvious. But it must be pointed out that apart from analysis of technological efficiency of investment we should carry out economic one. It is possible to achieve economic effects higher by 50 % compared to traditional or visual strategy by using qualitative 6th strategy of soil nitrogen enrichment (50 kg ha⁻¹ prior to sowing and 41 kg ha⁻¹ at top dressing in NH_4^+ form).

4. Conclusion

Economic efficiency analysis of N-min method application in wheat top dressings showed that:
1. Wheat yields increase with increased rate of soil-applied nitrogen
2. Yield increase trend is more expressive only with 6th strategy (9.5 kg of wheat per 1 kg of soil applied nitrogen) compared to traditional (8.8 kg) and visual methods (9.3 kg)
3. Economic efficiency of nitrogen top dressing (η) for the all observed strategies including traditional and visual method decreases with

increased soil nitrogen application starting from minimal rates (130 kg of total N, Table 1).

4. References

Chang, F. R.(1994): Optimal growth and recursive utility: phase, diagram analysis. JOTA 80, *425-439*

Jukic D., Scitovski, R.(1997): Existence of optimal solution for exponential model feast squares. Journal of Computational and Applied Mathematics 78, 311-328.

Petraa B. (1997): Economic aspects of Wheat processing in Eastern Croatia. First Croatian Congress of Cereal Technologists with International Participation, "Bra{no-kruh '97", Opatija, 1997, 77-83.

Rakitowski, D.A (1985): Handbook of Nonlinear Regresion Models. M. Deker, New York, 1990. Scitovski, S. Kosanovi}, rate of change in economics research, Economics analysis XIX, 65-75.

Werman, J. Sharpf. H., J. (1986), The N-min method an aid to integrating vrious objectives of nitrogen fertilization. Z. Pflanzenernahr, Bodenk 147, 128-440.

Chapter 15

Application of N-min method at top-dressing of winter wheat

I. Juric[1], I. Zugec[1], B. Petrac[1], M. Crnac[2], B. Stipesevic[1]
[1] Faculty of Agriculture Osijek, University of J.J Strossmayer Osijek, Croatia
[2] Faculty of Economy Osijek, University of J.J Strossmayer Osijek, Croatia

Key words: wheat, N fertilization, N-min method

Abstract: Uniform N nutrition of wheat may lead to over or under fertilization which can cause leaching or environmental pollution. A field trial was performed to study the effect of preseeding N fertilization for winter wheat related to mineral N ($NO_3 + NH_4$) in the soil and the amount of N for top-dressing . Results showed that with the use of N-min method, the quality of N fertilizer at top dressing has decreased regarding the visual impression and foliar analysis.

1. Introduction

Traditional uniform wheat fertilization by nitrogen during vegetation usually leads to over or under-fertilization. Overfertilization can cause increased NO_3 leaching and unnecessary expenditure on N, while underfertilization can cause considerable yield decrease. Related to N nutrition of wheat, the efficient fertilization recommendation method should be considered. The main goal must be to provide enough nitrogen for the wheat physiological processes. Also, leaching and environmental pollution must be avoided. This can be achieved by monitoring the changes of mineral nitrogen in the soil at different depths, according to root-system development. All of this must be done during vegetation and with N-min method.

2. Materials and Methods

A field trial with winter wheat was carried out in 1996-1998 at Feričanci, Osijek. The soil was classified as a pseugogley, with a pH (H_2O) of 5.2, and P_2O_5 and K_2O contents of , 8.5 and 14.4 mg $100g^{-1}$ of soil, respectively. The experiment was designed in split plots, N prior to sowing (0, 30, 60, 90, 120 kg ha^{-1}) was considered as main plots and criteria of top-dressing as sub-plots. Criteria were as follows: 1. visual impression of winter wheat status, 2. according to foliar analysis, 3. NO_3-N content, 4. NO_3-N + 30% NH_4-N, 5. NO_3-N + 60% NH_4-N, 6. NO_3-N + 100% NH_4-N.

The amount of mineral nitrogen ($NO_3 + NH_4$) at tillering (Zadoks 25) was determined at the depths of 0-30 and 30-60 cm, and the first top-dressing was realized according to these quantities; and at forking (Zadoks 30), the amount of mineral nitrogen ($NO_3 + NH_4$) was additionally determined at the layer of 60-90 cm, and the second top-dressing was applied accordingly.

The aim of this study was to evaluate the effect of preseeding nitrogen fertilization for winter wheat related to mineral nitrogen ($NO_3 + NH_4$) in the soil and the amount of N for winter wheat top-dressings.

3. Results and Discussion

The amount of mineral nitrogen in the soil is changing continuously. Differences are especially significant between the years. At tillering stage a small amount of NO_3-N was found in the topsoil (0-30 cm). The amount of NH_4-N was higher. During these two years the ratio between NO_3-N and NH_4-N was 1:2.5 at the depth 0-30 cm, and 1:1.3 at the depth 30-60 cm. But, this ratio can change according to NO_3-N.

The amount of NO_3-N and NH_4-N at both depths (0-30 and 30-60 cm) did not dependent on preseeding nitrogen quantity.

Later, prior to forking, the mineral N has increased in ratio 1:0.22:0.21 for the depths 0-30, 30-60 and 60-90 cm respectively. In both years at the topsoil (0-30 cm) NH_4-N was higher than NO_3-N, and the ratio was 1:0.7. At the next depth (30-60 cm) there were no differences, but at the lowest depth (60-90 cm) the ratio was 1:0.42.

At the forking phenophase, mineral nitrogen at different depths of soil did not dependent on the amount of preseeding nitrogen fertilization. One of differences in the mineral nitrogen of the soil during the years was the practice of using different amounts of fertilizers at top-dressing. According to N-min method it is possible to reduce the quantity of nitrogen at top-dressing. Even the smallest quantity of nitrogen was used when total mineral nitrogen (NO_3-N + 100% NH_4-N) was considered. But, relatively more nitrogen was used according to the visual impressions and foliar analyses.

The wheat yield was more dependent upon climatic conditions than on the quantity of nitrogen and criteria of top-dressing.

Table 1. Mineral nitrogen (NO_3 + NH_4) at the depths of 0-30 and 30-60 cm at tillering phenophase (Zadoks 25) in 1997 year

Preseeding fertilization (kg ha^{-1} N)	0-30 cm		30-60 cm	
	NO_3-N	NH_4-N	NO_3-N	NH_4-N
	(mg 100 g^{-1} of dry soil)		(mg 100 g^{-1} of dry soil)	
0	0.23	0.84	0.35	0.78
30	0.27	0.61	0.22	0.75
60	0.35	0.50	0.29	0.61
90	0.16	0.44	0.26	0.84
120	0.24	0.90	0.25	0.75
x	0.25	0.65	0.27	0.75

Table 2. Mineral nitrogen (NO_3 + NH_4) at the depths of 0-30 and 30-60 cm at forking phenophase (Zadoks 30) in 1997 year

Preseeding fertilization (kg ha^{-1} N)	0-30 cm		30-60 cm		60-90 cm	
	NO_3-N	NH_4-N	NO_3-N	NH_4-N	NO_3-N	NH_4-N
	(mg 100 g^{-1} of dry soil)		(mg 100 g^{-1} of dry soil)		(mg 100 g^{-1} of dry soil)	
0	0.53	1.38	0.18	0.24	0.11	0.64
30	1.07	1.76	0.32	0.37	0.02	0.50
60	1.18	1.81	0.66	0.55	0.04	1.23
90	1.17	2.38	0.51	0.30	0.02	0.66
120	2.09	2.35	0.31	0.43	0.14	0.62
X	1.21	1.94	0.40	0.38	0.07	0.73

Table 3. Effect of different methods of N top-dressing on amount of N in kg ha^{-1}, 1997

Preseeding fertilization (kg ha^{-1} N)	Top-dressing criteria							According to N-min method (3-6)
	1	2	3	4	5	6	x	
0	121.5	108.0	62.1	45.9	24.3	67.0	64.8	49.8
30	121.5	108.0	86.4	86.4	78.3	54.0	89.1	76.2
60	121.5	108.0	99.9	102.8	102.6	89.1	104.0	98.6
90	121.5	108.0	99.9	108.0	108.0	108.0	108.0	106.0
120	121.5	108.0	108.0	108.0	108.0	108.0	110.2	108.0
X	121.5	108.0	91.3	90.2	84.2	77.2	95.4	87.7

Table 4. Wheat grain yield according to criteria of N top-dressing in 1997, t ha^{-1}

Preseeding fertilization (kg ha^{-1}N)	Top-dressing criteria (B)						x	According to N-min method (3-6)
	1	2	3	4	5	6		
0	2.65	3.30	3.36	3.57	3.20	3.20	3.21	3.38
30	3.42	3.71	3.75	3.70	3.47	3.62	3.61	3.63
60	3.98	4.23	3.79	3.84	3.83	3.69	3.98	3.78
90	3.88	4.14	3.80	3.95	3.76	3.94	3.91	3.86
120	3.90	4.00	4.27	4.62	4.62	3.76	4.11	4.31
X	3.57	3.88	3.78	3.94	3.78	3.64	3.76	3.79

LSD A $_{(P=5\%)}$ = 0.64
$_{(P=1\%)}$ = 0.80

Table 5. Mineral nitrogen (NO_3 + NH_4) at the depths of 0-30 and 30-60 cm at tillering phenophase (Zadoks 25) in 1998. year

Preseeding fertilization (kg ha^{-1} N)	0-30 cm		30-60 cm	
	NO_3-N (mg 100 g^{-1} of dry soil)	NH_4-N	NO_3-N (mg 100 g^{-1} of dry soil)	NH_4-N
0	0.12	0.27	0.49	0.31
30	0.14	0.29	0.54	0.26
60	0.12	0.32	0.72	0.41
90	0.12	0.27	0.66	0.38
120	0.14	0.39	0.60	0.49
x	0.13	0.31	0.60	0.37

Table 6. Mineral nitrogen (NO_3 + NH_4) at the depths of 0-30 and 30-60 cm at forking phenophase (Zadoks 30) in 1998. year

Preseeding fertilization (kg ha^{-1} N)	0-30 cm		30-60 cm		60-90 cm	
	NO_3-N (mg 100 g^{-1} of dry soil)	NH_4-N	NO_3-N (mg 100 g^{-1} of dry soil)	NH_4-N	NO_3-N (mg 100 g^{-1} of dry soil)	NH_4-N
0	2.30	2.75	0.63	0.77	0.44	0.50
30	2.52	2.96	0.69	0.84	0.50	0.61
60	2.43	2.83	0.66	0.80	0.40	0.48
90	2.94	3.41	0.43	0.52	0.41	0.50
120	2.80	3.27	0.54	0.65	0.59	0.72
x	2.60	3.04	0.59	0.71	0.47	0.56

Table 7. Effect of different methods of N top-dressing on the amount of N in kg ha^{-1}, 1998

Preseeding fertilization (kg ha^{-1} N)	Top-dressing criteria (B)						x	According to N-min method (3-6)
	1	2	3	4	5	6		
0	94.5	94.5	67.5	59.4	55.1	43.3	69.0	56.3
30	94.5	94.5	91.8	89.1	86.4	83.7	90.0	87.7
60	94.5	94.5	102.6	102.6	99.9	97.2	98.5	100.6
90	94.5	94.5	108.0	108.0	108.0	108.0	103.5	108.0
120	94.5	94.5	108.0	108.0	108.0	108.0	103.5	108.0
X	94.5	94.5	95.5	93.4	91.5	88.0	92.9	92.1

Table 8. Wheat grain yield according to criteria of N top-dressing in 1998, t/ha

Preseeding appl. (A) N kg ha⁻¹	Top-dressing criteria (B)							According to N-min method (3-6)
	1	2	3	4	5	6	x	
0	6.16	6.05	6.03	5.86	5.48	5.79	5.89	5.79
30	6.16	6.21	6.28	5.91	6.29	5.92	6.13	6.10
60	6.42	6.57	6.50	6.50	6.28	6.26	6.42	6.38
90	6.74	6.74	6.57	6.43	6.13	6.78	6.49	6.48
120	6.47	6.47	6.47	6.88	6.71	6.61	6.66	6.67
X	6.41	6.41	6.37	6.32	6.18	6.27	6.32	6.28

LSD	A	B	AB
P=5%	0.73	0.15	0.45
P=1%	1.03	0.20	0.34

4. Conclusion

The amount of mineral N in the soil doesn't strongly depend on preseeding fertilization with this element.

Prior to forking of wheat, mineral N was increased at the top soil in the ratio 1:0.22:0.21 (0-30, 30-60 and 60-90 cm).

With the application of N-min method, the quantity of N fertilizer in top-dressing has decreased concerning the visual impression and foliar analysis.

The smallest amount of N in top-dressing was used concerning the total quantity of NO_3-N and NH_4-N in the soil.

Between the years, the level of wheat yield was dependent upon climatic conditions.

5. References

Janzen, H.H. (1990). Deposition of nitrogen into rhizosphere by wheat roots. Soil Biol. Biochem. 22:1155-1160

Knapp, W.R. and Knapp, J.S. (1978) Response of winter wheat to date of planting and fall fertilization. Agron. J. 70: 1048-1053

Maxwell, T.M., Kissel, D.E., Wagger, M.G., Whitney, D.A., Carbera, M.L. and Moser, H.C. (1984) Optimum spacing of preplant bonds N and P fertilizer for winter wheat. Agron. J. 76: 234-247

Mitchel, R.D.J, Hassink, J. and Webb, J. (1998) Nutrient limitations, utilization and transformation, ESA, Short communications, Volume II: 7-8

Mitchel, R.D.J, Webb, J. and Harrison, R. (1998) Residual effects of crop residues on soil mineral N and N leaching. ESA, Short communications, Volume II: 9-10

Mitchell, R.D.I., Hassink, I and Webb (1998) Nutrient limitations and transformations. 5. Congress ESA - Nitra, the Slovak Republic, 28. June - 2. July 1998.

Murphy, D.V., Willison, T.W., Baker, J.C. and Gulding, K.W.T (1998) Assessing the potential for nitrogen loss in agricultural soils. ESA, Short communications, Volume II: 17-18

Soper, R.J., Racz, G.J. and Fehr, P.I. (1971) Nitrate nitrogen in the soil; as a means of predicting the fertilizar nitrogen requirements of barley. Can. J. Soil. Sci. 51: 45-49

Wehermann, J. and Scharpf, H.J. (1986) The N-min method on aid to integrating various objectives of nitrogen fertilization. 2. Pflanzenernähr. Bodenk. 149: 428-440

Wehermann, J. and Scharpf, F. (1979) Der Mineral stickstoffgehalt des Bodens als Masstab fur den Stickstoffdüngerbedarf (N-min Methode). Plant and Soil 52: 109-126

Chapter 16

Nitrogen utilization in spring barley genotypes

A. Öztürk, Ö. Çağlar, Ş. Akten

University of Atatürk, Faculty of Agriculture, Department of Agronomy, 25240 Erzurum, Turkey

Key words: N uptake efficiency, N translocation efficiency, agronomic efficiency, physiological efficiency, spring barley

Abstract: Plant characters related to the utilization of nitrogen were investigated in fifteen barley genotypes. The results showed that differences among the genotypes for all characters were significant. Grain yield, total N yield and grain protein content of genotypes ranged between 2186 and 3222 kg.ha^{-1}, 72.7 and 102.3 kg.ha^{-1}, 12.01 and 13.27 %, respectively. Tokak 157/37 had the highest uptake (1.28 kg N uptake/kg N applied), translocation (72.1 %) and agronomic (40.3 kg grain/kg N applied) efficiency of N. The highest physiological efficiency was obtained from cv. 1515 (32.5 kg grain/kg N uptake). Thus, the potential for developing superior, N efficient genotypes appears to exist.

1. Introduction

Today, there is a need to develop plant genotypes that assure the highest yields with the lowest inputs. Therefore, researchers have been trying to develop cereal genotypes which can take up more N from the soil, transfers more of the N to the grain. Efficiency in uptake and utilization of N in the production of grain requires that those processes associated with absorption, translocation, assimilation and redistribution of N operate effectively (Moll et al., 1982). Earlier studies showed that important variations exist among barley genotypes in terms of grain protein content, grain N %, straw N %, N yield, N uptake efficiency and N translocation efficiency (Grant et al., 1991; Tillman et al., 1991; Bulman and Smith, 1994). Thus, the potential for developing superior, N-efficient genotypes appears to exist. The major objective of the present study was to determine the differences in the characters related to the utilization of N by barley in the Erzurum region of Turkey.

2. Materials and Methods

The field experiment was carried out in 1996 and 1997. Fifteen barley genotypes were tested in an experiment with a randomized complete block design with three replications. Seed was sown with plot drill at rate of 400 seeds m^{-2}. Each plot was 1.2 by 6.0 m with six rows. All plots were fertilized with 80 kg N ha^{-1}. Half of the N (ammonium nitrate, 26 %) was applied at sowing, and the other half applied at the beginning of stem elongation. Plants were irrigated twice, once at stem elongation and the other at heading. 2,4-D was used to control broadleaf weeds. At maturity, 4 m of the four center rows of each plot were harvested. Plant samples were then separated into grain and straw (stem, leaf and chaff included), oven-dried for 24 h at 80 °C. The N concentration in the grain and straw was determined by the Kjeldahl method.

Total precipitation and average temperatures in April, May, June, July and August were 30.6, 39.7, 17.2, 24.3, 16.7 mm and 3.8, 11.6, 13.8, 20.1 and 19.3 °C in 1996; 40.7, 66.1, 32.0, 3.7, 6.4 mm and 3.1, 11.7, 14.7, 18.3 and 19.5 °C in 1997 respectively. These figures suggest that the amount of precipitation and average temperatures were more suitable in 1997 than in 1996.The soil was clay loam with available P$_2$O$_5$ and K$_2$O of 39 and 601 kg.ha^{-1}

respectively. Soil pH was 7.6 and the organic matter content was 1.8 %.

Grain protein content (GPC), N yield, N uptake efficiency (NUE), N translocation efficiency (NTE), agronomic efficiency (AE), physiological efficiency (PE) were estimated (Anderson, 1985; Grant et al., 1991; Gonzales-Ponce et al., 1993).

3. Results and Discussion

The results of variance analysis showed that

differences among genotypes were significant for the characters investigated (Tables 1 and 2). More suitable climatic conditions in 1997 increased grain, straw and grain N yield, NTE, AE and PE (Gonzales-Ponce et al.,1993)

Yields and N concentrations

Averaged over the data for both years, the grain yield of the genotypes ranged between 2186 and 3222 kg.ha^{-1}. FAO 69180, Bilge 2 and Iris genotypes had the lowest yields

Table 1. The yields and N percentages of barley genotypes

Variable	Grain yield (kg ha^{-1})	Straw yield (kg ha^{-1})	Grain N (%)	Straw N (%)	Grain N yield (kg ha^{-1})	Straw N yield (kg ha^{-1})
Years (Y)						
1996	2590	5496	2.29	0.54	59.3	29.6
1997	2893	5784	2.14	0.49	61.8	28.2
Mean	2742	5640	2.22	0.52	60.5	28.9
Genotypes (G)						
26	2554 efg	5121 bc	2.13 cde	0.49 bcd	54.4 fg	24.7 fg
Brems	3180 a	5659 ab	2.20 a-e	0.52 bcd	69.7 ab	29.5 bcd
Tokak 157/37	3222 a	5602 ab	2.30 ab	0.51 bcd	73.8 a	28.5 c-f
Afyon Kılıç	3042 ab	6029 a	2.27 ab	0.54 ab	68.8 ab	32.3 abc
Yeşilköy 5703	2595 ef	5611 ab	2.31 a	0.54 ab	59.6 def	29.2 bcd
FAO 69180	2186 h	4734 c	2.30 ab	0.48 cde	50.2 g	22.5 g
1502	2823 cd	5621 ab	2.11 de	0.59 a	59.3 def	32.9 ab
1511	2935 bc	5778 ab	2.17 be	0.59 a	63.5 cd	33.8 a
1515	3111 ab	5943 a	2.09 e	0.52 bcd	65.0 bc	31.0 a-d
1522	2610 ef	5873 ab	2.18 e	0.53 bc	56.9 ef	30.7 a-d
1524	2565 efg	5845 ab	2.27 ab	0.43 e	58.1 def	25.1 efg
1527	2750 cde	6114 a	2.12 cde	0.47 de	58.3 def	28.7 cde
Cytris	2685 def	5817 ab	2.24 a-d	0.51 bcd	60.2 cde	29.5 bcd
Iris	2499 fg	5518 ab	2.25 abc	0.51 bcd	55.9 ef	27.9 def
Bilge-2	2369 gh	5335 abc	2.31 a	0.51 bcd	54.6 efg	27.0 def
LSD	195.8	698.1	0.12	0.05	5.00	3.60
C.V. (%)	4.65	8.05	3.59	7.27	5.39	8.05
Y	P<0.001	P<0.01	P<0.001	P<0.001	P<0.001	P<0.01
G	P<0.001	P<0.001	P<0.001	P<0.001	P<0.001	P<0.001
YxG	P<0.001	NS.	NS.	NS.	P<0.001	P<0.01

*The means with the same letter within each variable are not significantly different (Duncan's multiple range test), NS = not significant.

Table 2. Total N yield, grain protein content (GPC), N uptake efficiency (NUE), N translocation efficiency (NTE), agronomic efficiency (AE) and physiological efficiency (PE) of barley genotypes

Variable	Total N yield (kg ha^{-1})	GPC (% Nx5.75) (%)	NUE (kg N uptake / kg N applied)	NTE (%)	AE (kg grain / kg N applied)	PE (kg grain / kg N uptake)
Years (Y)						
1996	88.8	13.20	1.11	66.8	32.4	29.2
1997	89.8	12.30	1.12	68.7	36.2	32.2
Mean	89.3	12.75	1.12	67.8	34.3	30.7
Genotypes (G)						
26	79.1 gh	12.23 cd	0.99 gh	68.6 bc	32.0 efg	32.3 a
Brems	99.1 ab	12.64 a-d	1.24 ab	70.2 ab	39.8 a	32.1 ab
Tokak 157/37	102.3 a	13.20 ab	1.28 a	72.1 a	40.3 a	31.5 abc
Afyon Kılıç	101.1 a	13.05 ab	1.27 a	68.1 be	38.0 ab	30.1 cd
Yeşilköy 5703	87.3 def	13.25 ab	1.09 def	68.4 bcd	32.4 ef	29.9 cd
FAO 69180	72.7 h	13.23 ab	0.91 h	69.0 bc	27.3 h	30.0 cd
1502	92.2 bcd	12.12 cd	1.15 bcd	64.3 f	35.3 cd	30.7 a-d
1511	97.2 ab	12.47 bcd	1.22 ab	65.3 def	36.7 bc	30.3 bcd
1515	96.0 abc	12.01 d	1.20 abc	67.8 b-e	38.9 ab	32.5 a
1522	87.6 def	12.54 a-d	1.10 def	65.0 ef	32.6 ef	29.8 cd
1524	83.2 efg	13.06 ab	1.04 efg	69.8 abc	32.1 efg	30.9 a-d
1527	87.0 def	12.20 cd	1.09 def	67.0 b-f	34.4 cde	31.7 abc
Cytris	89.6 cde	12.89 abc	1.12 cde	67.2 b-f	33.6 def	30.0 cd
Iris	83.8 efg	13.12 ab	1.05 efg	66.7 c-f	31.3 fg	29.8 cd
Bilge-2	81.5 fg	13.27 a	1.02 fg	66.9 b-f	29.6 gh	29.1 d
LSD	6.50	0.69	0.08	2.86	2.45	1.74
C.V. (%)	4.73	3.50	4.78	2.74	4.64	3.68
Y	NS.	P<0.001	NS.	P<0.001	P<0.001	P<0.001
G	P<0.001	P<0.001	P<0.001	P<0.001	P<0.001	P<0.001
YxG	P<0.001	NS.	P<0.001	P<0.01	P<0.001	P<0.01

* The means with the same letters within each variable are not significantly different (Duncan's multiple range test), NS= not significant.

while Tokak 157/37, Brems and 1515 genotypes had the highest yields (Table 1). Results showed that high yielding genotypes had also high NUE and PE values (Isfan,1990). There were significant differences in straw yield between genotypes. The straw yield varied from 4734 (FAO 69180) to 6114 (Cytris) kg.ha^{-1}. Grain N % varied from 2.09 % in the 1515 genotype to 2.31 % in the Yeşilköy 5703 and Bilge 2 genotypes. Similar results were reported by Isfan et al.(1991). As shown in Table 1, there was a significant variation in straw N % among genotypes. 1524 genotype had the lowest straw N % (0.43 %) while 1502 and 1511 genotypes had the highest straw N % (0.59 %). Grain N yield, straw N yield and total N yield of genotypes ranged between 50.2

(FAO 69180) and 73.8 (Tokak 157/37), 22.5 (FAO 69180) and 33.8 (1511), 72.7 (FAO 69180) and 102.3 (Tokak 157/37) kg.ha^{-1} respectively. Grain N yield was higher in 1996 while straw N yield was higher in 1997.

GPC varied between years (Table 2). It was higher in 1996, possibly due to insufficient rainfall during the growth stage. On an average of both years 1515 had the lowest GPC (12.01 %), Bilge-2 had the highest GPC (13.27 %). Genetic variation was also reported in GPC by Grant et al., (1991) and Bulman and Smith, (1994).

Uptake and Translocation efficiency of N

Genotypic differences in NUE were observed (Table 2). Tokak 157/37 had the highest NUE

(1.28 kg N uptake/kg N applied) and FAO 69180 the lowest (0.91 kg N uptake/kg N applied). Afyon Kılıç, Brems and 1515 genotypes were the next highest in NUE, while 26 and Bilge-2 were next lowest. Genotypic variation was reported for barley in NUE (Tillman et al., 1991). NTE is often regarded as a measure of retranslocation efficiency of N from vegetative plant parts to the grain. NTE was lower in 1996 than in 1997. This was possibly due to high temperatures that occured during the grain filling and insufficient rainfall received during the growth stage, all of which made the transport of N from the straw to the grain very difficult. Proof of this was a high residual content of N found in the straw (Gonzales-Ponce et al.,1993). There was a significant variation in NTE among genotypes (Table 2). NTE of genotypes ranged between 64.3 % (1502) and 72.1 % (Tokak 157/37). Genetic variation exist for NTE in barley (Grant et al.,1991; Tillman et al., 1991).

Agronomic and Physiological Efficiency

Significant differences among the genotypes were found for AE and PE (Table 2). Although the year x genotype interaction was significant for AE, genotypes FAO 69180 (27.3 kg grain/kg N applied) and Tokak 157/37 (40.3 kg grain/kg N applied) had the lowest and highest values. AE was higher in 1997. Similar results were reported by Anderson (1985). The highest (32.5 kg grain/kg N uptake) and the lowest (29.1 kg grain/kg N uptake) PE values were obtained from 1515 and Bilge-2 respectively. Similar variations of PE were also found by Isfan (1990) in spring barley. PE varied with years. It was higher in 1997 than in 1996.

The results showed that genotypes had N uptake of 1.07 % of the total dry matter produced. Grain yield increased depending on increases in N uptake. Although some genotypes took up more N, they tended to store

a less of the N taken up (1502, 1511, 1522). A reverse situation was observed in some genotypes (26, FAO 69180). The genotypes such as Tokak 157/37 and Brems were found to be superior to others with regards to NUE and NTE. Tokak 157/37, Brems and 1515 genotypes showed a better performance in the ability to convert the fertilizer N into yield. Thus, the potential for developing superior, N-efficient genotypes seemed to exist.

Acknowledgements

We thank Dr. Faik Kantar for critical reading of the text.

4. References

Anderson WK (1985) Differences in response of winter cereal varieties to applied nitrogen in the field. II. Some factors associated with differences response. Field Crop Res. 11: 369-385.0

Bulman P and Smith D (1994) Post-heading nitrogen uptake, and partitioning in spring barley. Crop Sci. 34: 977-984.

Gonzales-Ponce R Salas ML and Mason SC (1993) Nitrogen use efficiency by winter barley under different climatic conditions. Journal of Plant Nutrition 16: 1249-1261.

Grant CA Gauer LE Gehl DT and Bailey LD (1991) Protein production and nitrogen utilization by barley cultivars in response to nitrogen fertilizer under varying moisture conditions. Can. J. Plant Sci. 71: 997-1009.

Isfan D (1990) Nitrogen physiological efficiency index in some selected spring barley cultivars. Journal of Plant Nutrition 13:907-914.

Isfan D Cserni I and Tabi M (1991) Genetic variation of the physiological efficiency index of nitrogen in triticale. Journal of Plant Nutrition 14: 1381-1390.

Moll RH Kamprath EJ and Jackson WA (1982) Analysis and interpretation of factors which contribute to efficiency of nitrogen utilization. Agron. J. 74: 562-564.

Tillman BA Pan WL and Ullrich SE (1991) Nitrogen use by northern-adapted barley genotypes under no-till. Agron. J. 83: 194-201.

Chapter 17

The effect of nitrogen doses on nitrogen uptake and translocation in spring barley genotypes

Ö Çağlar, A. Öztürk

University of Atatürk, Faculty of Agriculture, Department of Agronomy, 25240 Erzurum, Turkey

Key words: N uptake, N translocation efficiency, spring barley

Abstract: In this research we investigated the effects of N doses on the yield, N uptake and N translocation in genotypes of spring barley. The research was done in Erzurum, in 1996 and 1997. Three barley genotypes (Tokak 157/37, Cytris, and 1515) were used with five doses of N (0, 20, 40, 60 and 80 kg ha^{-1}). Significant differences were observed between genotypes, in the grain N concentration, grain protein concentration, grain N yield, total N uptake and N translocation efficiency. The highest grain yield (2706 kg ha^{-1}), grain N % (1.98), grain N yield (54 kg ha^{-1}), total N uptake (80.9 kg ha^{-1}) and N translocation efficiency (66.9 %) were obtained from Tokak 157/37. N doses had a significant influence on the characters investigated. Grain yield, straw yield, concentrations of grain N, straw N, and grain protein , grain N yield, straw N yield and total N uptake all increased with increasing doses of N whereas N translocation efficiency decreased.

1. Introduction

The effectiveness with which N is used by cereals has become increasingly important because of the increased costs of manufacture and distribution of N fertilizer. Research on the N-metabolism of cereals indicated that environmental conditions significantly influence N uptake, translocation and storage as well as the yield of plants (Hamid and Sarwar, 1976; Whitfield et al., 1989). These physiological processes are affected by N fertilization. Field studies indicate that the N concentration, [N], of grains and straw, the total N uptake (Kucey and Schaalje, 1986; Nedel et al., 1993), and protein concentration (Lauer and Partridge, 1990; Bulman et al.,1994) all increase with increasing application of N. Conversely, the efficiencies of N uptake and translocation decreased (Bole and Pittman, 1980; Bulman and Smith, 1994). However, it is stressed that more work be carried out with a larger number of genotypes and levels of N

application to reveal obvious differences in these parameters. Since N fertilization is important in cereal production in Turkey, it is necessary to determine a deliberate policy for the use of nitrogen fertilizer.

2. Materials and Methods

This research was carried out in Erzurum, in 1996 and 1997 using three barley genotypes Tokak 157/37, Cytris, and 1515) and five levels of application of N (0, 20, 40, 60, 80 kg ha^{-1}) in a factorial design, with three replications. Tokak 157/37 is a 2-row genotype and is facultative, Cytris and 1515 are 2-row, spring genotypes. Ammonium nitrate, (35 % N) was used as a source of nitrogen. Each plot was 6.0 m long and 1.2 m wide; and contained six plant rows at 0.2 m spacing. Sowing was made with a plot drill at a rate of 400 seeds m^{-2}. All nitrogen was applied at sowing. Irrigation was applied at the beginning of stem elongation and at heading. Weeds were controlled using 2,4-D. At maturity, a 4 m length of the four centre rows of each plot was harvested. Plant samples were taken and separated into grain and straw (stems, leaves and chaff included). These were oven-dried for 24 h at 80 °C. Grain [N] and straw [N] were determined by the Kjeldahl method.

The climatic conditions in the second year were more favourable for barley than in the first. The total precipitation in April, May, June, July and August was 30.6, 39.7, 17.2, 24.3, 16.7mm respectively in 1996 and 40.7, 66.1, 32.0, 3.87, 6.4 mm in 1997. Average temperatures over the same periods were 3.8, 11.6, 13.8, 20.1 and 19.3 °C respectively in 1996 and 3.1, 11.7, 14.7, 18.3 and 19.5 °C in 1997. Available P_2O_5 and K_2O contents of the clay loam soil were 39 and 617 kg ha^{-1} respectively. Soil pH was 7.7 and the organic matter content was 1.8 %.

Grain protein content (GPC), grain N yield, straw N yield, total N uptake (TNU) and N translocation efficiency (NTE) were examined following Grant et al. (1991) and Gonzales-Ponce et al. (1993).

3. Results and Discussion

Effect of Years Analysis of variance showed statistically significant differences between years for the characters investigated (except straw yield, straw [N], straw N yield and TNU). More favourable climatic conditions increased grain yield, grain N yield and NTE and decreased grain [N] and grain protein concentration in 1997 (Table 1). This is consistent with the findings of Grant et al. (1991). In this study, only statistically significant interaction was found between genotype and dose for straw yield, no significant interaction was determined in the other characters.

Effect of Genotypes Averaging genotype data across years, significant differences were observed between genotypes in grain [N], GPC, grai"1n N yield, TNU and NTE (Table 1). The highest grain [N] (1.98 %) and GPC (11.41 %) were obtained from Tokak 157/37. The genotype 1515 had the lowest levels. Brunori et

al. (1980) reported that high yielding genotypes with high protein concentration synthesised protein at a higher rate and for a longer period than did lower yielding genotypes. Grant et al. (1991) reported significant variation in these characters among barley genotypes. In our work, the ranking of genotypes for grain N yield and total N uptake was similar to that for grain [N] and GPC (Table 1). The highest grain N yield and TNU were obtained from Tokak 157/37 and the lowest from genotype 1515.

Bulman and Smith (1994) reported significant variations in spring barley genotypes in both grain N yield and TNU. Dhugga and Waines (1989) investigating the cause of such variations pointed out that there was insufficient storage capacity in low grain- and N-yielding genotypes to allow significant uptake of N after the flowering period. In genotypes with higher yield potential, the amount of N needed is greater and TNU should be at least 1.2 % of biological yield for a good yield (Stanford and Hunter, 1973). Genetic variation was also reported in NTE (Grant et al., 1991; Bulman and Smith, 1994). We found the highest NTE in Tokak 157/37 (66.9 %). Canvin (1976) and Brunori et al. (1980) considered that N translocation to the grains is mostly determined by their storage capacity. Tokak 157/37 was capable of translocating large amounts of the N taken up to the grains and, so, had a higher TNU than in other genotypes. The other genotypes did not show a relation between N uptake and their capacity for N translocation.

Effect of N applications Analysis of variance showed that all parameters investigated were significantly influenced by N doses (Table 1). Grain yields increased with N application up to 60 kg N ha^{-1}. Differences between the effects of the application of 60 and 80 kg N ha^{-1} were statistically non-significant for all the parameters examined. Increases in N doses up to the optimum level increased grain yield by increasing spike and spikelet numbers per unit area (data not shown) as was reported also by Kucey and Schaalje (1986) and by Nedel et al (1993). The effect of N doses to

Table 1. The effects of N doses on yields, N percentages, grain protein content (GPC), total N uptake (TNU) and N translocation efficiency (NTE) in barley genotypes*

Variable	Grain yield (kg ha⁻¹)	Straw yield (kg ha⁻¹)	Grain N (%)	Straw N (%)	Grain N yield (kg ha⁻¹)	Straw N yield (kg ha⁻¹)	GPC (%)	TNU (kg ha⁻¹)	NTE (%)
Years (Y)									
<1996	2506	5555	1.99	0.49	50.3	27.5	11.42	77.7	64.8
1997	2815	5575	1.85	0.47	52.4	26.5	10.62	78.9	66.5
Mean	2661	5565	1.92	0.48	51.3	27.0	11.02	78~	65.7
Genotypes (G)									
Tokak	2706	5633	1.98a	0.48	54.0a	26.9	11.41a	80.9a	66.9a
Cytris	2647	5625	1.91b	0.49	50.9b	27.8	10.99b	78.7ab	64.9b
1515	2630	5437	1.85c	0.47	49.1b	26.3	10.65c	75.3b	65.2b
Ndoses (N) (kg ha⁻¹)									
0	2057d	4427d	1.75c	0.41b	35.9d	17.7d	10.05c	53.5d	66.9a
20	2345c	5011c	1.81c	043ab	42.5c	21.7c	10.42c	64.2c	66.2ab
40	2776b	5654b	1.91b	0.49ab	52.8b	27.5b	10.97b	80.4b	65.8ab
60	3128a	6348a	2.03a	0.52ab	63.2a	33.1a	11.65a	96.3a	65.4ab
80	2997a	6385a	2.09a	0.55a	62.3a	35.0a	12.02a	97.3a	64.1 b
Y	P<0.001	NS.	P<0.001	NS.	P<0.05	NS.	P<0.001	NS.	P<0.05

* The means with the same letter within each variable are not significantly different (Duncan's multiple range test), NS= statistically non-significant.

increase straw yield, grain N yield, straw N yield, and TNU matched that on grain yield (Table 1). Increases in the application of N decreased NTE (Table 1). This is consistent with Schjorring et al.(1989) and with Kucey's finding (1987) that increased application of N was associated with increased amounts of N not translocated from the straw by maturity. It is well known that over- application of N can result in lodging and a severe reduction in both assimilation and the translocation of N to the grain (Canvin, 1976), however, that did not happen in this study.

Taken over both years there were positive correlations between TNU and all of the following: grain filling period (r = 0.237*), plant height (r =0.518**), spikes per m² (r =0.696**), kernels per spike (r = 0.568**), grain yield (r =0.903**), straw yield (r = 0.914**), GPC (r = 0.741**), grain [N] (r = 0.739**), grain N yield (r =0.979**), straw [N] (r = 0,797**), and straw N yield (r =0.947**). Similarly there were positive correlations between NTE and grain filling period (r = 0.240*), 1000-kernel weight (r = 0.249*), and harvest index (r = 0.594**). In contrast there were negative correlations between NTE and plant height (r = -0.273**), straw yield (r = 0.314**), kernels per spike (r = - 0.264*), straw N % (r = -0.663**), straw N yield (r = - 0.550**), TNU (r = -0.269**).

Acknowledgements

We thank Dr. Faik Kantar for critical reading of the text.

4. References

Bole JB and Pittman UJ (1980) Spring soil water, precipitation and nitrogen fertilizer: Effect on barley grain protein content and nitrogen yield. Can. J. Soil Sci. 60:471-477.

Brunori A Axmann H Figueroa A and Micke A (1980) Kinetics of nitrogen and dry mater accumulation in the developing seed of some varieties and mutant lines of *Triticum aestivum* Z. Pflanzenzücht 84:201-218.

Bulman P and Smith DL (1994) Post-heading nitrogen uptake, retranslocation and partitioning in spring barley. Crop Sci. 34: 977-984.

Bulman P Zarkadas CG and Smith DL (1994) Nitrogen fertilizer affects amino acid composition and quality of spring barley grain Crop Sci. 34:1341-1346.

Canvin DT (1976) Interrelationships between carbohydrate and nitrogen metabolism In genetic improvement of seed proteins. pp. 172-190. National Academy of Sciences, Washington DC.

Dhugga KS and Waines JG (1989). Analysis of nitrogen accumulation and use in bread and durum wheat. Crop Sci. 29:1232-1239.

Gonzalez-Ponce, B Salas ML and Mason SC (1993) Nitrogen use efficiency by winter barley under

74

different climatic conditions. Journal of Plant Nutrition 16:1249-1261.

Grant CA Gauer LE Gehi DT and Bailey LD (1991) Protein production and nitrogen utilization by barley cultivars in response to nitrogen fertilizer under varying moisture conditions. Can. J.. Plant Sci. 71:997-1009.

Hamid A and Sarwar G (1976) Effect of split application on N uptake by wheat from ^{15}N labelled ammonium nitrate and urea. Expl. Agric. 12:189-193.

Kucey RMN and Schaalje GB (1986) Comparison of nitrogen fertilizer methods for irrigated barley in the Northern Great Plains. Argon J. 78:1091-1094.

Kucey RMN (1987) Nitrogen fertilizer application practices for barley production under south western Canada prairie conditions. Commun. Sci. Plant Anal. 18: 753-769.

Lauer JG and Partridge JR (1990) Planting date and nitrogen rate effects on spring malting barley. Agron. J. 82:1083-1088.

Nedel JL Ullrich SE Clancy LA and Pan WL (1993) Barley semi-dwarf and standard isotype yield and malting quality response to nitrogen. Crop Sci. 33:258-268.

Schjorring JK Nielsen NE Jensen HE and Gottshau A (1989) nitrogen losses from field grown spring barley plants as affected by rate of nitrogen application. Plant and Soil 116:167-175.

Singh Y Singh R and Sekhon GS (1977) Uptake of primary nutrients by dryland wheat as influenced by N fertilization in relation to soil type, profile water storage and rainfall. 3. Indian Soc. Soil Sci. 25:175-181.

Stanford G and Hunter AS (1973). Nitrogen requirements of winter wheat (*Triticum aestivum* L.) varieties "Blueboy" and "Redcoat". Agron J. 65:442-447.

Whiffleld DM Smith CJ Gyles OA and Wright OH (1989) Effects of irrigation, nitrogen and gypsum on yield, nitrogen accumulation and water use. Field Crop Res. 20: 261-267.

2- Crop quality – nutrient management by foliar fertilization

Chapter 18

The effect of KNO₃ application on the yield and fruit quality of olive

U. Dikmelik[1], G. Püskülcü[1], M. Altuğ[1], M.E.İrget[2]
[1]Olive Research Institute, 35100 Bornova/İzmir/Turkey
[2]Ege Uni. Fac. Agric. Soil Sci. Dept. 35100 Bornova/İzmir/Turkey

Key words: olive, foliar application, potassium nitrate, quality, table olive, oil

Abstract: Seventy two percent of olive soils in Turkey are low in potassium and thirty two percent of the olive trees in the same region are insufficient in K nutrition. Moreover, 52 % of these soils are high in Ca. Potassium application to these olive groves is quite rare and generally in the form of compound fertilizers.

This research was established to see the effect of foliar application of K on some quality properties of oil and table variety characteristics of olives. KNO₃ at 4 % concentration was applied to the foliage twice at 20 days of interval, first after fruit set and second after pit hardening. Oil percentage and some table characteristics were studied at green ripe and maturity periods. At these growth periods, leaf and fruit K contents and yield were also measured.

Treatments positively affected the fruit size, hundred fruit weight, fresh weight and pulp/pit ratio. A slight increase was measured in oil percentage and no effect was determined on yield. Fruit K content increased however was not as significant as the increases in leaf.

1. Introduction

Seventy two percent of olive soils in Turkey are insufficient in available potassium and it was found that 32 % of olive trees are deficient in potassium. The high available calcium content in 52 % of these soils makes it difficult for the olive trees to utilize fully the soil available potassium in soil (Dikmelik,1995). In these olive groves the use of potassium fertilizers is limited, and it is generally in the from of compound fertilizers mostly, as 15-15-15 and 20-20-0 (Dikmelik, 1995; Anaç and Çolakoğlu, 1994). In important olive growing areas such as Marmara and Akhisar regions, foliar application of potassium is recommended due to soil characteristics in these areas (Genç et al., 1991).

In this study the effect of foliar application of potassium on yield and some quality characteristics of table and oil olives were studied during the fruit development period. In addition, potassium contents of olive leaves and fruits were evaluated.

2. Materials and Methods

The study was conducted in the experimental orchard of Olive Research Institute Kemalpaşa-İzmir, Turkey. The experimental trees (cv Memecik) were 23 years old and severely affiliated to alternate bearing. The fruits of this variety are utilized both as table olive and for oil extraction. Research soils were loamy in texture, slightly alkaline, poor in organic matter and P, medium in available K and Mg, rich in Ca and CaCO₃. Leaf N, P, K and Mg content were of quite insufficient levels (Püskülcü and Aksalman, 1983) before the KNO₃ application.

Experimental plots were arranged in randomized parcel design as 5 replicates each having; 1) Control + Foliar water spray 2) Control (non-treated) 3) NPK (from soil) 4) NPK + 4 % KNO₃ (foliar) 5) 4 % KNO₃ (foliar). KNO₃ at 4 % concentration was applied to the foliage twice at 20 days of interval, first after fruit set and second after pit hardening. NPK fertilization consisted of 1.750 kg

$(NH_4)_2SO_4$; 0.8 kg $(NH_4)_2HPO_4$ and 1 kg K_2SO_4 per tree. The study was performed on two bearing years.

Fruit and leaf samples were taken during pit hardening (August), green ripening (October) and maturity (December). Fruit pulp analyses were carried out with fresh samples. Also, leaf and pulp samples were analysed for K by flamephotometer and oil content analysed by soxhlet extraction method. Reducing sugars were analysed through the spectrofotometric method (Ross, 1959). Yield responses were determined at two bearing years of trees and the yield was calculated in cubic meter (Pastor, 1983).

4.   Results and Discussion

Foliar application of potassium increased fruit size, hundred fruit weight, fresh weight and pulp/pit ratio. These increases were observed more clearly during the period from August to October. It was found that the oil content of green-ripe olives was slightly increased, while potassium application didn't have an important effect on the oil content of the final product.

It was observed that potassium content of leaves was considerably increased with foliar applications. This increase was higher than the increase which was brought about by potassium application through soil. Moreover, the potassium content in the pulp was determined to increase only in trees which were foliar treated with potassium.

Foliar application of potassium is recommended for important olive growing areas in Turkey due to present soil characteristics of these areas. In a study carried out on olives, a 4 % KNO_3 treatment combined with urea during the pit hardening stage proved to exert positive effects on table olive quality (Pastor, 1983). Similar results were obtained in another research work where application were started before flowering and continued until the end of August (Tan, 1995). In the present study the foliar applications of KNO_3 which are carried out from the first rapid growth period of fruit (July) till maturity (end of September) were found to have a positive effects on the quality of table olives and to improve the leaf potassium content as well. However, these applications were found to be inefficient during the second rapid growth period of the fruit at which the potassium demand of the fruit is the highest. As a result, the application of potassium seems to be necessary until the fruit ripening (maturity) period.

5. References

Dikmelik. Ü., 1995. Nutrient Status of Olive Orchards in Turkey and Their Fertilization. İlhan Akalan Soil and Environment Symposium.Sep.27-29.

Anaç, D., Çolakoğlu, H., 1994. Current K Fertilization Use in Turkey. Potash Review. IPI. 3/94:1-12.

Genç, Ç., ve Ark., 1991. Marmara Bölgesi Sofralık Zeytinliklerin Beslenme Durumunun Tesbiti. Bahçe D. 20 (1-2): 49-58

Püskülcü. G., Aksalman. A., 1989. The Rules of Taking Leaf and Soil Samples and Fertilizer Recommandation on Olive Trees. Olive Res. Inst. No:44.

Pastor. M., 1983. Plantation Density. Int. Course on Fert. and Inten. of Olive Cultivation. Cordoba, April.

Androulakis. I., Perica. S., Loupassaki. M.H., 1991. Effect of Summer Application of N and K on the Development of Olive Fruits. 8th Cons. of Eur. Coop. Res. Network on Olives. September, İzmir.

Tan. M., 1995. Research About the Effect of Prunning and Leaf Fertilization on the Fruit Yield and Quality of Edremit Variety. E. U. Fac. Agr. Horticulture Dept. (pH D. Thesis) Bornova, İzmir - Turkey.

Ross. F. A., 1959. Dinitrophenol Method for Reducing Sugars. In: Potato Processing. AVI Publ. Co. Connecticut. 169-170.

Table 1. The effect of foliar application of KNO_3 on some fruit quality characteristics.

1st BEARING YEAR	AUGUST (Pit Hardening)						
TREATMENT	Width (cm)	Length (cm)	Weight of 100 fruit (g)	Fresh Weight 100 g	Dry Weight 100 g	Dry Mat. (%)	Pulp Pit
1.Control + Water Spray	11.2	17.7	144.7	64.2	21.8	34.1	1.80
2 Control (Non-Treated)	11.7	18.1	141.5	65.3	21.0	32.2	1.88
3. NPK (From Soil)	11.5	17.9	148.1	65.0	20.3	31.3	1.88
4. NPK +% 4 KNO_3	11.3	17.7	142.7	64.1	22.0	34.4	1.82
5. % 4 KNO_3 (Foliar)	12.0	18.4	161.0	68.4	19.7	28.9	2.20
2ND BEARING YEAR							
1. Control + Water Spray	12.4 ab	18.7	158.4	67.9 b	16.1	23.7	2.10 b
2.Control (Non-Treated)	12.2 b	18.6	151.2	68.1 b	15.7	23.1	2.14 b
3. NPK (From Soil)	12.0 b	18.8	156.5	67.5 b	15.5	23.0	2.06 b
4. NPK +% 4 KNO_3	12.3 b	18.4	155.7	68.1 b	15.7	23.0	2.13 b
5. % 4 KNO_3 (Foliar)	12.9 a	19.1	174.6	70.4 a	16.3	23.1	2.40 a

	1st BEARING YEAR	OCTOBER (Green Ripenesse)							
	Width (cm)	Length (cm)	Weight of 100 fruit (g)	Fresh Weight 100 g	Dry Weight 100 g	Dry Mat. (%)	Pulp Pit	Oil (d.m. %)	Red. Sug. (%)
1	14.2 b	20.0	254.5	80.6	25.6	31.8	4.18	45.68	2.85
2	14.5 b	20.1	260.0	79.9	27.9	34.9	4.00	47.30	2.60
3	14.8 ab	20.1	284.3	80.7	26.8	33.3	4.18	49.50	2.84
	15.0 ab	20.5	284.4	80.9	25.1	30.9	4.28	50.22	2.48
5	15.7 a	21.3	319.1	82.4	26.9	32.6	4.80	52.16	2.75
	2ND BEARING YEAR								
1	13.7 b	19.7	219.7	74.1	29.5	41.4	2.96	37.74 ab	3.43
2	13.8 b	19.9	237.5	76.7	26.7	36.6	3.34	30.74 bc	3.80
3	13.7 b	19.8	231.3	75.4	27.7	37.7	3.14	31.98 bc	3.50
4	15.3 a	19.7	226.2	74.7	30.5	41.8	3.10	28.20 c	4.35
5	15.1 a	20.6	280.1	78.6	28.7	37.7	3.90	38.02 a	4.01

	1st BEARING YEAR		DECEMBER (Maturity)							
	Width (cm)	Length (cm)	Weight of 100 fruit (g)	Fresh Weight 100 g	Dry Weight 100 g	Dry Mat. (%)	Pulp Pit	Oil (d.m. %)	Red. Sug. (%)	Yield (m^3)
1	12.3	17.6	203.1	75.8	41.1	54.2	3.16	54.4	4.35	1.69
2	12.2	17.4	204.4	76.2	45.0	59.0	3.24	51.0	4.19	1.51
3	12.3	17.8	214.5	75.7	39.1	51.8	3.20	57.6	4.11	1.72
4	12.0	17.3	220.8	76.3	42.8	56.2	3.24	50.8	4.35	1.57
5	12.1	17.8	243.6	76.8	42.3	55.0	3.56	50.0	3.70	1.71
	2ND BEARING YEAR									
1	14.1	19.6	217.5	74.3	35.2	47.6	2.90	47.8	2.00 a	2.47
2	14.0	20.4	208.7	73.8	34.2	46.3	2.84	51.0	1.74 ab	2.15
3	13.8	20.6	217.4	73.4	37.3	50.1	2.80	48.8	1.29 bc	2.07
4	15.1	20.8	329.4	76.1	36.6	47.5	3.36	53.9	1.20 c	2.35
5	15.9	20.5	278.7	77.7	34.8	46.0	3.54	52.9	1.70 abc	2.14

Table 2 . The potassium contents of olive leaves and fruit pulps following foliar application of KNO_3

LEAVES						
TREATMENT	1ST BEARING YEAR			2ND BEARING YEAR		
	August	October	December	August	October	December
1- Control + Water Spray	0.93 b*	0.77 b	0.62 ab	0.75 b	0.56	0.61 b
2- Control (Non-Treated)	0.93 b	0.76 b	0.57 b	0.74 b	0.59	0.68 b
3- NPK (From Soil)	0.90 b	0.77 b	0.58 b	0.77 b	0.60	0.70 b
4- NPK +% 4 KNO_3	0.99 b	0.83 b	0.64 a	0.81 b	0.72	0.82 a
5- % 4 KNO_3 (Foliar)	1.14 a	0.97 a	0.64 a	1.01 a	0.71	0.81 a
PULPS						
TREATMENT	1ST BEARING YEAR			2ND BEARING YEAR		
	August	October	December	August	October	December
1- Control + Water Spray	2,38	2,05	1,62	2,46 bc	1,76	1,56
2- Control (Non-Treated)	2,30	1,90	1,56	2,39 c	1,62	1,62
3- NPK (From Soil)	2,28	2,04	1,83	2,63 ab	1,67	1,54
4- NPK +% 4 KNO_3	2,23	2,16	1,66	2,54 abc	1,56	1,49
5- % 4 KNO_3 (Foliar)	2,56	2,15	1,87	2,68 a	1,62	1,65

*$p \leq 0.05$

Chapter 19

Effects of foliar potassium nitrate and calcium nitrate application on nutrient content and fruit quality of fig

M.E. İrget[1], Ş. Aydın[2], M. Oktay[1], M. Tutam[1], U. Aksoy [3], M. Nalbant[1]
[1]*Ege University Faculty of Agriculture Department of Soil Science Bornova-Izmir*
[2]*Fig Research Institute Erbeyli-Aydın*
[3]*Ege University Faculty of Agriculture Department of Horticulture Bornova-Izmir*

Key words: Ficus carica, fig, fruit quality, plant nutrients, potassium nitrate, calcium nitrate

Abstract: The research work is carried out to investigate the effects of foliar potassium and calcium nitrate applications on nutrient content and fruit quality of figs. The trial was performed as three replications on Sarılop trees, the main cultivar for commercial drying. The foliar fertilisation was made twice on July 10 and 25, 1997, and the program was carried out as potassium nitrate at 3.0 % level, calcium nitrate at 1.5 and 3.0 % levels and a combined application of 2.0 % potassium nitrate and 2.0 % calcium nitrate. The results put forth an increase of K and Ca contents of leaf lamina and petiole. Applications exerted marked effect on fruit width, length, shape and width of the ostiolar opening. Similar consequences appeared on texture and sugar fractions as fructose, α-glucose, β-glucose and galactose of dried fruits.

1. Introduction

The fig fruit even if accepted as an under-utilised species all over the world is an important crop in the Mediterranean countries and can be consumed as fresh or dried fruit. In Turkey, the major fig producer in the world, most of the production comes from the western part of the country. About 75 % of the national production are supplied as dried fruit from the Big and Small Meander Basins of Aydın, Izmir and Denizli provinces (Anonymous, 1995). The fig plantations in the Region are 99 % composed of a unique variety, Sarilop (=Calimyrna), the standard variety for drying (Kabasakal, 1990). Due to its intensity, the fig production plays an important role in the Region economy.

Sarilop variety is known all over the world with its high quality as the dried fig. The climatic conditions prevailing in the Aegean Region have an important impact on crop quality. The major attributes that determine the quality and price of dried fig fruit are listed as size, colour, taste, flavour, texture and ratio of defects. In the Aegean region, during the last 25 to 30 years, fig plantations are eradicated from the lowlands and established in the less fertile leached mountainous areas which consequently increased the importance fertilisation practices. In this respect, potassium and calcium applications play a marked role in terms of yield and quality.

Previous researchwork carried out on fig reveal that calcium and potassium fertilisation has an important impact on fruit quality and that potassium/calcium balance is the main decisive factor (Aksoy and Anac, 1994; Aksoy and Akyüz, 1993; Eryuce et al., 1996). Potassium is known to effect through water use, whereas, calcium overtakes a significant role in the cell wall structure in the form of calcium-pectate. In plant nutrition, calcium is known to create consequent physiological disorders due to its slow translocation as a plant nutrient (Mengel and Kirkby, 1982).

This experiment was designed to test the effects of foliar applied potassium and calcium on plant nutrients, fruit characteristics and composition of Sarilop under conditions prevailing in Aydin province.

2. Materials and Methods

The experiment was carried out in a 15 years old Sarilop fig orchard in Erbeyli village of Aydın province. The soil properties of the experimental orchard are given in Table 1.

The trial was set up according to randomised blocks design as three replicates and three trees per replicate (Atanasiu, 1968). Including the control parcel, five variables were tested: 1.5% Ca $(NO_3)_2$, 3 % Ca $(NO_3)_2$, 3 % KNO_3 and 2 % KNO_3 + 2 % Ca $(NO_3)_2$. The fertilisers were applied twice on July 10, 1997 and July 25, 1997. The leaves that had the first fruit in its axil (generally the third basal most) were sampled for nutrient analysis. After wet ashing K and Ca analysis were done separately in the lamina and the petiole (Kabasakal, 1983; Kacar, 1972). Fresh fruit samples were taken during the intensive harvest period and analysed for fruit width, length, index, neck length, ostiole width and fruit weight. In dried fruit samples, sugar fractions, fruit size, texture and colour was determined. Texture (1-3 scale) and colour (1-5 scale) were analysed by panel tests (Aksoy, et al., 1987). Sugar fraction, fructose, α-glucose, β-glucose and galactose, were quantified by gas chromatography (Hakerlerler et al., 1994). The potassium and calcium contents of the fruit samples were also determined (Kabasakal, 1983; Kacar, 1972). The data obtained was statistically analysed by TARIST program (Açıkgöz et al., 1993).

3. Results and Discussion

Results appearing from calcium and potassium nitrate applications are given in Tables 2-4. The effect on leaf lamina and petiole and fruit K and Ca contents was statistically significant. Different levels of applied Ca increased Ca contents of various plant parts compared to the untreated control parcel. K applications, on the other hand, tended to increase K contents and decreased Ca levels. A trial carried out on fig in the Aegean Region had revealed confirming results. Foliar application of Ca $(NO_3)_2$ at 0.25 % and KNO_3 at 0.50 % levels increased potassium content and decreased Ca due to their antagonistic relationship (Eryuce et al, 1996; Aydemir and Ince, 1988). In some fig varieties, it is reported that Ca applications increase the leaf and fruit Ca levels (Aksoy et al., 1992). $Ca(NO_3)_2$ treatments advanced fruit-Ca levels in tomatoes, as well (Kılınç and Tuna, 1997). Significant positive effects of potassium applications on K levels of grape (Cokuysal, 1990), citrus (Jones and Burns, 1972: Jones et al., 1967) and peach (Kenworthy, 1965) are reported. The findings on fig are in accordance with the outcome of these researches.

Calcium is an element known for its vital role as a cell wall component and in fruit development and ripening. Fruit firmness and thus post-harvest losses are generally closely related with the fruit Ca content. Due to its deficiency, loose and juicy areas may form on fruits (Aktas, 1991). Calcium applications are also known to regulate the activities of the plant growth substances mainly of indole acetic acid (IAA) that is formed in the apex and that plays a regulatory role in cell division and thus enhance growth rate (Bergmann, 1992; Marschner, 1974; Marschner, 1995).

The fruit size of fig fruits was highly affected by the foliar applications (Table 3). All the treatments were found to be statistically different than the control at 1 % level. The most oblate fruits were harvested in trees treated with Ca $(NO_3)_2$ at 3 % level. Fruit size was increased by all of the treatments. Untreated control trees had the lowest average fresh fruit weight with 92.1 g. Neck length is an important parameter in terms of fresh harvest since longer neck may ease hand picking. This is more important in varieties that have a short neck like Sarilop. Even though the effect of foliar applications were not statistically significant, the shortest neck values were found in control fruits.

A wide ostiolar opening (eye) is accepted as an important characteristic in fig production since it enables the entrance of pathogens and their vectors. In all of the trials on Sarilop, a common target is to obtain a closed or narrower opening. In this trial, all of the applied solutions provided a narrower ostiolar opening compared to the control (Table 3). Aksoy et al. (1992) report that the size of the ostiole is highly correlated with the fruit size so having a smaller opening together with a bigger fruit size obtained by the tested foliar applications as an additional value.

Sarılop is known to possess the most desired dried fruit attributes among the fig varieties. The major characteristics can be mentioned as fruit size, light colour and soft texture. The effect of the applications on fruit size of the fresh and dried fruits did not prove to be statistically significant, however, the biggest dried fruits (18.85 g) were obtained in parcels applied with 3 % KNO_3 (Table 4). The K content of the petiole was significantly correlated with the average dried fruit weight (r=0.625*). The smallest sizes were recorded in 1.5 % $Ca(NO_3)_2$ (15.73 g) and control (15.96 g) trees. In respect to dried fruit colour, similar findings appeared. Fruit colour was darker in 1.5 % $Ca(NO_3)_2$ applied and control fruits, on the other hand, lighter colours were brought about by 3 % KNO_3 . Fruit texture was softer (higher index value) in 2 % KNO_3 + 2 % $Ca(NO_3)_2$ and 3 % KNO_3 applications, whereas, harder in untreated fruits or in fruits treated with $Ca(NO_3)_2$ only. The correlation coefficients of leaf lamina and fruit K contents with texture were significant at 1 % level (r= 0.719 ** and 0.963 **, respectively). Similar relationships with the colour index (leaf lamina K x colour r= 0.598*; fruit K x colour r=0.630*) were significant at 5 % levels.

Fig fruits are rich in sugars and soluble solids that are mostly composed of sugars. One of the sugar fractions that receive the highest attention is fructose due to its comparatively higher sweetness. Hakerlerler et al. (1994) cite the sugar composition of fig fruits belonging to various cultivars as 31.6 % fructose, 19.6 % α-glucose and 22.1 % β-glucose and 1.0 % sorbit. The effect of all tested applications on fruit sugar composition was found significant at 1 % level. KNO_3 applications at 3 % level increased fructose, α-glucose, β-glucose and galactose compared to control as 45.3, 19.9, 18.1 and 1.8 %, respectively. The control fruits ranked the lowest in this respect. 2 % KNO_3 + 2 % Ca $(NO_3)_2$ and 3 % KNO_3 applications were grouped together in terms of α-glucose, β-glucose and galactose contents. Increasing K levels in the leaf lamina advanced fructose (r= 0.624*) and α-glucose (r=0.525*) contents. Petiole-K was found to be highly correlated with fructose (r= 0.550*) and fruit K with galactose (r=0.662**).

It can be concluded that potassium applications at tested concentrations enhanced dried fig quality through its marked positive effects on colour and texture and total sugar and fructose contents. Calcium applications, on the other hand, resulted in harder and darker fruits and thus producing inferior fruit quality. The petiole-Ca was negatively correlated with fruit K (r=-0.565*) so had a negative impact through its antagonistic effect on potassium. The positive effect of calcium was to narrow the ostiolar opening. Although positive responses related to fruit quality were obtained by 2 % KNO_3 + 2 % Ca $(NO_3)_2$ treatment, generally leaf burn was accompanying, so it is recommended that this combination need to be tested further at lower concentrations.

4. References

Acıkgoz N, Akkaş M.E., Moghaddam A and Ozcan K (1993) TARIST: A package program of statistics and quantitative genetics for PC (Turkish), Computer Applications Symp., Konya-Turkey, 133.

Aksoy U and Akyuz D (1993) Changes in K, Ca and Mg contents in different parts of the fig fruit during development, VIII th Int. Coll. for the Optimization of Plant Nutrition, 309-312.

Aksoy U and Anac D (1994) The effect of calcium chloride application on fruit quality and mineral content of fig, Acta Horticulturae, 368: 754-762.

Aksoy U, Anac D and Gul N (1992) Relationships between fresh and dried fruit quality criteria of the fig fruit, 1st National Horticultural Congress (Turkish). Izmir-Turkey.

Aksoy U, Anac D, Hakerlerler H and Düzbastılar M (1987) Nutritional Status of fig orchards around Germencik and relationships between the analysed plant nutrients and yield and quality (Turkish), TARİŞ Res. and Dev. Directorate, Bornova-Izmir.

84

Aktaş M (1991) Plant Nutrition and Soil Fertility (Turkish), Ankara Univ. Fac. of Agric. Publ. 1202, Ankara.

Anonymous (1995) State Statistics Institute, TR Prime Ministry State Statistics Institute 2091.

Atanasiu N (1968) Field Experimental Design (Turkish), Ege Univ. Publ. 140.

Aydemir O and İnce F (1988) Plant Nutrition (Turkish), Dicle Univ. Publ. No. 2.

Bergmann W (1992) Nutrient Disorders of Plants. Gustav Fisher Verlag, Jena-Stutgart.

Çokuysal, B (1990) The effect of GA and foliar fertilization on leaf nutrients and crop quality (Turkish) , MSc. Thesis, Ege Univ. Dept. of Soil Sci., İzmir-Turkey.

Eryüce N. Çokuysal B, Çolakoğlu H. and Aydın Ş (1996) The effects of different nitrogen levels and foliar fertilization on the leaf and fruit nutrient contents of fig.IX th Int. Coll. For the Optimization of Plant Nutrition, Czech Rep., 301-305.

Hakerlerler H, Aksoy U, Saatçı N, Üçdemir and Hepaksoy S (1994) The leaf and fruit nutrient content of some fig cultivars and clones and their relationships with the fruit sugar fractions (Turkish), Jour. Ege Univ. Fac. Agric. 31(1): 73-80.

Jones W.W. and Burns, R.M. (1972) Correction of the K deficiency in grapefruit, Calif. Citrograph, 231-277.

Jones, W.W., Embleton T.W., Gorber M.J. and Gree C.B. (1967) Creasing of Orange Fruit, Hilgardia.

Kabasakal A (1990) Fig Growing (Turkish) TAV Publ. 20- Yalova, Turkey.

Kabasakal A (1983) Researches on seasonal variation of some mineral nutrients in Fig cv. Sarılop and the relationships between soil, plant, shoot and fruit properties (Turkish), Ege Univ. Fac. Agric. Ph. D. Thesis.

Kacar B (1972) Chemical Analysis of Plant and Soil II. Plant Analysis, Ankara Univ. Fac. Agric. Publ. 453.

Kenworthy A.L., (1965) Deciduous and Small Fruits, Potassium Council Fertilizer, 41:32.

Kılınç R and Tuna L (1997) The effects of soil or foliar applied calcium nitrate on yield and quality of processing tomato (Turkish), Hasat, 12: 33-38.

Marschner, H (1974) Calcium Nutrition of Higher Plants, Neth. J.Agric. Sci.22: 275-282.

Marschner H (1995) Mineral Nutrition of Higher Plants, Academic Press Inc., London.

Mengel K and Kirkby E.A., (1982) Principles of Plant Nutrition, Int. Potash Inst. Switzerland.

Table 1. Soil properties of the experimental plot

Soil depth	pH	CaCO$_3$ (%)	T. Salt (%)	Organic matter(%)	Total N (%)	Sand (%)	Silt (%)	Clay (%)	Texture
0-30 cm	6.69	0.5	0.03	1.99	0.101	36.16	42.72	21.12	Loamy
30-60cm	6.03	0.5	0.03	1.08	0.064	60.16	22.72	17.12	Sandy loam

Soil depth	P	K	Ca	Available (mg kg^{-1}) Mg	Na	Fe	Zn	Mn	Cu
0-30 cm	1.12	152	1600	340	18	17.4	0.24	9.8	0.96
30-60cm	0.25	96	1480	380	24	8.4	0.60	4.4	3.06

Table 2. Effect of KNO$_3$ and Ca (NO$_3$)$_2$ applications on leaf lamina petiole and fruit K and Ca contents

	Fruit K (%)	Ca (mg kg^{-1})	Lamina (%) K	Ca	Petiole(%) K	Ca
Control	0.780ab	2878b	0.980b	3.573bc	0.800bc	4.273b
1.5% Ca (NO$_3$)$_2$	0.757b	3330a	0.853b	4.023c	0.703c	4.747a
3.0% Ca (NO$_3$)$_2$	0.740b	3539a	0.910b	3.967b	0.910b	4.677a
3%KNO$_3$	0.823a	2951b	1.300a	3.463a	1.113a	3.810b
2%KNO$_3$+ 2% Ca (NO$_3$)$_2$	0.837a	3189ab	1.007b	3.777bc	0.830bc	4.203ab
LSD (5%)	0.058*	370*	0.248**	0.375*	0.375*	0.524

Table 3. Effect of KNO$_3$ and Ca(NO$_3$)$_2$ applications on fresh fruit quality parameters

	Width (mm)	Length (mm)	Fresh Fruit Characteristics Weight (g)	Shape index (W/L)	Neck length (mm)	Ostiole width (mm)
Control	47.8b	37.7ab	92.1	1.26c	7.91	8.437a
1.5 % Ca(NO$_3$)$_2$	52.8a	38.9a	101.1	1.32abc	10.70	7.133b
3.0% Ca(NO$_3$)$_2$	51.8a	36.1b	96.5	1.38a	11.16	6.560b
3% KNO$_3$	53.7a	39.3a	99.4	1.30bc	9.92	7.143b
2% KNO$_3$ + 2% Ca(NO$_3$)$_2$	53.9a	38.3a	97.5	1.35ab	10.45	7.347b
LSD (5%)	3.82**	1.89**	ns	0.062**	ns	0.871*

Table 4. Effect of KNO$_3$ and Ca(NO$_3$)$_2$ applications on dried fruit quality parameters and sugar fractions

	Texture Index (1=hard 3=soft)	Colour Index (1=very dark 5=very soft)	Weight (g)	Fructose (%)	α-glucose (%)	β-glucose (%)	Galactose (%)
Control	1.26bc	2.60	15.96	16.77d	5.31c	6.70d	0.53b
1.5%Ca(NO$_3$)$_2$	1.23bc	2.60	15.73	30.57bc	12.75abc	15.69b	1.13ab
3.0%Ca(NO$_3$)$_2$	1.13c	2.67	17.41	26.48c	10.74ba	12.49c	0.76b
3% KNO$_3$	1.73a	3.47	18.85	45.34a	19.92a	18.12a	1.59a
2% KNO$_3$ + 2% Ca(NO$_3$)$_2$	1.53a	3.23	18.21	33.53b	16.98ab	17.30ab	1.51a
LSD (5%)	0.33**	ns	ns	4.89**	8.77**	2.40**	0.70**

a, b, c :Mean values (average of replications) in each column represented by different letters are statistically different (*p< 0.05 *; ** p<0.01)/

Chapter 20

Influence of foliar fertilization on quality of pear (*pyrus communis* l.) cv. 'williams'

M. Hudina, F. Štampar

University of Ljubljana, Biotechnical Faculty, Agronomy Department, Institute for Fruit growing, Viticulture and Vegetable growing, Jamnikarjeva 101, 1000 Ljubljana, Slovenia

Key words: pear; *Pyrus communis* L.; sugar; fructose; glucose; sucrose; sorbitol; organic acid; quality; nutrition

Abstract: The influence of foliar fertilization on quality of pears (*Pyrus communis* L.) cv. 'Williams' was observed. The contents of sugars (glucose, fructose and sucrose), sorbitol and organic acids (citric, malic, shikimic and fumaric) were determinated with HPLC analyses and contents of mineral elements in pear fruits were determinated in Phosyn Laboratories. Spraying of foliar fertilization Hascon M 10 AD caused the increase of soluble solids by 0,6 °Brix in comparison to the control. Content of total titratable acids was higher at the control, and the same applies for the pH juice of pears. The control showed the contents of fructose to be by 0,13 g kg^{-1} higher and that of sucrose to be by 0,24 g kg^{-1} higher than at the foliar fertilization. Content of sorbitol and glucose was higher at foliar fertilization. In fruits, sprayed with foliar fertilizer the content of citric acid was higher by 0,2 g kg^{-1} and fumaric acid by 0,066 mg kg^{-1}, the content of which is scarce in fruits at picking. Content of shikimic acid was in both treatments the same (0,3 g kg-1). Metabolism of malic acid was more active at foliar fertilization. Foliar fertilization brought about higher content of K in fruits, and also better absorbency of Mg and P in fruits. The control showed too high level of Mg and P in fruits. Quantity of N was in both treatments the same.

1. Introduction

Sugars, alcohol sugars, organic acids and vitamins serve as indicators of metabolic activities in fruits, indicating the changes in qualitative compound of fruits. Changes in contents of organic acids, sugars and alcohol ratio can result in the changes of flavour and firmness of the fruits (Doyon et al., 1991).

Sugar content in fruits is connected with technological measures being applied in orchards (training system, nutrition, assimilation area, irrigation). It is substantial to increase the quality of fruits and yields with control of the sugar metabolism during the development and growth of fruits. Time of the cell expansion which starts shortly before the June fruit drop, begins with the sugar accumulation. Fruits can obtain sucrose through two possible ways, either the sucrose is being transported to fruits from other parts of plant or it is being synthesised from the existent sorbitol. Changes in sorbitol concentration, glucose, fructose and sucrose in mesocarp of the growing fruits were studied by Berüter (1985), who determined the domination of sorbitol some two weeks after the blooming in the fruits, however, its content drops quickly, and after that it remains the same from the end of the June fruit drop till the fruit ripeness. Fructose and sucrose demonstrate even layering during the entire fruit growth period, whereas the glucose content becomes lower in July. Decreased glucose content in July is connected with the starch accumulation being at peak in August, then it decreases again, whereas the glucose content increases. Pentose - phosphate cycle activity decreases substantially during the ripening. But in the ripe fruits the absorption of the main translocation sugar sorbitol which serves to fructose synthesis is enabled by the enzyme sorbitol dehydrogenase not found in young fruits.

Organic acids are synthesised from the stored carbon hydrates and amino acids by means of carbon dioxide assimilation and the enzyme PEP (phosphoenol piruvat carboxilase). Organic acids in fruits can either be freely dissolved in cell juice or in form of salts, esters and glycozides (Arfaioli and Bosetto, 1993). Malic and citric acids are the most frequently found in pears, but other organic acids are traced in smaller quantities (quinic, glycolic, shikimic, glyceric, mucic, succinic, lactic, acetic) (Morvai and Molnar-Perl, 1992; Arfaioli and Bosetto, 1993). Malic acid together with sugars is the key substance for the assimilation process, and citric acids are important in the Krebs cycle; one of their tasks is the reduction of catalytic activities of some metals and the protection of the vitamin C against oxidation (Arfaioli and Bosetto, 1993). Metabolism of malic acid is very active and its content during the fruits ripening period could be reduced even by 50% (Seymour et all., 1993).

The aim of our study was to determine the influence of foliar fertilization on the content of sugars (glucose, fructose and sucrose), sorbitol and organic acids (malic, citric, shikimic and fumaric) from the fruit set to the ripeness in pears cv. 'Williams' and on the content of individual elements (nutrients) in ripe fruits.

2. Materials and Methods

The sugar content (glucose, fructose and sucrose), sorbitol and organic acids (citric, malic, shikimic and fumaric) were studied from June 1, 1997 until the harvesting (August 17, 1997) by the cultivar 'Williams' on the quince MA. The experiment covered two treatments repeated four times: control (C) and foliar fertilization (F). Foliar fertilization was applied 5 times in 8 - 10 day - interval starting on 24th May, 1997, with Hascon M 10 AD make

containing 21% P, 28 % K, B, Mn and Mo. Samples to determine contents of sugar, sorbitol and organic acids were prepared out of three fruits in 4 repetitions, firstly by homogenisation with manual blender (Braun), then with Ultra-Turrax T-25 (Ika - Labortechnik). 10 g of mashed fruit was dissolved with bidistillated water up to 40 ml and centrifuged at 6000 rotation/min for 15 minutes. Prior to the injection in the column the samples were filtrated through 0,45 µm Minisart filtre (RC-25, Sartorious). For each HPLC analysis of sugars, sorbitol and organic acids 20 µl of sample were used. The HPLC system of the TSP manufacturer (Thermo Separation Products) was used. Sugars and sorbitol were analysed in the column Aminex - HPX 87C with flow of 0,6 ml/min and at 85 °C. For mobile phase bidistillated water was used and RI detector for identification. Organic acids were analysed in the column Aminex - HPX 87H with flow of 0,6 ml/min and at 65 °C. For mobile phase 4 mM sulphuric acid (H_2SO_4) was used and UV detector with wavelength at 210 nm for identification. Soluble solids was determined in the juice with the refractometer (Kübler) at 20 °C. Fluka Chemical (New York, NY, USA) standards were applied for sugars, sorbitol and organic acids.

3. Results and Discussion

Contents of sucrose, fructose, glucose and sorbitol in the pear fruits cv. 'Williams' from the fruit set till the harvesting for the control and foliar fertilization are shown in Fig. 1 to 4. Content of sucrose and fructose during the growing period increase in both treatments, whereas the content of glucose and sorbitol decreases. Upon application of foliar fertilization the increase of the glucose content was notified in the treatment with foliar fertilization comparing to the control. At the same time the content of sorbitol was higher in the case of foliar fertilization than in the case of control. During the harvesting the control showed the content of fructose to be by 0.13gkg[-1] and the content of sucrose to be by

0.24 g kg⁻¹ higher than in the case of foliar fertilization.

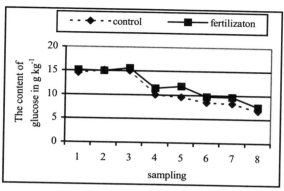

Figure 1. The content of glucose in g.kg⁻¹.

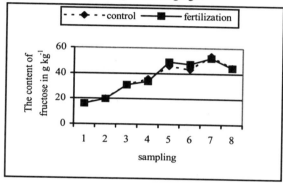

Figure 2. The content of fructose in g kg⁻¹

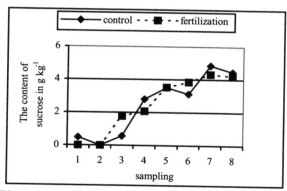

Figure 3. The content of sucrose in g kg⁻¹

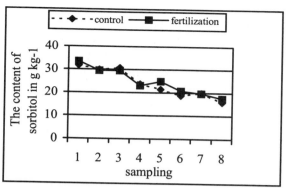

Figure 4. The content of sorbitol in g kg⁻¹

Table 1. The content of malic and citric acids (g kg⁻¹)

Date of sampling	malic		citric	
	C	F	C	F
1 June	3.13	2.90	4.04	3.07
16 June	3.60	3.29	2.67	2.00
29 June	3.51	2.81	0.29	0.25
13 July	1.90	0.83	0.90	0.46
27 July	0.26	0.00	1.39	0.89
3 August	0.31	0.00	1.18	2.05
10 August	0.23	0.43	1.91	1.17
17 August	0.26	0.15	1.39	1.59

Table 2. Total soluble solids (Brix) and titratable acids (use of NaOH (ml)).

date of sampling	soluble solids		titratable acids	
	C	F	C	F
1 June	9.13	9.52	/	/
16 June	8.55	8.82	4.23	4.22
29 June	10.19	9.97	4.58	3.81
13 July	10.73	10.98	3.81	3.50
27 July	11.53	11.97	3.76	3.26
3 August	11.79	11.94	3.22	3.33
10 August	12.57	12.67	3.19	2.70
17 August	12.55	13.14	3.26	3.22

Table 1 gives the contents of malic and citric acids from the fruit set till the harvesting, and the Table 2 shows values for total soluble solids and titratable acid expressed in the use of 0.1 N NaOH. Content of malic and citric acids during the growing period decreases in both treatments. Content of malic acid is higher during the harvesting in the case of control if compared to that of foliar fertilization which is probably due to higher content of Ca in fruits as the control showed 5.5 mg Ca/100 g of fresh fruit, whereas foliar fertilization demonstrated 4.9 mg Ca/100 g of fresh fruit. Higher content of Ca in fruits preserves the malic acid, as there is the reverse proportion between the Ca content and assimilation of malic acid. (Kovács and Djedjro, 1994). During the harvesting the fruits sprayed with foliar fertilizer had by 0.2 gkg⁻¹ more citric acid. Also, the results obtained showed by 0.066 mg kg⁻¹ more fumaric acid which during the ripening period can be traced in fruits in very small quantities. Content of shikimic acid showed the same value in both treatments (0.3 g kg^{-1}). During the growing period and the harvesting the content of total soluble solids was higher in the case of foliar fertilization treatment due to the foliar spraying with the fertilizer having a high percentage of K. The ripe fruits in the case of foliar fertilization contained 78 mg K/100 g of fresh fruit, and in the case of control the value was 74 mg K/100 g of fresh fruit. Also Marcelle (1995) determines positive correlation between the K content and soluble sugars. Foliar fertilization treatment shows less titratable acids than in the case of control. Foliar fertilization resulted in better absorption of Mg and P in fruits, as the cotrol showed too high levels of P and Mg content in the case of control. Content of nitrogen in fruits during the harvesting was the same in both treatments (45 mg N/100 g of fresh fruit).

It was determined that foliar fertilization results in the increase of content of glucose, sorbitol and soluble solids, content of K in fruits and in the decrease of malic acid and titratable acids. At the same time, foliar fertilization enables better absorption of P and Mg in fruits.

4. References

Arfaioli P and Bosetto M (1993) Time changes of free organic acid content in seven Italian pear (Pyrus communis) varieties with different ripening times. Agr. Med. 123:224-230.

Berüter J (1985) Sugar accumulation and changes in the activities of related enzymes during development of the apple fruit. Journal of Plant Physiology 121:331-341.

Doyon G, Gaudreau G, St-Gelais D, Beaulieu Y, Randall CJ (1991) Stimultaneous HPLC determination of organic acids, sugars and alcohols. Can. Inst. Sci. Technol. J 1/2:87-94.

Kovacs E and Djedjro GA (1994) Changes in organic acids of fruits after different treatments. Acta Horticulturae 368: 251-261.

Marcelle RD (1995) Mineral nutrition and fruit quality. Acta Horticulturae 383: 219-226

Morvai M and Molnár-Perl I (1992) Stimultaneous Gas Chromatographic Quantitation of Sugars and Acids in Citrus Fruits, Pears, Bananas, Apples and Tomatoes. Chromatographia 9/10: 502-504.

Seymour GB, Taylor JE, Tucker GA (1993) Biochemistry of Fruit Ripening. London, Chapman & Hall, 325-346.

Chapter 21

Influence of foliar fertilization on yield quantity and quality of apple (*Malus domestica* borkh.)

F. Štampar, M. Hudina, K. Dolenc, V. Usenik

University of Ljubljana, Biotechnical Faculty, Agronomy Department, Institute for Fruit growing, Viticulture and Vegetable growing, Jamnikarjeva 101, 1000 Ljubljana, Slovenia

Key words: apple; *Malus domestica* Borkh.; yield; sugar; organic acid; quality; nutrition

Abstract: The influence of foliar fertilization with Zn, B, P, Ca (Phosyn programmeme) on yield, quality and quantity was studied on cvs. 'Elstar', 'Jonagold' and 'Golden Delicious'. Foliar fertilization improved the yield increase up to 30% and the share of the first class. The quality was determined by the contents of sugars (glucose, fructose and sucrose) and alcohol sugar sorbitol and organic acids (citric, malic, shikimic and fumaric) with HPLC (High Performance Liquid Chromatography) analyses and contents of mineral elements (Ca, K, Mg, N) in the fruits of apple trees. Contents of individual sugars and sorbitol of all cultivars were higher in Phosyn treatment than the control. Also, citric and malic acid contents were higher, whereas differences in shikimic and fumaric acids were less. Fruit mineral elements showed that foliar fertilization caused better quantities and relations among elements of individual cultivars and higher contents of soluble solids with regard to control treatment. Complex foliar fertilization had positive impact on quantity and quality of fruits at different cultivars of apple tree.

1. Introduction

Nutrition of fruit trees is very important due to the decisive influence it has on the quality and quantity of the yield. In certain periods fruit trees require substantially larger quantities of nutrients than available from the soil. This supply can be worsened either by water deficit or water surplus. Occasionally ,fruit trees demand specific nutrients due to poor absorption from the soil which can be compensated with foliar fertilization.

In fruit orchards the foliar application of Ca is quite important, since the latter, when the water regime is unfavourable, influences the water status in cells and regulates the balance of cell solution, contributes to the electrostatic water equalization and enables the partition and length growth of cells. Increased share of the physiological and storage diseases of apples is correlated to lesser Ca content and higher

contents of N, P, K and Mg in the fruits (Glenn and Poovaiah, 1985; Raese and Staiff, 1990).

Potassium (K) takes part in the protein synthesis and regulates the osmotic water condition in the plant cells and influences the stomata regulation and net assimilation of CO_2. Because of its effect on the cell expansion potassium is of vital importance to increase the fruit growth, as well as in the carbohydrates storage, thus effecting the quality, pigmentation and organic acids accumulation. Phosphorus (P) is engaged in the regulation of many enzyme processes and energy transport mechanizm which effects the synthesis of sugars and alcohol esters through the ATP activity. Foliar application of boron (B) in autumn or just before blooming is of great significance in spring periods with lower air temperatures, as it increases cell expansion in flowers on which frost has a break influence and therefore inhibits the flower drop (Crassweler et al, 1981). Boron

91

deficiency in apple fruits causes flatness and formation of corky tissue, whereas, high B in fruits particularly after the summer application decreases cracking of the epidermis of the fruits in cv. 'Elstar' (Zude et al, 1997). Zinc (Zn) in plants is engaged in the synthesis of proteins and the plant hormone auxin (IAA).

Nutritive condition of fruit plants is monitored through leaf and fruit analysis. According to Werth (1995), the following contents of nutrients are required in fruits if normal storage is to be achieved: for Ca 4 - 5 mg/100 g of fresh fruit, K 100 - 130 mg/100 g of fresh fruit, Mg 4 - 5,5 mg/100 g of fresh fruit, N 35 - 50 mg/100 g of fresh fruit. Rate of potassium to calcium plays an important role in fruit nutrition and is to be below 30.

2. Materials and Methods

A macro experiment (4ha) of foliar applied nutrients was accomplished in 1997 in the Hmezad - Sadjarstvo Mirosan d. d. orchards on cv. 'Elstar'(E), 'Jonagold' (J) and 'Golden Delicious' (G) to examine the effect of Zn, B, P and Ca on quantity and quality of yield. Cultivars were planted in one row planting system with the planting distance of 4 x 1,6 m, with cv. 'Jonagold' and 'Elstar' and of 3,5 x 1 m with cv. 'Golden Delicious'. The experiment was set up as two treatments: control (C) and foliar nutrient application according to the Phosyn (F) (Zintrac – first application before leafing and second before leaf drop, Bortrac – two times before and once after blooming, Seniphos – four times in the period of fruit cell division, Stopit – four times in the period of fruit enlargement) programmeme in two repetitions (2 x 20 trees). The following parameters were studied: number of flower buds, fruit number and fruit weight per quality class/tree and yield coefficient. After harvesting 10 fruits were chosen at random from each application and then analysed for each element. Soluble solids were determined using 50 fruits (TSS%). Sugars and organic acids were determined by HPLC (High Performance Liquid Chromatography) according to the Hudina and Štampar (1998) method.

3. Results and Discussion

The year of 1997 was climatically rather unfavourable due to the April temperatures being below 0 °C several times (from April 8 to 10 the minimum temperatures 5 cm above the ground were -10, -8,8 and -4,7 °C, those from April 12 to 14: -3,8, -4,4, -8,5 °C, whereas in the period from April 16 to 19, the records showed -6,4, -4,8, -1,5, -2,7 °C and 24th, 25th April -6,5 and -5,2 °C). Lower temperatures caused severe frost mainly with 'Jonagold' and partly with cv. 'Elstar' and 'Golden Delicious'. In addition, the growing period (July, August) was cooler and humid than the average of many years. Tables 1 and 2 show the number of flower buds, number and weight of fruits/tree per different quality class, yield coefficient and yield/ha.

Table 1. Mean number of flower clusters and fruits per tree per different quality class and yield coefficient of apple cultivars in different treatments; Mirosan, 1997.

Treat-ments	No. of flower clusters	No. fruits per tree		Yield coefficient
		1st classs	2nd class	
E – C	170.2	128.3	21.9	0.88
E – F	168.9	185.1	61.4	1.57
J – C	174.5	71.2	3.1	0.43
J – F	189.9	111.6	4.8	0.61
G – C	187.7	122.6	18.9	0.75
G – F	187.7	143.2	18.9	0.86

Table 2. Mean yield per tree per different quality classes, percentage of 1st quality fruits and yield per hectar of various apple cultivars in different treatments; Mirosan, 1997.

Treat-ments	Yield per tree kg		% of 1st quality	Yield per hectar in (kg)
	1st class	2nd class		
E – C	17.1	2.0	89.5	26740
E – F	21.6	5.8	78.8	38360
J – C	15.3	0.3	98.1	21840
J – F	23.4	0.5	97.9	33460
G – C	20.7	1.8	92.0	58500
G – F	23.6	2.0	92.2	66560

Results of foliar application of Zn during the budding which stimulates the synthesis of proteins and auxins and the auxin activity (Faust, 1989), of B prior to blooming and immediately after it which effects the increased cell division in a blossom and better fructification

(Crassweler et al, 1981), of Phosphorus regulating the enzyme processes and cell numbers (Morin et al, 1994), of Ca which effects the stability of the cell wall (Glenn in Poovaiah, 1985) and consequently the storage capacity and of Mn with its impact on the chlorophyll synthesis (Faust, 1989) and the green colour of the fruit, show a notable impact on the yield coefficient, and above all on the quantity and quality of yield. With respect to the low temperatures in April and the repeated frost, the effect of the added boron is quite notable also in larger number and weight of fruits/tree with cv. 'Elstar', 'Jonagold' and 'Golden Delicious'.

Table 3. Content of different elements in mg/100 g fresh fruit and soluble solids (SS%) content of various apple cultivars in different treatments; Mirosan, 1997.

Treat ment	Ca	K	Mg	N	K/Ca	SS%
E – C	3.6	133	5.3	53	36.9	11.5
E – F	4.7	128	5.9	35	27.2	12.0
J – C	3.2	91	4.0	31	28.4	13.2
J – F	2.8	95	4.4	36	33.9	13.8
G – C	3.6	114	4.8	30	31.6	13.0
G – F	3.4	121	4.7	44	35.5	13.0

Calcium content which should be between 4 to 5 mg/100 g of fresh fruit (Werth, 1995), was obtained in cv. 'Elstar' with the treatment Phosyn (Table 3). Potassium deficiency in fruits was noticed in cv. 'Jonagold', whereas in cv. 'Elstar' the content of K – which is the antagonist of Ca - noted to be high in the control. The content of Mg was somewhat higher in cv. 'Elstar' (treatment Phosyn), whereas in the rest of the cultivars it was within the standard quantities. Similar to the findings of K, the high N was determined in cv.'Elstar' (control), which is most likely due to the low content of Ca. Rate of K to Ca which is assumed to be an important criteria in tree nutrition should be below 30, however such a level was not achieved for cv. 'Elstar' (control), 'Jonagold' (Phosyn) and cv. 'Golden Delicious' (both treatments). Soluble solids was higher in the phosyn treatment than in the case of control for all cultivars, the only exception being cv. 'Golden Delicious' where the same quantity was measured in both treatments. With respect to the five times

application of Ca (Stopit), higher fruit Ca contents were expected. However, such contents were distinctly recorded only when a tree had a normal yield (cv. 'Elstar' - Phosyn) and when the ratio between K/Ca was smaller than 30. Should the ratio be above 30, then K demonstrates negative impact on Ca absorption in plant which cannot be recovered by foliar application of Ca. Application of diverse nutrients had a positive impact on the content of soluble solids.

Table 4. Sucrose, glucose, fructose and sorbitol of fresh fruit (g kg^{-1}) of various apple cultivars in different treatments; Mirosan, 1997.

Treatment	Sucrose	Glucose	fructose	sorbitol
E – C	33.93	15.35	47.02	2.67
E – F	38.71	15.42	51.14	3.49
J – C	38.88	21.62	63.45	5.06
J – F	41.04	22.28	63.31	6.38
G – C	34.54	22.80	65.94	3.44
G – F	39.37	23.53	66.24	4.59

Glucose, fructose, sucrose and sorbitol contents was higher in all cultivars when treated with Phosyn (Table 4).

Table 5. Citric and malic (g kg^{-1}), shikimic and fumaric acid contents (mg kg^{-1}) of fresh fruits of various apple cultivars in different treatments; Mirosan, 1997.

Treatment	citric	malic	shikimic	fumaric
E – C	0.047	8.56	4.38	0.72
E – F	0.341	8.78	4.54	0.63
J – C	0.483	11.52	8.48	0.76
J – F	0.127	11.71	7.90	0.67
G – C	0.290	7.35	5.13	0.58
G – F	0.652	8.72	5.59	0.69

Malic and citric acid contents which are the most important organic acids in the apple fruits, were higher in all cultivars in the case of the Phosyn treatment. Shikimic and fumaric acids are found in substantially lower quantities than the citric and malic acids and their share to total acids remain modest. Despite unfavourable weather conditions foliar fertilization with Zn, B, P, Ca had very positive influences on the quantity and quality of the apple fruits in cv. 'Elstar', 'Jonagold' and 'Golden Delicious'.

The application of the foliar fertilizer Phosyn proved to be very efficient in terms of the increasing yield. From ecological point of view foliar fertilization is more acceptable due to

94

smaller quantities of nutrients provided for immediate consumption by a plant. Experiment results of one year indicate positive trend in the direction of a more economical apple production.

4. References

Crassweller RM, Ferree DC and Stang EJ, (1981) Effects of overtree misting for bloom delay on pollination, fruit set, and nutrient element concentration of 'Golden Delicious' apple tree. J. Amer. Soc. Hort. Sci. 106(1): 53-56.

Faust M (1989) Physiology of Temperate Zone Fruit Trees. John Willey & Sons, New York 338 p.

Glenn GM, Poovaiah BW (1985) Cuticular permeability to Calcium compounds in 'Golden Delicious' apple fruit. J. Amer. Soc. Hort. Sci. 110(2):192-195.

Hudina M, Štampar F (1998) Influence of foliar fertilization on quality of pear (*Pyrus communis* L.) cv. 'Williams'. Unpublished data.

Morin F, Fortin JA, Hamel C, Granger RL, Smith DL (1994) Apple rootstock response to vesicular-arbuscular mycorrhizal fungi in a high phosphorus Soil. J. Amer. Soc. Hort. Sci. 119:578-583.

Raesse J, Staiff D (1990) Fruit Calcium, Quality and Disordes of Apples (*Malus domestica* Borkh) and Pears (*Pyrus communis*) Influenced by Fertilizers. In: Plant Nutrition-Physiology and Applications, Kluwer Academic Publishers 619-623.

Werth K (1995) Farbe & Qualität der Südtiroler Apfelsorten.- Vog, Bozen, 88 p.

Zude M, Alexander A, Lüdders P (1997) Einfluβ von bor-sommerspritzung auf den borgehalt und die lagerungseigenschaften der apfelsorte 'Elstar'. Erwerbsobstbau 39: 62-64.

Chapter 22

The recovery of citrus from iron chlorosis using different foliar applications. Effects on fruit quality.

M. Pestana[1], D.A. Gonçalves[1], A. De Varennes[2], E.A. Faria[1]

[1]U.C.T.A. – Universidade do Algarve, Campus de Gambelas, 8000 Faro – Portugal.

[2] Instituto Superior de Agronomia, Depto de Química Agrícola e Ambiental, Tapada da Ajuda, 1300 Lisboa - Portugal

Key words: Citrus; chlorophyll density; foliar applications; fruit quality.

Abstract: The response of Encore trees grown on a calcareous soil to different foliar applications to offset iron deficiency. Four treatments were tested: distilled water (control); iron (II) sulphate (500 mg Fe.L^{-1}); sulphuric acid (0.5 mM H_2SO_4) and Fe-EDDHA (120 mg Fe.L^{-1}). The recovery from iron chlorosis was evaluated with the SPAD apparatus and the values converted to total chlorophyll density. We also evaluated effects of the treatments on some physical and chemical characteristics of the fruit.Chlorophyll density in the leaves, and the total sugar content of the fruits, were greater in all experimental treatments, compared with control, but with no significant differences between treatments. The concentration of citric acid decreased in the treated plants. The greatest diameter and fresh weight of fruits were obtained in the treatment with iron chelate. Foliar applications of iron sulphate or sulphuric acid led to values of these parameters that were intermediate between those of the control and the iron chelate treatments.

1. Introduction

The quality of citrus fruits depends on several parameters, namely fruit diameter, acidity, and the sugar content. Fruit maturation is related to seasonal variation of some juice components. For example, during maturation, the percentage of citric acid decreased from 2.2% to 0.5% (v/v of juice) and inversely, sugar concentration, expressed as total soluble solids, increased from 9° to 12° brix (El-Kassas, 1984; Spiegel-Roy and Goldschmidt, 1996).

It is known that fruit acidity varies with citrus species, and even between cultivars. This parameter is affected by plant nutrient status and climate (El-Kassas, 1984; Spiegel-Roy and Goldschmidt, 1996). The synthesis of citric acid is dependent on the aconitase enzyme, which catalyses the isomerization of citrate to isocitrate (Marschner, 1997).

The correction of iron chlorosis in trees is normally achieved by the application of iron chelates, like Fe-EDDHA, to the soil (Legaz et al., 1992). In calcareous soils, iron is rapidly immobilised and this practice has to be repeated every year. The cost of this operation may reach 60% of total fertiliser costs. Environmental impacts are also expected; for instance, synthetic chelation agents may decrease the absorption of some other metals, like manganese, copper and nickel (Wallace et al., 1992). Foliar sprays with acids, may release iron immobilised within the plant, and provide an alternative treatment (Sahu et al., 1987; Tagliavini et al., 1995). The aim of this work was to evaluate the recovery of citrus trees established on a limestone soil, and suffering from iron chlorosis using different foliar applications, and to study the effects of the treatments on fruit quality. The compounds we tested may act as a source of iron (iron sulphate and iron chelate) or may mobilise the iron pool in chlorotic leaves (sulphuric acid).

2. Materials and Methods

Sixteen Encore trees (*Citrus reticulata* Blanco), grafted on *Citrus aurantium* L., were randomly selected from a grove established on a calcareous soil (total $CaCO_3$= 47%; active lime=12%; $pH(H_2O)$=7.8) located in Almancil (Algarve, Portugal). In the beginning of the experiment all trees were very similar in shape and size, and showed a moderate chlorosis (chlorophyll density averaging 279.30 ± 4.82 $mg.m^{-2}$).

Four different foliar sprays were tested: distilled water (control); iron sulphate (500 mg $Fe.L^{-1}$), sulphuric acid (0.5 mM H_2SO_4) and Fe-EDDHA (120 mg $Fe.L^{-1}$). For each treatment, four trees were sprayed, generally every fortnight, from the 15 August 1997 (when fruit diameter had a diameter of approximately 30 mm) to 14 December 1997. The crop was harvested on the 29 December. No foliar applications were possible in October due to adverse weather conditions. Treatments were applied to at least four branches per tree, selected to have different orientations.

Chlorophyll density was determined just before the next spray application, in all the branches selected, using the SPAD apparatus. Each reading consisted of the mean value for 5 leaves. Three readings were made per branch. In all at least 36 values were made per treatment on each occasion. Chlorophyll density was determined using the equation:

$$Y= 6.8352 * X - 53.846$$

where Y is the chlorophyll density ($mg.m^{-2}$) and X is the SPAD reading (r^2=0.93; p=0.001; n=25). This equation was obtained in a separate essay, in which discs were cut from the leaves and the chlorophyll density determined according to Abadia and Abadia (1993).

To study the effect of foliar application on fruit quality, at least 10 fruits at a uniform stage of maturity were collected from each of the selected trees. Fresh weight, diameter, juice content (expressed in ml or as a percentage of whole fruit weight), total sugar concentration (expressed by degree brix), and citric acid concentration (expressed as v/v of juice) were determined for each fruit. All these analysis were carried out according to A.O.A.C methods (1975).

To determine if fruit diameter could be used to estimate fruit fresh weight on the tree, the correlation between these parameters was tested.

The effects of treatments were evaluated with analysis of variance (ANOVA) and Duncan Multiple Range Test (DMRT) at 5%.

3. Results

The recovery from iron chlorosis was determined by changes in the chlorophyll density of the leaves. Foliar application of iron or sulphuric acid led to similar values of chlorophyll density, which were greater than those of the control (Fig. 1). Since no foliar applications were done in October, the upward trend in the chlorophyll density observed in the experimental treatments was reserved. Nonetheless, at the end of the experiment, the recovery from iron chlorosis was visible in all treatments, with the exception of control. The fresh weight per fruit was related to fruit diameter by the equation:

$$Y = 3.6942 * X - 134.15$$

where Y is the fresh fruit weight and X is the fruit diameter (r^2=0.97; n=290; p=0.0001).

Fresh weight and fruit diameter increased in all experimental treatments, compared with controls, but the most effective was with iron chelate (Table 1). The percentage of juice relative to whole fresh weight was similar in all treatments. However, the treatments with iron chelate or with sulphuric acid led to greater volume of juice.

The concentration of citric acid in the fruit decreased, and the total sugar concentration increased, in all experimental treatments, compared with control (Table 1). However, no significant differences were observed between experimental treatments.

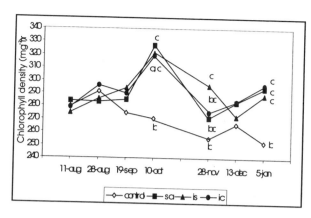

Figure 1. Variation of total chlorophyll density (mg.m^{-2}) in leaves of orange trees during the experiment. Treatments were: control (c), iron sulphate (is), sulphuric acid (sa) and iron chelate (ic). In each date, means followed by the same letter are not significantly different at 5% (Duncan test). Statistical analysis is shown only for the dates in which significant differences were obtained.

Table 1. The effect of treatments on fresh weight (FW), diameter (D), juice content (JC), total sugar content (TSC), and citric acid concentration (CAC) of Encore fruits. Means in a column followed by the same letter are not significantly different at 5% (Duncan test).

Treatment	FW (g)	D (mm)	JC (ml)	JC (% FW)	TSC (° brix)	CAC (% v/v J)
Control	66 c	55 c	32.5 c	46.8 a	8.45 b	2.0 a
Iron sulphate	87 b	60 b	39.2 bc	48.2 a	9.02 a	1.7 b
Iron chelate	107 a	65 a	50.7 a	47.2 a	9.22 a	1.5 b
Sulphuric acid	89 b	60 b	46.2 ba	47.6 a	9.16 a	1.4 b

4. Discussion

All experimental treatments resulted in a similar chlorophyll density within the leaves at the end of the experiment. Each of the sprays was effective in re-greening chlorotic leaves. Foliar application of iron likely increased the available iron pool within the leaves. Results of the sulphuric acid sprays appeared to have improved the availability of leaf iron leading to increased chlorophyll synthesis (Sahu *et al.*, 1987) consistent with the conclusions of Mengel (1995). However, selection of a commercial spray product needs also to consider effects on fruit quality.

The fresh weight of the fruit, their diameter and juice content (ml) were greatest following treatment with iron chelate. This is consistent with the view that iron in a chelated form is more stable and available than in the ionic form. All experimental treatments improved total sugar and citric acid concentrations over controls, as also reported by El-Kassas (1984) for limes.

The only measure of fruit quality that was not affected by treatments was the percentage of juice relative to fresh weight. It seems reasonable to assume that this parameter will depend more on soil water than on iron status.

Fruit diameter was strongly correlated with fruit fresh weight, and can provide a non-destructive method to estimate fruit development on the three.

In this experiment the treatments were applied during fruit growth and maturation. The impacts of offsetting iron chlorosis, prior to fruit formation, on subsequent quality await further investigations.

Acknowledgements

This work was supported by the EC program AIR3 -CT94 - 1973, and by the Praxis XXI program, project 3/3.2/HORT/2160/95.

5. References

Abadia, J; Abadia, A (1993) Iron and plant pigments. *In*: Iron chelation in plants and soil microorganisms. Academic Press. pp. 327-343.

AOAC (1975) Association of Official Agriculture Chemists. Official Methods of Analysis. 12th. Ed. Washington, D.C.

El-Kassas, SE (1984) Effect of iron nutrition on the growth, yield, fruit quality, and leaf composition of balady lime trees grown on sandy calcareous soils. J. Plant Nutr. 7: 301-311.

Legaz, F; Serna, MD; Primo-Millo, E; Martin, B (1992) Leaf spray and soil application of Fe-chelates to Navelina orange trees. Proc. Int. Soc. Citriculture 2: 613-617.

Marschner, H (1997) Mineral nutrition of higher plants. 2[a] ed. Academic Press. pp. 313-324.

Mengel, K (1995) Iron availability in plant tissues – iron chlorosis on soil calcareous *In* Iron nutrition in soils and plants. Kluwer Academic Publishers. Netherlands: pp. 389-397.

Sahu, MP; Sharma, DD; Jain, GL; Singh, HG (1987) Effects of growth substances, sequestrene 138-Fe and sulphuric acid on iron chlorosis of garden peas (*Pisum sativum L.*). J. Hortic. Sci. 62: 391-394.

Spiegel-Roy, P and Goldschmidt, EE (1996) Biology of Citrus. Cambridge University press. pp. 107-113.

Tagliavini, M; Scudellari, D; Marangoni, B; Toselli, M (1995) Acid-spray regreening of kiwifruit leaves affected by lime-induced iron chlorosis. *In*: Iron nutrition in soils and plants. Kluwer Academic Publishers, Netherlands. pp. 191-195.

Wallace, A; Wallace, GA; Cha , JW (1992) Some modifications in trace metal toxicities and deficiencies in plants resulting from interactions with other elements and chelating agents.- The special case of iron. J. Plant Nutr. 15: 1589-1598.

Chapter 23

The effect of foliar fertilization with KNO_3 on quality of dry wines

H. Kalkan[1], M.E. İrget[2], A. Altındişli[2], S. Kara[2], M. Oktay[2]

[1] Ege University, Food Engineering Department, Bornova, İzmir, Turkey.
[2] Ege University, Faculty of Agriculture, Bornova, İzmir, Turkey.

Key words: fertilization, KNO_3, wine quality

Abstract: The goal of foliar fertilization with KNO_3 is to obtain efficiency and dominancy of a determined quality criteria in grapes and respectively in wines. This study was carried out to determine the effect of foliar fertilization with different KNO_3 doses (control, 1%, 2%, 3%) on quality of white (Colombard) and red (Carignane) wines. Considering the chemical and organoleptical results, it could be concluded that the effects of foliar fertilization with KNO_3 on wine quality is significant in the range of 1 % KNO_3 for Colombard and 2 % KNO_3 for Carignane.

1. Introduction

World production of wine has steadily risen over recent years. Despite the fact, surplus of high quality wines is not found. There are a lot of environmental and viticultural inputs which are related to wine quality. The main factors are grape composition (soil and water; competition; canopy management; climate; vine growth; rate of maturation), harvesting time, vinification technique (temperature of fermentation; used quantity of SO_2; duration of pomase contact; stabilizing agent types), aging time and place of aging (Jackon and Lombard, 1993). In investigations on Cabernet sauvignon (Acetituno et al., 1987) and Chenin blanc (Jacson and Lombard, 1993) fertilization, results showed insignificant effect of N on the quality of wines. Studies on N for Riesling (Spayol et al., 1993) and on P and K for Foch grapes in British Colombia showed that N and K increased yield (Meneely and Battarbee, 1976; Jackon and Lombard, 1993). In some cases K increased must pH however did not affect soluble solids (Berg et al., 1979; Soyer et al., 1992; Jackon and Lombard., 1993). Potassium is associated with grape composition

(Amen, 1973; Cox et al., 1977) and wine quality and its present appearance is linked with acids and pH (Wong and Caputi,1966; Boulton 1980; Haba et al,1997). Musts which contain high amounts of K tend to have high pH and high malate, although during vinification malate may drop, pH may increase even further. Generally, tartarates are not affected by K content (Amerin, 1958; Hale.1977).

This study was carried out to determine the effects of foliar fertilization with KNO_3 on the wine quality of two grape varieties.

2. Materials and Methods

Ten year old Carignane and Colombard grape varieties were foliar fertilized with KNO_3 at 0, 1 %, 2 % and 3 % concentrations on June 21 and on July 5 in 1996. The experiment was performed in randomised complete block design with 3 replicates in a vineyard with a 3x2.5 planting distance. Grape samples were hand picked and processed into wines at Food Engineering Department, Ege University. Since the study was carried out to determine the effect of foliar fertilization (control, 1 %, 2 %, 3 %) on wine quality of each grape variety, grapes were divided into four groups. According to fertilization programme, wine production was considered in triplicate. For each grape variety the propagation of fermentation was done by pure yeast culture (1.5 % *S. cerevisiae var. elipsoideus*). Before inoculation (12 hours), as an antioxidant an antimicrobial agent was used, 5 % SO_2 solution (30 mg kg^{-1} for white must and 25 mg kg^{-1} for red must). Fermentation was carried out at 25 ± 1°C. The other parameters of white and red wine productions were done according to the traditional procedures for each category of wine (white and red; dry wines). After racking 2 times, filtration and bottling, chemical and organoleptical analyses were performed. For complete evaluation of wine quality, density (aerometrically), total soluble solids (TSS), pH, total acidity of must (Ough and Amerin. 1987) total, bound and free SO_2, alchohol content, volatile acidity (Akman, 1962), total acidity (Ough and Amerine,1987) were performed. Organoleptical analyses were conducted according to OIV (Office International du Vin) tests. Statistical analyses were conducted with SPSS program by applying Anova and Duncan test.

3. Results and Discussion

Must analyses of Carignane and Colombard grapes are shown in Table 1 and Table 2, respectively.

According to the results of density concerning both grapes, there was a significant difference between control and the treated samples. As seen from the tables, the highest value of density was reached in the musts of 2 % KNO_3 treated grapes for white and red wines, respectively (1.079 and 1.078 gr/cm^3). Treatments with KNO_3 increased the leaf K and P contents and augmented the grape sugar content by 1.5 -2.0 % compared with the control (Jackon and Lombard, 1993).

The pH values of KNO_3 applied grapes were higher in both varieties according to the controls. This increase was more significant in 2 % KNO_3 and 3 % KNO_3 treated grape musts.

Boulton (Amen, 1973) reports an inverse relationship between the free hydrogen ion concentration and the potassium content for Australian red musts and wines of two vintage.

TSS values of KNO_3 applied Carignane and Colombard musts were higher than controls, excluding the 3%K treatment. The highest value of TSS was reached in 2 % KNO_3 treatment

The titratable acidity for both types of musts were quite lower than their corresponding controls.

Chemical analyses of Carignane and Colombard wines are shown in Tables 3 and Tables 4, respectively.

Related to the SO_2 results of both grape wines, the lowest total SO_2 value was measured in the control treatment. Furthermore, the lowest free SO_2 value of Carignane wine was in 2 % KNO_3 treated samples. In the case of Colombard wines, there was no difference in free SO_2 values compared to the control. As seen from Table 3, the volatile acidity of Carignane had the lowest value 0.43 in 2 % KNO_3 treated samples but at the same time titratable acidity of the same sample had the highest value. Regarding the volatile acidity of Colombard wines, there was no significant difference between treatments. Titratable acidity had the highest value in 2 % KNO_3 treated samples.

Alcohol content of wines showed that there was no significant difference between the control and treated samples of Colombard wines but for the Carignane the highest value of alcohol content was not in the control but in the 1 % KNO_3 treated samples. Unvolatile acidity values of both wines showed the same correlation with the titratable acidity values. For both wine types, the unvolatile acidity values of the controls were lower than that of treated samples. In Colombard, 2 % KNO_3 resulted in a quite high value. Similar situation was seen in Carignane variety. The berry K results showed that there was no difference between the control and treated samples.

Table 1. Must analyses of Carignane grapes

Treatment	Density (g cm^{-3})	PH	TSS (%)	Titratable acidity (g L^{-1} tartaric acid)
Control	1.071c	3.853b	18.127d	6.172a
1 % KNO$_3$	1.076b	3.853c	18.635b	6.172a
2 % KNO$_3$	1.078a	3.873a	18.882a	6.132b
3 % KNO$_3$	1.071c	3.873a	18.362c	5.422c

$P < 0,05$

Table 2. Must analyses of Colombart grapes

Treatment	Density (g cm^{-3})	pH	TSS (%)	Titratable acidity (g L^{-1} tartaric acid)
Control	1.071c	3.653c	18.015d	6.205a
1 % KNO$_3$	1.077b	3.675a	18.500b	6.163c
2 % KNO$_3$	1.079a	3.713b	18.588a	6.185ab
3 % KNO$_3$	1.076c	3.665b	18.338c	6.178bc

$P < 0.05$

Table 3. Chemical analyses of Carignane wines

Treatment	Total SO$_2$	Free SO$_2$	Bound SO$_2$	K	Titratable acidity	Unvolatile acidity	Volatile acidity	Alcohol
	mg kg^{-1}				g L^{-1} (tartaric acid)			%
Control	72.25d	39.00b	33.25d	841n.s	5.05b	4.98b	0.20d	9.87ab
1 % KNO$_3$	154.50a	57.75a	96.75a	876n.s	5.38a	5.28a	0.32b	10.63a
2 % KNO$_3$	93.75b	25.25c	68.50b	886n.s	5.45a	5.32a	0.43a	9.60b
3 % KNO$_3$	84.00c	27.50c	56.50c	892n.s	5.48a	5.40a	0.23c	9.48b

$P < 0,05$, n.s: non signficant

Table 4. Chemical analyses of Colombard wines.

Treatment	Total SO$_2$	Free SO$_2$	Bound SO$_2$	K	Titratable acidity	Unvolatile acidity	Volatile acidity	Alcohol
	mg kg^{-1}				g L^{-1} (tartaric acid)			%
Control	75.00b	22.50n.s	55.00b	502n.s	7.84ba	7.75a	0.31n.s	10.40n.s
1 % KNO$_3$	150.00a	26.25n.s	123.70a	536n.s	7.77ba	7.68ba	0.29n.s	10.57n.s
2 % KNO$_3$	137.00a	24.25n.s	113.25a	544n.s	8.33a	8.24a	0.27n.s	10.43n.s
3 % KNO$_3$	82.50b	23.25n.s	59.25b	552n.s	7.25b	7.08b	0.28n.s	10.70n.s

$P < 0,05$, n.s: non signficant

Table 5. Organoleptical results of Carignane and Colombard wines (OIV).

Treatment	Carignane wine	Colombard wine
Control	18.7[a]	17.7[ab]
1 % KNO$_3$	18.8[a]	17.7[ab]
2 % KNO$_3$	18.9[a]	18.0[a]
3 % KNO$_3$	18.0[b]	17.4[b]

$P < 0.05$

Organoleptical results of both wine degustation are shown on Table 5.

As seen from Table 5 both wines are in the category of quality wines. According to the results of statistical analyses, there was no significant difference between control and, 1 % KNO_3; 2 % KNO_3 samples of Carignane wines. In the Colombard wines, the highest value was reached at 2 % KNO_3 treatment.

The organoleptical results concerning the wines of Colombard, the best result was reached definitely by 2 % KNO_3 treatment.

4. Conclusion

Considering the must and wine results of this research, it could be concluded that the effect of foliar KNO_3 treatments on Carignane and Colombard grapes is significant in the range of 1 % KNO_3 and 2 % KNO_3 for wine quality. For all chemical and organolleptical results, it could be evaluated that up to 2 % KNO_3 application for Colombard and up to 1 % KNO_3 for Carignane have positive effects on wine quality.

5. References

Acetituno, C., J. Merida, J.L. Gonzales, M. Medina, 1987, Effect of Nitrogen Fertilizers on Nutrients, Acids and Sugar in Leaves of Vitis vinifera "Pedro Ximerez", Anales de Edafolofia y Agrobiologia 46 (7-8), 951-961.

Akman, A. V. 1962. Şarap analiz metodları. Ankara Üniv. Ziraat Fak. Yayınları, No:33, Ankara.

Amen, R. J. 1973. Minerals as nutrients. Food Prod. Dev. 7(7): 32-42.

Amerine, M. A. 1958. Composition of wines II. Inorganic . Adv. Food Res. 8: 133-224.

Berg, H. W., Min Akiyoshi, M. A. Amerine, 1979, Potassium and Sodium Content of California Wines Am. J. Enol. Vitic. 30(1), 55-57.

Boulton, R., 1980, The Relationship Between total Acidity, Titratable Acidity and pH in Wine, Am. J. Enol. Vitic., 31(1), 76-80.

Cox, R. J., R. R. Eitenmiller, and J. J. Powers, 1977. Mineral content of some California wines. Food Sci. 42: 879-850.

Haba, M. A., Mulet and A. Berna, 1997, Stability in Wine Defferentiation of Two Close Viticultural Zones, Am. J. Enol. Vitic., 48(3), 285-290.

Hale, C.R. 1977, Relation Between Potassium ,Malate And Tartarate Contents of Grape Berries, Am. J. Enol. Vitic., 16(1), 9-19.

Jackon, D.I.and P. B. Lombard, 1993, Environmental and Management Practices Affecting Grape Composition and Wine Quality - A review, Am. J. Enol. Vitic., 44(4), 409-430.

Meneely, G. R. and H. R. Battarbee, 1976. Sodium and potassium. Nutr. Rev. 34: 225-235.

Ough, C. S. and M. A. Amerine, 1987. Methods for analysis of musts and wines. Univ. of California. 2nd ed. John Wiler and Sons.

Soyer, J. P.; J.Delas; C. Molot; B. Mocquot, 1992, Vineyard Cultivation Techniques, Potassium Status and Grape Quality, In Proceeding, Second Congress of the European Society for Agronomy, Warwick University 23-28 August, 308-309.

Spayol, S. E.; R. L. Wample, R. G. Stevens, R. G. Evans, A. K. Kavakami, 1993, Nitrogen Fertilization of White Riesling in Washington: Effect on Petiole Nutrient Concentration, Yield, Yield Components and Vegetative Growth, Am. J. Enol. Vitic., 44(4), 378-387.

Wong, G. and A. Caputi, 1966. A new indicator for total acid determination in wines. Am. J. Enol. Vitic. 17: 174-177.

Chapter 24

Effect of foliar applied KNO_3 on yield, quality and leaf nutrients of Carignane and Colombard wine grapes

A. Altındişli[1], M. E. İrget[1], H. Kalkan[2], S. Kara[1], M. Oktay[1]

[1] Ege University, Faculty of Agriculture, Bornova, İzmir / Turkey;
[2] Ege University, Food Engineering Department, Bornova, İzmir / Turkey

Key words: fertilization, potassium nitrate, grape quality

Abstract: KNO_3 foliar applications given in different doses effected the yield, 100 berry weight, TSS, acidity, pH, TSS/acidity, N and K content of leaf blade and petiole in Carignane and Colombard wine grapes. Results showed that 1 and 2 % KNO_3 applications increased the yield and hundred berry weight compared to control and had positive effect on TSS. Because of leaf blights caused by 3 % KNO_3 application, certain problems appeared despite an increase in yield and some other parameters according to control. Foliar KNO_3 applications caused to increase N and K contents of leaf blade and petiole. It was concluded that the doses of 1 and 2 % can be suggested for agricultural practise.

1. Introduction

Roots are the primary organs where plants obtain their nutrients. Existence of factors which limit the availability of nutrients in the soil decrease the expected benefit from fertilizers. Under these conditions, nutrients can be provided to plants by foliar applications.

At present, KNO_3 is a significant fertilizer applicable to leaves to achieve increases in yield and quality in different plants. KNO_3 is composed of 13% N and 46% K_2O. It is very soluble in water and commonly used in vineyards after blooming. Potassium has been recognized as a factor of considerable influence on grapes and consequently on wine by affecting pH, color, fermentation processes and ultimately, the flavor and clarity of the bottled wine (Hepner and Bravdo, 1985).

Grape yield, berry size, TSS and acidity have great importances in terms of wine production in wine-grape varieties. Despite the fact that it can be changed according to the type of wine to be produced, the ratio of TSS/acidity is an important criteria besides the quantity.

The present trial was conducted to determine the effects of foliar applied KNO_3 doses on yield, quality and leaf nutrient contents in Carignane and Colombard wine-grape varieties which have economic importance.

2. Materials and Methods

The trial was carried out in a Carignane and Colombard vineyard in Menderes, İzmir through the growing season of 1996. The 10 year old vineyard was established with rooted cuttings at a distance of 3 x 2.5 m and trained as double cordon trellis system. Soil characteristics of the experimental area are given in Table 1.

Table 1. Results of soil characteristics

Soil Depth	pH	CaCO3 (%)	Salt (%)	O. M. (%)	Sand (%)	Silt (%)	Clay (%)	Texture
0-30 cm	7.45	0.5	<0.03	1.34	62.88	28.00	9.12	Sandy-loam
30-60 cm	7.44	0.5	< 0.03	0.54	58.88	28.00	13.12	Sandy-loam

Soil Depth	Total N %	Available (mg kg^{-1})								
		P	K	Ca	Mg	Na	Fe	Zn	Mn	Cu
0–30 cm	0.084	2.76	130	2400	130	25	4.69	0.88	2.86	1.39
30–60 cm	0.075	1.90	120	1980	120	20	4.50	0.82	2.44	1.35

The trial was designed in randomized blocks with 4 replications. Three different KNO_3 doses of 1, 2 and 3 %, and control were the treatments. First application was made in 21.6.1996, the second one 15 days after. Leaf samples were taken from each replication after the applications at the beginning of ripening. The leaves opposite to the first fruit cluster on each cane were sampled (Cooke and Carson, 1961; Beyer, 1962; Levy, 1968).

Ten vines were chosen among the thirty treated vines in each replication and harvested in 19.9.1996. Besides, two hundred berries were collected from each replication according to Amerine and Cruess (Amerine and Cruess, 1960) method. The yield (kg/vine), hundred berry weight (g), TSS (%), acidity (g/1) and pH were determined. Nitrogen and K contents of leaf blade + petioles were measured after wet digesting (Christense et al., 1994; Kacar, 1995). Statistical analyses of the data were made by SAS program and Duncan test were used for evaluation (SAS, 1989).

3. Results and Discussion

Results related to each variety are separately given in Tables 2 and 3.

Yield: All of the three KNO_3 doses significantly increased the yield in both varieties according to control. In both varieties, the highest yield was obtained from the application of 2 % KNO_3, followed by 1 % and 3 % KNO_3 doses. It is concluded that the effects of KNO_3 applications on the yield could be due to the increases in berry weights. Likewise in a trial conducted in Seedless Sultanina in Aegean Region, two applications of 1 % KNO_3 at the beginning of ripening and at the maturation, caused increases in hundred berry weights (Çokuysal, 1990). It is considered that the decrease in the yield of the highest dose of 3 % KNO_3 may be related to the partial blights of the leaf margins. In this regard, the results of leaf analysis showed that low petiole NO_3 concentration is accompanied by higher yields (Spayd et al., 1993).

Hundred berry weight: It is clear that all KNO_3 doses increased hundred berry weight in both varieties compared to control. The largest berry was measured in 2 % KNO_3 dose followed by 1 % 3 % and control respectively.

Total soluble solids: All KNO_3 doses increased TSS in both varieties compared to the

Table 2. Effect of different KNO_3 doses on yield, some quality and leaf nutrient contents in Colombard grapes.

	Yield[1] (kg/vine)	100 berry[1] (g)	TSS[2] (%)	Acidity[2] (g/l)	pH[2]	TSS / Acidity[1]	N (%) Petiole[2]	N (%) Blade[2]	K (%) Petiole[2]	K (%) Blade[1]
Control	9.20 b	175 d	19.4	7.27	3.00	2.66 a	0.455	1.902	0.760	0.315 c
1%KNO₃	11.37 a	206 b	19.7	8.13	3.13	2.42 c	0.518	1.990	0.850	0.473 b
2%KNO₃	11.40 a	211 a	19.8	8.53	3.16	2.32 d	0.550	2.045	1.020	0.665 a
3%KNO₃	11.37 a	182 c	19.6	7.97	3.03	2.45 b	0.565	2.108	1.353	0.718 a

1: p < 0.05. 2 : non significant

control, however, results were not statistically significant. The highest value in terms of TSS was determined in 2 % treatment. In 3 % application reductions were seen. TSS is one

deacidification may be required (Jackson and Lombard. 1993).

pH: Treatments did not statistically effect pH levels in two varieties. pH levels varied

Table 3. Effect of different KNO_3 doses on yield. some quality and leaf nutrients in Carignane grapes

	Yield[1] (kg/vine)	100 berry[1] (g)	TSS[2] (%)	Acidity[2] (g/l)	pH[2]	TSS / Acidity[1]	N (%) Petiole[2]	Blade[2]	K (%) Petiole[2]	Blade[1]
Control	8.40 c	164 d	17.7	5.66	3.33	3.12 a	0.530	1.840	0.680	0.413 c
1%KNO$_3$	10.27 b	199 b	18.6	6.40	3.40	2.90 b	0.570	2.053	1.115	0.643 b
2%KNO$_3$	12.00 a	201 a	19.2	6.16	3.43	3.11 a	0.610	2.060	1.275	0.722ab
3%KNO$_3$	9.27 b	171 c	18.2	6.43	3.40	2.83 c	0.650	2.047	1.305	0.800 a

1: $p < 0.05$. 2 : non significant

of the most important criteria in wine-grape varieties in terms of wine quality. Lafon (1955) states that 1 % foliar KNO_3 application increased bunch weight, SS + TSS values per vine in a study conducted in three different locations of France in terms of crop and quality criteria (Lafon and Couillaud. 1955). In a similar study (Rose, 1980), important increases were reported due to four application doses of KNO_3 on Thomson seedless in terms of fruit setting, crop quantity per vine, average bunch weight and brix. In this present study, the increases in TSS achieved in all applications excluding the control might be due to the positive effect of K, however, the decrease in the highest dose might be from N. Therefore, mutual effects of K and N should be taken into consideration in terms of high quality production when K and N were applied together. Likewise, in a study conducted by Christensen et al. (1994) on four different grape varieties including the Colombard, decreases in TSS has been reported under extreme nitrogen fertilization (Christense et al.. 1994).

Acidity: No statistically important effect of KNO_3 doses was found on acidity in two varieties. Failla et al. (1996) weekly sprayed %0.75 KNO_3 to Chardonnay and Cabernet Sauvignon cultivars starting from fruit set to the beginning of ripening and determined no important effects on acidity (Failla and et al., 1996). In another study made on Shiraz cultivar, a direct relation between potassium and tartrate was not found in ripe grapes (Hale, 1977). Wine with too much acid (10 g/l TA equivalents and above) is tart to the taste and

between 3.0 and 3.13 in Colombard, whereas 3.33 and 3.43 in Carignane. It is suggested that a pH level above 3.60 in wine may cause problems. High pH levels increase the relative activity of micro-organisms such as bacteria, lower the colour intensity in red wines, bind more SO_2 and reduce the free SO_2 content, and can shorten the ability of wine to age (Jackson and Lombard. 1993).

*TSS/acidity :*This ratio known as maturity index was statistically affected by the treatments in both varieties. The highest ratios were obtained from the control parcels of the two varieties.

Leaf N and K contents: It was observed that KNO_3 applications in two varieties increased both leaf blade and petiole nitrogen content but results were not statistically important. Christensen et al. (1994) report that NH_4NO_3 soil applications increase nitrogen content of leaf blades. The effects of foliar KNO_3 applications on K content of blades were not statistically significant in both varieties. However, applications increased leaf petiole K importantly. The highest K content in the leaf blade occurred at the dose of 3 %, followed by 2 % and 1 % in two varieties.

Present results notifying that foliar KNO_3 fertilization increase N and K contents of the leaves are in agreement with many similar studies conducted in vines (Soyer et al., 1992), grapefruits (Jones and Burns. 1972), Valencia oranges (Jones and et al.. 1967), tomatoes and cotton (Angelick et al.. 1970; Howard and Gmathmey. 1995). In long-term studies conducted in the vines of Bordeaux region,

France, it has been reported that KNO_3 applications increased K contents of leaf petiole, grape berries, must and wine (Soyer et al., 1992).

4. Conclusion

In conclusion, 1 and 2 % foliar applications of KNO_3 suggested to increase the yield, hundred berry weight and TSS and to improve leaf N and K contents, however, higher rates are to be avoided due to NO_3 accumulations in the leaves.

5. References

Amerine. M.A. and M.V. Cruess.1960. The Technology of Wine Making. The Avi Publishing Comp.. Inc.. Westport. Connecticut. U.S.A.. 709 pp.

Angelick . M. . M. Boreket and U. Montag.. 1970. The influence of potassium nitrate foliar spray on tomatoes. Field Extension Service. Ministry . of Agriculture (Hebrew).

Beyers . E. .1962 . Diagnostic Leaf Analyses for Decidious Fruit . South African Journal of Agricultural Sci. 5 (2):315-329 .

Christensen. L.P.. M. L. Bianchi. W. L. Peacock. and D. J. Hirschfelt. 1994. Effect of Nitrogen Fertilizer Timing and Rate on Inorganic Nitrogen Status. Fruit Composition. and Yield of Grapevines. Am. J. Enol. Vitic.. Vol. 45 (4): 377-387.

Cooke . J. A. and C.J. Carson .. 1961. California Vineyards Respond to Potash When Needed . Better Crops with Plant Foods . 45(3): 2

Çokuysal . B.. 1990. Sultani Çekirdeksiz Üzümde Gibberellik Asit ve Yaprak gübrelemesinin Yaprak Besin Elementleri ve Ürün Kalitesi Üzerine Etkileri . Ege Üni. Fen Bilim . Ens . Toprak Anabilim Dalı. Bornova- İzmir .

Failla.. O.. A. Scienza. and L. Brancadoro. 1996. Effects of Nutrient Spray Applications on Malic and Tartaric Acid Levels in Grapevine Berry. Journal of Plant Nutrition. 19 (1): 41-50.

Hale. C. R.. 1977. Relation Betweeen Potassium and the Malate and Tartrate Contentets of Grape Berries. Vitis. 16: 9-19

Hepner. Y.. and B. Bravdo. 1985. Effect of Crop Level and Drip Irrigation Scheduling on the Potassium Status of Cabernet Sauvignon and Carignane Vines and Its Influence on Must and Wine Composition and Quality. Am. J. Enol. Vitic. Vol 36 (2): 140-147.

Howard . D. D.. . and C. O. Gmathmey .1995 . Influnce of Surfactants on Potassium uptake and Yield Response of Cotton to Foliar Potassium Nitrate . Journal of Plant Nutrition. 12 (12): 2669-2680 .

Jackson. D. I. and P.B. Lombard. 1993. Environmental and Management Practices Affecting Grape Composition and Wine Quality- A Reviev. Am. J. Enol. Vitic. Vol. 44 (4): 409-430.

Jones . W . W .and R. M. Burns . 1972 . Correction of K Deficiency in Grapefruit. Calif. Citrograph. pp. 231-277.

Jones . W .W. .T. W. Emblleton .M. J. . Gamber . and C. B. Gree . 1967 . Creasing of Orange Fruit . Hilgardia. 38 (6) : 271.

Kacar . B . . 1972 . Bitki ve Toprağın Kimyasal Analizleri . II: Bitki Analizleri. A.Ü. Zir. Fak.Yayın No: 453. Uygulama Klavuzu.155. Ankara.

Kacar. B..1995. Bitki ve Toprağın Kimyasal Analizleri. III: Toprak Analizleri. A.Ü. Zir. Fak. Eğitim. Araştırma ve Geliştirme Vakfı Yayınları:3. Ankara.

Lafon . J. and P. Couillaud . 1955 .Extrait de la Revue . La Potasse p .265 .

Levy . J. F. .1968. L' Application du Diagnostic Foliaire la Determination de Vesions Alimentaires des Vines . Le Controle de la Fertilization de Planes Cultiv'ees. (II. Collog . Eur . Medit . Semilla . 1968) :295-305 .

Rose. J. .1980 . Effect of Supplemental Foliar and Drip Irrigation Applications of KNO_3 on Grapes. Haifa Chemicals Ltd. Israel.

SAS Institute Inc: SAS User's Guide. 6.03 Edition. 1989. Cary NC.

Spayd. S. E.. R. L. Wample. R.G. Stevens. R. G Evans. and A. K. Kawakami. 1993. Nitrogen Fertilization of White Riesling in Washington: Effects on Petiole Nutrient Concentration. Yield. Yield Component. and Vegetative Growth. Am. J. Enol. Vitic.. Vol.44 (4): 378-386.

Soyer. J.P.. J. Delas. J.. Molot. B. Mocout. 1992. Vineyard Cultivation Technigues. Potassium Status and Grape Quality. In: Proceeding Second Congress of Europan Society for Agronomy. 308-309.

Chapter 25

Effect of foliar applied Epsom Salt on sugar beet quality

A. Kristek[1], V. Kovacevic[2], E. Uebel[3] and Rastija M.[2]

[1]*Sugar Beet Breeding Institute, M. Divalda 320 HR-31000 Osijek, Croatia*
[2]*Faculty of Agriculture, Trg sv. Trojstva 3 HR-31000 Osijek, Croatia*
[3]*International Potash Institute (IPI), c/o Kali und Salz GmbH, D-34111 Kassel, Germany*

Key words: epsom salt, sucrose, sugar beet

Abstract: Low sucrose contents (average of 14.03 % in the period of 1991-1995) and satisfied root yields (37.5 t ha^{-1} respectively) characterize sugar beet growing in Crotia. The objective of this study was to examine the possibility to improve sugar beet, especially sucrose contents, by Epsom Salt foliar fertilization. Survey of a 3-year experience was shown in this study. Recognized (trial A) and experimental (trial B) sugar beet genotypes were grown in three experimental sites situated in a large farm for two growing seasons (1995 and 1996) in four replications with a plot size of 10 m^2. Two replications were fertilized with Epsom salt (ESF treatment), 5% (w/v) MgSO$_4$.7H$_2$O solution at rate of 400 l ha^{-1} twice at 10 day intervals in June, while the others were used as control. Results were shown as averages of three localities for each level of fertilization treatment. Sucrose contents were determined polarometrically, root K and Na by flamephotometre and root amino-nitrogen (a-N) colorimetrically. Depending on the growing season, sucrose ranged from 12.14 to 15.92 %, root yield from 38.1 to 58.0 t ha^{-1}, root K from 4.03 to 4.10 mmol 100 g^{-1} and a-N from 1.72 to 2.37 mmol 100 g^{-1}.

1. Introduction

Low sucrose contents (average of 15.55% in the period of 1986-1990 and 14.03% in the period of 1991-1995) and satisfied root yields (43.7 and 37.5 t ha^{-1} respectively) characterize sugar beet growing in Croatia. Average harvested area of sugar beet in Croatia for 30-year period of 1961-1990 was 24 000 ha year^{-1} (Kristek et al. 1998a).

The objective of this study was to examine the possibility to improve sugar beet quality, especially sucrose contents by Epsom salt foliar fertilization. Kristek et al., (1997 and 1998b) report their experiences on Epsom Salt foliar applications. Survey of 3-year experience was shown in this study.

2. Materials and Methods

For 1995 and 1996 growing seasons, recognized (trial A) and experimental (trial B) sugar beet genotypes were grown in four replications of plot size = 10 m^2 and in three localities of a large farm. Two replications were fertilized with Epsom salt (ESF treatment), 5% (w/v) MgSO$_4$.7H$_2$O solution at rate of 400 l ha^{-1} twice at 10-day interval in June, while the others were used as control. Sugar beet was sown at the beginning of April and it was harvested at mid of October. Results were shown as averages of three localities for each fertilization treatment and Trial A and B comprised 78 and 140 results for the year 1995 and 102 and 90 for the year 1996. In third year of testing, sugar beet hybrids were grown in eight replications (four ESF treatments and four controls) according to previous years of testing (A), but also small farmers (three trials, five replications of each treatment) were included in testing (B): the results for the 1997 are the averages of 228 and 15 individual data, for the A and B, respectively.

Table 1. Influence of Epsom salt foliar fertilization on sugar beet properties

Fertiliza tion	Experiment A						Experiment B					
	Suc.	Yield (t/ha)	mmol/100g root				Suc.	Yield (t/ha)		mmol/100g root		
	(%)	Root	Suc.	K	Na	a-N	(%)	Root	Su	K	Na	a-N
						(1995)						
Control	12.25	33.4	3.4	3.95	2.37	1.91	11.80	40.6	4.0	4.26	2.48	2.38
Mg	12.50	33.8	3.5	4.00	2.36	1.99	12.00	44.4	4.3	4.06	2.52	2.45
5%	0.19	ns	ns	ns	ns	ns	0.14	1.63		0.09	ns	0.06
1%	0.24						0.18	2.15	0.2	0.11		0.08
						(1996)						
Control	14.95	53.1	6.9	3.90	0.93	1.63	14.80	46.3	5.8	4.20	1.10	1.92
Mg	15.13	52.9	6.9	3.80	0.98	1.49	14.96	46.4	5.9	4.22	1.02	1.82
5%	0.11	ns	ns	ns	ns	0.05	0.07	ns	ns	ns	0.04	0.05
1%	0.14					0.06	0.09				0.05	0.06
						(1997)						
Control	15.18	55.2	7.1	3.83	1.32	2.48	16.20	59.0	8.2	4.29	0.79	2.41
Mg	15.44	57.4	7.5	3.95	1.29	2.45	16.87	60.2	8.8	4.32	0.61	2.12
5%	0.08	0.8	0.1	0.04	ns	ns	0.41	0.4	0.2	ns	0.06	0.15
1%	0.10	1.1	0.2	0.05			0.60	0.6	0.3		0.08	0.22

* Suc = sucrose; a-N = alfa amino nitrogen; ns = non significant differences on tested level of statistical significance (LSD 5% and LSD 1%)

Table 2. Weather characteristics for three growing seasons (1995, 1996 and 1997) and long-term means (the data of Osijek Weather Bureau)

Month	Rainfall (mm) and mean air-temperatures (°C)							
	1995		1996		1997		1961-1990	
	mm	°C	mm	°C	mm	°C	mm	°C
March	47	5.8	41	2.9	20	6.7	47	6.2
April	44	11.5	66	11.2	54	7.8	57	11.2
May	84	15.4	76	18.4	36	17.7	65	16.0
June	98	18.5	30	20.9	96	20.5	87	19.2
July	2	23.6	96	19.6	98	20.7	69	20.5
August	96	20.1	115	21.0	57	21.4	63	20.3
September	107	15.2	129	13.6	59	16.5	51	16.8
October	2	12.2	61	12.3	61	12.3	52	11.3
Total (mm)	480		614		481		491	
Mean (°C)		15.3		15.0		15.5		15.1

Sucrose contents were determined polarometrically, root K and Na by flamephotometre and root amino-nitrogen (a-N) colorimetrically.

3. Results and Discussion

Drought and heat stresses characterized the growing season from June 30 to August 9 in 1995: 26 mm of rainfall and average 21.2°C of air-temperature (Osijek Waether Bureau). In the 1996 and 1997 growing seasons, weather conditions were normal (Table 2). Degree of the individual growing season favorability for sugar beet growing is possible to estimate based on rainfall and air-temperatures in July: 2 mm and 23.6°C (1995), 96 mm and 19.6 °C (1996), 98 mm and 20.7 °C (1997).

In general, sucrose contents increased significantly by the ESF as follows; (for the 1995, 1996 and 1997, respectively) 0.25%, 0.18% and 0.26% in A trial and 0.20%, 0.16% and 0.65% in B trial in 1995, 1996 and 1997,respectively. Accumulation of amino-N in sugar beet root is harmful because of disturbance of sucrose crystallization (Burba 1996). Significant influences of ESF on a-N decreasing were found in 1996 growing season in both trials, as well in the 1997 for the trial B only. However, K and Na status in sugar beet roots were less depended on ESF. In general, weather conditions, soil properties and genotypes influenced yield and sugar beet quality more than ESF. We would recommend ESF as a practice of crop management for sugar beet, especially under the conditions with moderate Mg and/or S supply.

4. Conclusion

Sugar beet yield and quality are under considerable influences of environmental factors, especially weather conditions. Also, improvement of these properties is possible by adequate fertilization including foliar spraying with magnesium sulfate under moderate Mg supplying soil conditions.

5. References

Burba M. (1996): Der schadliche Stickstoff als Kriterium der Rubenqualitat. Zuckerindustrie, 121 (3): 165-173.

Kristek A., Andres E., Kovacevic V., Liovic I., Rastija M. (1997): Response of sugar beet to foliar fertilization with Epsom salt (MgSO4.7H2O). In: Plant Nutrition – for sustainable food production and environment (Ando T. et al. Editors), Printed in Japan, Kluwer Academic Publishers, p. 939-940.

Kristek A., Rastija M., Magud Z. (1998a): Status and perspectives of sugar beet growing in Croatia. In: Short Communications, Volume I (M. Zima and M.L.-Bartosova, Editors), Fifth Congress of the European Society for Agronomy, 28 June – 2 July, 1998, Nitra, The Slovak Republic, p. 111-112.

Kristek A., Uebel E., Kovacevic V., Rastija M., Liovic I. (1998b): Influence of sugar beet fertilization with magnesium sulphate on root yields and quality. In: Book of abstracts, 6th European Magnesium Congress, May 13-16, 1998, Budapest, Hungary, p.76.

Chapter 26

Effect of micro-element application on mineral composition and yield of spinach (*spinacia oleracea*) grown in soils with different lime contents

A.Aydin[1], M. Turan[1], A. Dursun[2]

[1] *University of Atatürk, Faculty of Agriculture, Department of Soil Science, 25240- Erzurum/Turkey.*

[2] *University of Atatürk, Faculty of Agriculture, Department of Horticulture, 25240- Erzurum/Turkey.*

Key words: foliar fertilizer, macro and micro elements, spinach , lime soil

Abstract: This study was undertaken to determine the effects of soil or foliage applied microplex fertilizer (5.4 %Mg, 4% Mn, 4% Fe, 1.5 % Cu, 1.5 % Zn, 0.5 % B, 0.1 % Mo and 0.05 % Co) on dry matter and mineral content of spinach and uptake of plant nutrients from soil. Dry matter content of spinach increased with microplex application. It had been higher in calcareous soils than less calcareous soils. Mineral matter content (N, P, K, Ca, Fe, Mn, Zn and Cu) of plants increased depending on application doses. Mineral content of plants were higher in foliar application than that of soil application.

1. Introduction

Availability of micro nutrients in agricultural soils is controlled by several soil characteristics such as pH, organic matter and lime content etc. It is commonly known that micro nutrients should be sufficient in a soil for a high agricultural production.

High pH, lime and organic matter of soils affect the Fe, Mn, Cu, Zn and B contents negatively and limit plant growth in these soils. Therefore, foliar fertilizers containing micro elements are beneficial both for economical use of fertilizers and agricultural production.

The effect of fertilizers on yield of some plant species are reported by several authors (Aksoy, 1980; Aksoy and Danışman, 1983; Aydeniz et al., 1983; Zabunoğlu et al., 1984). They state that the impacts of foliar fertilizers on yield depend on soil properties, plant species, foliar content, foliar level, application time and occurrence.

2. Materials and Methods

Experimental Design A pot experiment was performed on six soils with different lime contents ranging from 1.7 to 26.4 %. Soils were sampled from 0-20 cm depth and placed in 2 kg pots. As a basal dressing 10 kg da^{-1} N, 10 kg da^{-1} P$_2$O$_5$ and 10 kg da^{-1} K$_2$O were added in the forms of urea, triple superphoshate and potassium sulphate. Six seeds were sown and thinned to two prior to germination. Four different levels (0, 50, 100 and 200 g da^{-1}) of Microplex foliar fertilizer (5.4 % Mg, 4 % Mn, 4 % Fe, 1.5 % Cu, 1.5 % Zn, 0.5 % B, 0.1 % Mo and 0.05 % Co) were applied in two methods, once to the soil and 4 times to the foliage with 15 day intervals during the growth period. The experiment was designed in randomized blocks, with 6 soils, 4 fertilizer levels, 3 application methods, soil and foliage, and 3 replications (144 pots). The plants were harvested 10 days after the fertilizer application and dried at 65 °C in order to determine dry matter weight. Plant samples were chemically analyzed and the data were statically evaluated.

Soil Analyses : Soil texture was determined following Bouyoucos hydrometer method (Demiralay, 1993); pH using glass pH meter of 1:2.5 soil-water mixture; organic matter by Smith-Weldon method; exchangeable cations, cation exchange capacity and plant available P by sodium bicarbonate blue-colour method (Sağlam, 1994); plant available Zn, Fe, Mn and Cu by DTPA method (Lindsay and Norwell, 1969).

Plant Analyses : Total nitrogen was determined following micro-Kjeldahl method; P, by Vanadomolibdoyellow - colour method spectrophotometrically; and K, Ca, Mg, Fe, Mn, Zn and Cu by using atomic absorption spectrophotometer (Bayraklı, 1987).

Statistical Analysis : The ANOVA was performed to determine the effects of the type of fertilizer application and doses on dry matter content and nutrient uptake (Düzgüneş et al., 1987).

3. Result and Discussion

Some chemical and physical properties of soils used are shown in Table 1. Organic matter of the soil samples were medium and high; Cu and Zn were sufficient; available P, Fe and Mn were generally not enough in most soils. Soil texture were medium and fine (Elgala et.al., 1986).

Dry Matter Content The effect of microplex treatments on dry matter contents of spinach are given in Table 2.

Table. 1. Some chemical and physical properties of soils

Soil Samp.	pH	O.M	CaCO$_3$	Ca+ Mg	K	Na	P$_2$O$_5$	Fe	Mn	Zn	Cu	Sand	Loam	Clay	Text. Class
	1:2.5	%	%	cmolkg^{-1}			kg da^{-1}	mg kg^{-1}				%			
1	7.3	2.3	1.7	28.8	2.2	0.1	6.9	5.9	5.3	1.45	1.68	35.7	38.4	25.9	L
2	7.9	2.2	3.2	38.9	2.1	0.3	4.8	1.6	4.2	1.16	2.06	23.4	44.9	31.7	CL
3	7.8	2.5	6.3	36.9	2.3	0.2	3.6	3.2	1.7	1.26	1.51	22.9	39.7	34.7	CL
4	8.1	3.4	12.6	37.6	2.4	1.0	2.3	2.9	3.1	0.85	1.63	15.3	42.5	42.2	SC
5	8.0	3.3	19.1	44.3	2.3	0.9	5.5	3.5	5.1	0.98	2.20	15.0	38.4	46.6	C
6	8.0	3.7	26.4	40.7	2.4	0.3	2.6	4.0	1.6	0.84	1.70	17.4	35.9	46.7	C

Table 2. Dry-matter content of spinach (g pot^{-1})

	Application Methods	Rates kg ha^{-1}	SOILS						Average (rates)**	Average (applica.)
			1	2	3	4	5	6		
Dry Matter Content (g pots^{-1})	Soil	0	2.83	2.58	2.49	2.81	2.32	2.55	2.60 c	
		0.5	3.14	2.86	2.98	2.92	2.73	2.80	2.86 b	
		1.0	3.32	3.03	3.10	3.15	2.94	2.90	3.08 a	
		2.0	3.53	3.43	3.08	3.21	3.14	3.11	3.18 a	
		Aver.	3.21	2.98	2.91	3.02	2.78	2.84		2.96a
	Foliage	0	2.82	2.52	2.50	2.77	2.38	2.58		
		0.5	2.97	2.76	2.84	2.89	2.61	2.84		
		1.0	3.23	2.86	2.98	3.21	3.12	3.08		
		2.0	3.17	2.85	3.07	3.17	3.14	3.27		
		Aver.	3.05	2.75	2.85	3.01	2.81	2.94		2.90 a
Average(soils) **			3.13 a	2.87 ab	2.88 ab	3.02 ab	2.80 b	2.89 ab		

** : Significant at P<0.01

Table 3. The effects of microplex on mineral contents of spinach

Soil Samp.	Applica. Meth.	Rates kg ha⁻¹	N %	P %	K %	Ca %	Mg, %	Fe mg kg⁻¹	Mn, mg kg⁻¹	Zn mg kg⁻¹	Cu mg kg⁻¹
1	Soil	0	2.34	0.344	8.31	0.735	0.445	198.9	72.8	164.2	13.9
		0.5	2.67	0.447	8.38	0.780	0.476	205.3	81.6	175.9	16.1
		1.0	2.75	0.458	8.50	0.812	0.516	224.1	88.2	186.2	19.0
		2.0	2.87	0.500	8.73	0.916	0.513	224.1	99.5	201.5	18.8
	Foliage	0	2.35	0.350	8.36	0.820	0.470	201.2	70.6	168.1	13.3
		0.5	2.86	0.510	9.04	0.880	0.492	262.8	105.7	219.8	20.7
		1.0	3.23	0.505	9.11	0.947	0.458	268.1	109.7	230.6	24.4
		2.0	3.42	0.530	9.27	0.991	0.514	329.9	134.7	261.8	25.6
2	Soil	0	2.74	0.312	8.61	0.989	0.614	150.5	69.9	137.0	14.8
		0.5	2.77	0.348	8.77	1.081	0.619	198.4	79.2	155.1	18.5
		1.0	2.99	0.314	8.84	1.096	0.623	257.9	103.1	168.7	19.7
		2.0	3.03	0.328	9.24	1.113	0.638	307.8	114.4	189.8	22.9
	Foliage	0	2.75	0.317	8.56	1.063	0.601	155.4	70.3	139.2	14.8
		0.5	3.00	0.378	8.84	1.038	0.610	219.7	98.2	165.8	20.9
		1.0	3.20	0.386	9.03	1.124	0.648	285.9	139.5	210.3	23.4
		2.0	3.56	0.394	9.26	1.132	0.655	340.7	143.3	232.9	22.2
3	Soil	0	2.58	0.364	7.77	0.872	0.538	139.7	66.6	143.1	20.2
		0.5	2.88	0.385	8.49	0.911	0.553	194.4	85.4	150.3	23.4
		1.0	2.95	0.387	9.11	1.056	0.562	234.9	100.6	163.4	19.3
		2.0	2.99	0.392	9.88	1.160	0.601	266.6	109.4	195.6	24.6
	Foliage	0	2.55	0.357	7.71	0.920	0.560	144.7	65.8	147.8	19.7
		0.5	2.95	0.378	8.77	0.978	0.576	208.0	93.7	162.0	25.9
		1.0	3.12	0.397	9.24	1.128	0.597	241.3	109.4	182.6	29.6
		2.0	3.17	0.409	9.84	1.197	0.618	302.8	124.2	225.6	34.4
3	Soil	0	2.58	0.364	7.77	0.872	0.538	139.7	66.6	143.1	20.2
		0.5	2.88	0.385	8.49	0.911	0.553	194.4	85.4	150.3	23.4
		1.0	2.95	0.387	9.11	1.056	0.562	234.9	100.6	163.4	19.3
		2.0	2.99	0.392	9.88	1.160	0.601	266.6	109.4	195.6	24.6
	Foliage	0	2.55	0.357	7.71	0.920	0.560	144.7	65.8	147.8	19.7
		0.5	2.95	0.378	8.77	0.978	0.576	208.0	93.7	162.0	25.9
		1.0	3.12	0.397	9.24	1.128	0.597	241.3	109.4	182.6	29.6
		2.0	3.17	0.709	9.84	1.197	0.618	302.8	124.2	225.6	34.4
4	Soil	0	2.71	0.320	8.33	1.021	0.530	178.4	90.2	127.0	15.7
		0.5	2.86	0.350	8.45	1.081	0.561	190.2	94.1	154.5	20.1
		1.0	2.97	0.366	9.12	1.036	0.538	214.4	94.1	164.2	19.3
		2.0	3.14	0.381	9.44	0.996	0.498	219.8	106.9	174.0	20.3
	Foliage	0	2.75	0.355	8.28	1.012	0.536	175.7	90.5	127.8	15.3
		0.5	3.08	0.380	8.85	0.997	0.490	254.7	106.8	174.2	22.2
		1.0	3.39	0.376	9.34	1.030	0.519	261.4	115.7	211.2	22.2
		2.0	3.35	0.386	9.41	1.086	0.546	275.2	120.7	210.6	23.4
5	Soil	0	2.73	0.344	8.10	0.993	0.446	134.1	50.2	120.6	13.9
		0.5	2.80	0.379	8.86	1.001	0.488	174.7	56.6	157.3	18.5
		1.0	2.97	0.391	8.92	1.117	0.518	204.9	67.7	192.6	20.2
		2.0	3.17	0.394	8.99	1.126	0.543	233.3	75.7	201.5	19.3
	Foliage	0	2.71	0.340	8.82	0.908	0.480	138.4	49.5	122.8	13.6
		0.5	3.01	0.371	8.94	1.013	0.477	222.5	85.7	172.4	20.5
		1.0	3.15	0.382	9.30	1.086	0.544	247.8	93.3	249.0	24.1
		2.0	3.37	0.396	9.32	1.201	0.538	297.7	107.1	247.1	25.9

114

Table 3 continued.

Soil Samp.	Applica. Meth.	Rates kg ha⁻¹	N %	P %	K %	Ca %	Mg, %	Fe mg kg⁻¹	Mn, mg kg⁻¹	Zn mg kg⁻¹	Cu mg kg⁻¹
6	Soil	0	2.53	0.360	8.26	0.788	0.410	163.9	77.7	157.5	14.8
		0.5	2.63	0.383	8.75	0.830	0.445	176.9	75.4	161.4	20.2
		1.0	2.69	0.420	8.84	0.916	0.442	195.6	89.2	192.9	21.0
		2.0	2.78	0.433	8.82	0.943	0.426	201.2	94.6	189.8	21.0
	Foliage	0	2.66	0.367	8.23	0.790	0.408	158.3	79.8	155.3	15.9
		0.5	2.91	0.431	8.60	0.881	0.440	273.5	102.0	203.8	21.4
		1.0	3.12	0.441	9.37	0.929	0.458	308.5	117.1	220.6	25.4
		2.0	3.18	0.470	9.51	0.996	0.461	292.6	124.6	234.2	26.6

The effect of application methods on dry matter contents of spinach were not significant. However, fertilizer rates and soil types significantly affected this parameter. The most effective rate was 2.0 kg ha⁻¹ and followed by 1.0 kg ha⁻¹, 0.5 kg ha⁻¹, respectively, as compared to that of the control.

The effects of rates and applications varied depending on different lime contents of the soils (Table 2). Foliar application of microplex was the most effective in soils with high lime content, but microplex soil application was effective in soils with low lime content.

The effects of microplex rates and applications on mineral contents of the plant were statistically significant at $p<0.05$. Mineral contents of the plant increased when microplex doses were increased. Mineral contents of the plant treated on foliage was higher than that of the soil application (Table 3).Similar results were reported by Aydeniz et. al., 1983 and Topçuoğlu et. al., 1997.

4. Conclusion

Dry matter contents of plants in both application methods increased with increasing microplex rates. Microplex soil application was more effective than the other application. But, application at foliage was more effective in the calcareous soil. Mineral contents of plants increased depending on high microplex application doses. It was higher at foliar application than that of the soil application.

5. References

Aksoy T (1980) Çeşitli yaprak gübrelerinin Orta Anadolu'da yetiştirilen buğday ve arpa bitkilerinin verimine etkisi. Türkiye Gübre Araş. Genel Müd. Yayın No: 78-1. 69-85.

Aksoy T and Danışman S (1983) Çeşitli Yaprak Gübrelerinin Mısır Bitkisinin Verimine Etkisi. A.Ü. Ziraat Fak. Yıllığı, Cilt: 33

Aydeniz A Danışman S Dinçer D ve Yıldız İ (1983) Yaprak Gübrelerinin Buğday, Arpa ve Fasulye Bitkilerinin Verim Düzeyine Etkisi. Köy İşleri ve Kooperatifler Bak., Topraksu Kartoğrafya Müd. Ankara.

Bayraklı F (1987) Toprak ve Bitki Analizleri 19 Mayıs Üniv. Yayınları, Yayın No: 17

Demiralay İ (1993) Toprak Fiziksel Analizleri. Atatürk Üni. Yayınları NO: 143. Erzurum.

Düzgüneş O Kesici T Kavuncu O and Gürbüz F (1987) Deneme Methodları. Ankara Üniv. Ziraat Fakültesi Yay., No:1021, Ankara.

Elgala AM İsmail A.S and Osman MU (1986) Critical levels of Iron , Mn and Zn in Egy.on Soils. J. of Plant Nut. 9(3-7):267-280.

Lindsay WL and Norwell WA (1969) Development of DTPA Soil Test for Zinc, Iron, Manganese and Cupper. Soil Sci. Soc. Amer Proc. Vol: 33, p:49-54.

Sağlam T (1994) Toprak ve Suyun Kimyasal Analiz Yöntemleri. Trakya Üni. Tekirdağ Ziraat Fak. Yayınları, No:189.

Topçuoğlu B Kütük C Demir K Özçoban M (1997) A. Sülfat ve A. Nitrat ile Gübrelenen Ispanak Bitkisine Yapraktan ve CaCl₂ Uygulamasının Verim İle Verim Fiziksel ve Kimyasal Bazı Kalite Özellikleri Üzerine Etkisi. Ankara Üniv. Ziraat Fak. Tarım Bil. Derg. Cilt: 3 Sayı : 3.

Zabunoğlu S Aksoy T Karaçal İ and Danışman S (1984) Yaprak Gübresinin Fasulye, Şeker Pancarı ve Armut'un Verimi Üzerinde Etkisi. A.Ü. Ziraat Fak. Yıl. Cilt:34. Fasikül 1-2-3-4'den ayrı basım (1986) Ankara .

Chapter 27

Effects of N-fertilizers on wheat grain protein through foliar application

M. Lotfollahi and M.J. .Malakouti

Soil and Water Research Institute., North Kargar Ave.P. O. Box: 14155-6185, Tehran, I. R., Iran.

Key words: wheat, protein, N fertilizer

Abstract: Nitrogen (N) is required in large quantities for crop production. Nitrogen deficiency is a major factor limiting productivity and quality (grain protein concentration = GPC) of wheat in some areas. The cereal crop is in need of various amounts of N at different stages of growth. It is important to use the N fertiliser efficiently by matching the N supply to the needs of the crop, both in terms of amount and timing of nutrient supply. Foliar application after anthesis rather than solid fertiliser is one of the effective methods for increasing the GPC of wheat. Spraying urea after anthesis is more efficient than at heading or at anthesis for increasing the rate of accumulation of protein in grain. Field experiments were conducted to study the effect of foliar N application after anthesis on GPC of wheat in the Fars province of Iran at 1997-98. With four rates of N applied after anthesis all the N treatments increased the GPC of wheat compared with the control.

1. Introduction

The cereal crop has different needs for N at different stages of growth. It is important to use the N fertiliser efficiently by matching the N supply to the needs of the crop, both in terms of amount and timing of nutrient supply because N fertilisers are expensive. Losses of N through processes such as leaching, which can be environmentally hazardous, can be limited by the efficient uptake and utilisation of N by the crop (Mason et al. 1972). A large number of field experiments in many countries has examined the relation between the time of application of N fertilisers to wheat and the yield and protein response produced. Late application of N fertiliser tends to increase GPC (Hunter and Stanford 1973, Hamid and Sarvar 1976, Strong 1982, Strong 1986, Recous et al. 1988, Smith et al. 1989, Cooper and Blakeney 1990). When N is applied before flowering (at boot) more N is assimilated into grain than when N is applied after flowering (Finney et al .

1957). However, application at flowering or later had the singular effect of increasing GPC. In contrast, application before flowering produced a grain yield response, thereby reducing its effectiveness to increase GPC. Thus a fertilisation strategy in which the N application is split between sowing and flowering would more likely result in grain of a higher protein concentration than the strategy in which the entire application is made at sowing. In previous pot experiment (Lotfollahi et al. 1997) placing N at a depth of 60 cm two weeks after flowering produced higher GPC compared with the control. The result also suggested that, as long as the N was not leached from the root zone, topsoil N application at sowing and subsoil N application after flowering had equivalent effects on GPC. However, it is clear that both depth and timing of N can also influence the yield and protein responses of wheat. Applying the N to the soil surface after flowering is not easy. Foliar application of N is one of the methods for increasing the GPC. For example, Strong (1982) reported that after anthesis foliar application rather than solid fertiliser was the more effective method. The experiment reported here compared treatments where the N was dusted at dough stage with a treatment where the mixture of water and insecticide was dusted. The experiment tests the hypothesis that foliar application of N after flowering (dough stage) will have improved the GPC.

2. Materials and Methods

The experiment was conducted at Fars province of Iran at 1997-98. Wheat (L.cv.Azadi) was grown in the field. The rate of base fertiliser was 200 kg ha^{-1} ammonium phosphate and 50 kg ha^{-1} urea. Dusting was done by airplane before dough stage (Zadok's 75) (19 June 1997). The experimental design was complete randomized with four N treatments (N0= no N, N1= 4 kg ha^{-1} urea, N2= 8 kg ha^{-1} urea, N3= 4 kg ha^{-1} ammonium sulfate) and three replications. The different N treatments were dissolved in a mixture of 24 liters of water and 1 liter of insecticide and dusted in one hectare. The concentration of N in the mixture was 0, 16, 32 and 16 percent for N0, N1, N2 and N3 respectively. Wheat plant samples were taken at maturity (Zadok's 92) above ground plant material was cut 50 to 100 cm of row, dried at 80°C, weighed and ground, grain was separated from the rest of the plant. Total nitrogen was determined in a sulphuric acid digest of the plant material by the method as described by Kjeldahl (Bremner 1965).

3. Results and Discussion

Foliar application of N after flowering increased the grain protein concentration of wheat compared with the control (Table 1).

The GPC of the plant was significantly higher with applying 8 kg ha^{-1} urea (N2) compared with control. Assuming that all of the N taken up after anthesis goes to the grain, a small amount of N taken up after anthesis (for example 5 kg ha^{-1}) can increase in GPC by about 1.5% (Lotfollahi 1996). The 3.7% increase in GPC that was obtained in this

experiment was remarkable, because the range of GPC commonly encountered in the field is 9-15%. This result is supported by the result of Sadaphal and Das (1966) who showed that when N was applied as a spray application after anthesis more N was assimilated into the grain than before anthesis. He reported for increasing the rate of accumulation of protein in grain spraying of urea (1 to 12%) after anthesis was more efficient than at heading or at anthesis. Copper and Blakeney (1990) reported that a foliar spray of urea increased the GPC of wheat about 2.9%.

Table 1: Effect of the N treatment on the concentration of N and grain protein at maturity.

Nitrogen Treatment	Grain (%) Nitrogen	Grain (%) Protein
N0= no N	1.54	8.95
N1= 4 kg ha^{-1} urea	1.79	10.41
N2= 8 kg ha^{-1} urea	2.19	12.68
N3= 4 kg ha^{-1} ammonium sulfate	1.87	10.86

4. Conclusion

From this study it can be concluded that foliar application of N after flowering (dough stage) significantly increased the GPC compared with the control. The foliar application of N to increase the GPC must be balanced with the amount of soil Nitrogen available. Further research to explain the relationship between N application and GPC of wheat is necessary.

Acknowledgement

The authors would like to gratefully acknowledge the generosity of ministry of Agriculture (REEO). The assistance of Dr. Abdolahi, Mr. Shahroghnia, Mr. Basiri and Mr. Aminpor are gratefully acknowledged.

5. References

Bremner J. M. (1965) Nitrogen availability indices. In Methods of Soil Analysis, Part 2 . (Ed. C. A Black.) pp. 1324-45. (American Society of Agronomy : Madison, WI.).

Cooper J. L. and Blakeney A. B. (1990) The effect of two forms of nitrogen fertiliser applied near anthesis on the grain quality of irrigated wheat. Australian Journal of Experimental Agriculture 30, 615-619.

Finney K. F., Meyer J. W. Smith F. W. and Fryer H. C. (1957) Effect of foliar spraying of Pawnee wheat with urea solutions on yield, protein content and protein quality. Agronomy Journal 49, 341-347.

Hamid A. and Sarwar G. (1976) Effect of split application on N uptake by wheat from N abeled ammonium nitrate and urea. Experimental Agricultual 12, 189-193.

Hunter A. A. and Stanford G. (1973) Protein content of winter wheat in relation to rate and time of nitrogen fertiliser application. Agronomy Journal. 65, 772-774.

Lotfollahi M. (1996) The effect of subsoil mineral nitrogen on grain protein concentration of wheat. Ph.D thesis, University of Adelaide.

Lotfollahi M., Alston A. M. and McDonald G. K. (1997) Effect of nitrogen fertiliser placement on grain protein concentration of wheat under different water regimes. Australian Journal of Agricultural Research 48, 241-250.

Mason M. G., Rowley A. M. and Quayle D. J. (1972) The fate of urea applied at various intervals after sowing of a wheat crop on a sandy soil in Western Australia. Australian Journal of Experimental Agriculture and Animal Husbandry. 12, 171-175.

Recous S., Machet J. M. and Mary B. (1988) The fate of labelled N urea and ammonium nitrate applied to a winter wheat crop. II. Plant uptake and N efficiency. Plant and Soil 112, 215-224.

Sadaphal M. N. and Das N. P. (1966) Effect of spraying urea on winter wheat (Triticum aestiuum) Agron. J. 58, 137-141.

Smith C. J., Freney J. R., Chapman S. L. and Galbally I. E. (1989) Fate of urea nitrogen applied to irrigated wheat at heading. Australian Journal of Agricultural Research 40, 951-963.

Strong W. M. (1982) Effect of late application of nitrogen on the yield and protein content of wheat. Australian Journal of Experimental Agriculture and Animal Husbandry 22, 54-61.

Strong W. M. (1986) Effect of nitrogen applications before sowing compared with the effects of split applications before and after sowing for irrigated wheat on the Darling Downs. Australian Journal of Agricultural Research 26, 201-7.

3-Crop quality – nutrient management under stress conditions

Chapter 28

Effect of saline conditions on nutritional status and fruit quality of satsuma mandarin cv. Owari

S. Hepaksoy, U. Aksoy, H. Z. Can, B. Okur, C. C. Kılıç, D. Anaç, S. Anaç
Ege University, Faculty of Agriculture, Bornova 35100 İzmir, Turkey

Key words: satsuma mandarin, saline conditions, nutritional status, fruit quality

Abstract: Citrus are grown in subtropical and tropical regions of the world. Mediterranean basin is the main region for citrus growing with irrigation. In the Mediterranean countries, due to the shortages of water, quantity and quality of irrigation water create problems. In Turkey, satsuma mandarins are mainly grown in the western part of the country. Gümüldür-Büyükalan, the main satsuma-growing region in Turkey is threatened by groundwater salination due to seawater intrusion. Citrus species are generally considered as sensitive to salt. Saline irrigation water reduces the vegetative and generative growth of trees, as well as yield and quality.In Büyükalan-Gümüldür/Turkey, 12 satsuma orchards were sampled in 1996 and 1997. The aim of the researchwork was to determine the effect of salinity on fruit quality and to find out its relationship with the nutritional status of the trees. As fruit quality parameters, average fruit weight, total soluble solids (TSS %) titratable acidity (as % citric acid) and Cl content of the fruit juice and crude fibre of the pulp were analysed. In addition, sugar composition was determined by gas chromatography. Sodium, potassium, calcium and chloride contents were analysed in leaf and irrigation water samples. Water samples were taken towards the end of the irrigation period. Leaf samples were taken during the fruit ripening period (end of October). In the statistical evaluation of the results, the effect of salinity on the analysed quality parameters were defined.

1. Introduction

In Turkey, citrus fruits with 1.4 million tons of production have an important position within total agricultural production and exportation. The Mediterranean region, the southern part of the country, contributes 86 % of the citrus production, however, in terms of mandarin production, the Mediterranean (55 %) is followed closely by the Aegean region (44 %). Due to earliness and high quality provided by the prevailing climatic conditions, satsuma mandarins are mainly produced in the Western Aegean region.

Salinity is known to affect yield, growth and quality of various crops. Citrus fruits known to be sensitive to saline conditions are threatened the most since the major growing areas are in the coastal regions. In previous researches carried out in Gümüldür-Büyükalan located in the central part of the production area, groundwater salination due to seawater intrusion was found to suppress tree growth of satsuma mandarins specifically the leaf area (Aksoy et al., 1997). The impact of saline irrigation water on Satsuma mandarins was also negative in respect to the gas exchange properties (Aksoy et al., 1998).

The present study aims to put forth the effects of groundwater salination in Gümüldür-Büyükalan region on fruit quality parameters.

2. Materials and Methods

The main and most intensive satsuma mandarin growing region in Turkey is around Gümüldür in the western part of the country. Büyükalan district, which is proximal to the seashore and threatened by salinity due to seawater intrusion was selected as the experimental site. A network of 6 east west transects parallel to the coast with a separation distance of 350 m. was worked out. To provide variation in terms of salinity, 2 orchards established by satsuma (Owari) mandarin budded on Trifoliate orange representing each transect were sampled. Irrigation water samples were taken in September 1996 and 1997 and analysed for Na, Cl and EC (US Salinity Staff, 1954). Leaf samples were collected during the fruit maturation period within the first week of November and analysed for their Na, K, Ca and Cl contents.

Fruit samples were collected from the same trees and total soluble solids (%) and titratable acidity (as citric acid %) contents of the juice and the mean fruit weight were analysed. The crude fibre content of the pulp was extracted and quantified according to AOAC (1984). The fruit pulp was lyophilised and analysed for the sugar fractions, sucrose, fructose, α and β glucose and galactose, by gas spectrophotometry (Carlo Erba Fractovap Model 2350) (Olano et al, 1986; Martinez-Castro et al., 1987). An ion/EC meter (Consort C733) determined fruit juice chlorine content.

3. Results and Discussion

The analysed parameters of the irrigation water showed that the EC (dS m^{-1}) values ranged between 0.60 and 2.4 dS m^{-1} in the selected orchards (Tables 1 and 2). Na contents, on the other hand, varied between 0.86 and 4.59 in 1996 and 0.96 and 4.60 in 1997 me l^{-1}. The minimum and maximum Cl values of the irrigation water 1.16-10.74 and 0.86-10.55 me l^{-1}, respectively.

The average Na content in the leaves was found between 88 and 740 mg kg^{-1}. The values obtained for 1996 were higher than 1997. The leaf Cl content ranged between 0.050 and 0.430 % (Tables 1 and 2).

The analysed fruit parameters were found to differ significantly according to the sampled orchards except titratable acidity (Table 3). Average fruit weight ranged between 83.5 and 126.6 grams in 1996 and 77.9 and 114.1 grams in 1997. In 1996, orchard 11 had the highest value and was grouped as different from the others which formed a single group altogether. In 1997, the variations in fruit weight values of orchards were higher and six statistical groups were formed. The total soluble solids content was highly affected in both years by the salinity parameters. Na, K, Cl and EC values of the irrigation water presented significant positive correlations with the total soluble solids. The range was between 8.0 and 11.1 %. In 1997, increasing leaf Cl contents resulted in higher accumulation of TSS. On the other hand, the amounts were affected negatively by the leaf K. The salinity parameters revealed significant correlations with the average fruit weight in 1997, whereas, more in 1996 with the TSS content.

In 1997, analysis of the pulp revealed that the sugars present were sucrose, fructose, β glucose, α glucose and galactose in the decreasing order (Table 4). The differences between the orchards were statistically significant in respect to β glucose content. Orchards possessing high levels of salinity in irrigation water had the lowest fructose content in their fruits. The correlation coefficients of the relationships between fructose content and EC (dS m^{-1}) (r=-0.549**), Na (r=-0.498*) and Cl (r=-0.512*) content of the irrigation water were statistically significant (Table 5).

Sugars are among the most significant compounds in the fruits. Although sugar concentrations of a particular species may vary with the variety and soil and climatic conditions, in citrus fruits, fructose is reported as the second most abundant sugar fraction after sucrose (Sugiyma et al., 1991). Binzel and Reuveni (1994) report that plant cells in culture accumulate reduced and non-reduced sugars to accommodate the more negative water potential of the saline stressed medium. The plants lower their osmotic potential by increasing its concentration through dry matter accumulation in order to establish the osmotic equilibrium and thus enhance water uptake (Greenway and

Table 1. Salinity parameters of water and leaf samples taken in 1996.

Orchards	Water					Leaf			
	EC (dS m⁻¹)	Na (me l⁻¹)	Cl (me l⁻¹)	K (me l⁻¹)	Ca+Mg (me l⁻¹)	Na (mg kg⁻¹)	Cl (%)	K (%)	Ca (%)
1	2.40	4.59	10.74	0.24	18.70	310	0.17	0.88	2.42
2	1.65	3.81	7.74	0.18	12.90	255	0.22	0.60	2.82
3	2.10	2.24	2.56	0.16	8.30	420	0.13	0.59	2.50
4	1.50	2.24	3.97	0.12	12.60	740	0.16	0.74	2.85
5	0.80	1.79	1.86	0.10	5.70	400	0.07	1.02	2.57
6	1.70	2.50	6.07	0.20	13.90	440	0.19	0.86	3.00
7	1.10	1.72	2.33	0.06	9.50	340	0.09	0.91	2.53
8	1.40	2.35	4.20	0.10	12.00	440	0.16	0.78	2.67
9	0.80	0.97	1.40	0.05	6.90	280	0.09	0.73	3.25
10	0.60	1.00	1.16	0.04	5.10	480	0.07	1.14	2.17
11	0.60	0.86	1.16	0.04	5.00	340	0.08	0.93	2.75
12	1.50	2.72	3.97	0.18	12.80	680	0.22	0.85	2.81
F						28.247**	19.850**	3.665*	9.243**
LSD (%5)						41.007	0.04	0.259	0.287

• No statistical analysis performed on water data.

Table 2. Salinity parameters of water and leaf samples taken in 1997.

Orchards	Water					Leaf			
	EC (dSm⁻¹)	Na (me l⁻¹)	Cl (me l⁻¹)	K (me l⁻¹)	Ca+Mg (me l⁻¹)	Na (mg kg⁻¹)	Cl (%)	K (%)	Ca (%)
1	2.40	4.60	10.55	0.22	19.80	220	0.43	0.05	4.98
2	2.15	4.82	10.41	0.20	15.50	242	0.25	0.75	3.12
3	1.10	2.22	2.19	0.16	7.90	176	0.08	0.62	3.93
4	1.70	2.37	3.76	0.13	14.20	308	0.11	0.63	3.78
5	0.80	1.95	1.90	0.10	6.10	220	0.07	1.20	3.21
6	1.41	2.64	4.52	0.21	12.20	220	0.32	0.81	3.73
7	1.00	1.80	1.81	0.07	8.50	132	0.06	0.94	3.90
8	1.10	2.14	3.14	0.09	8.70	242	0.14	0.91	4.08
9	0.72	0.99	1.00	0.06	6.60	132	0.07	0.78	3.81
10	0.80	1.03	1.19	0.06	7.50	176	0.07	1.19	3.78
11	0.70	0.96	0.86	0.05	6.00	88	0.05	1.07	3.90
12	1.50	2.73	3.97	0.18	12.80	220	0.13	1.94	3.69
F						44.727**	1167.6**	100.556**	26.064**
LSD (5%)						27.677	0.011	0.07	0.278

• No statistical analysis performed on water data.

Table 3. Fruit quality characteristics of the sampled satsuma mandarin orchards.

Orchards	Fruit weight (g)		TSS (%)		Titratable acidity(%)		Cl (ppm)		Crude fibre (%)	
	1996	997	1996	1997	1996	1997	1996	1997	1996	1997
1	83.5	77.9	10.100	11.1	1.389	1.510	124.6	102.0	0.91	0.70
2	89.7	85.6	9.000	11.0	1.389	1.533	96.5	91.2	0.93	0.68
3	87.2	83.1	9.500	10.7	1.408	1.418	84.6	87.8	0.86	0.57
4	100.3	98.1	9.650	10.1	1.626	1.340	105.4	101.6	0.87	0.60
5	88.3	80.8	9.400	10.3	1.338	1.431	61.2	97.95	0.79	0.51
6	89.7	91.1	8.100	10.5	1.632	1.465	100.8	79.3	0.73	0.44
7	90.0	87.6	9.850	10.9	1.376	1.152	72.3	88.5	0.69	0.39
8	94.5	92.5	10.000	9.15	1.389	1.476	75.2	88.6	0.62	0.32
9	95.3	94.0	10.350	9.15	1.389	1.439	57.5	90.8	0.61	0.43
10	93.5	90.9	9.500	8.65	1.632	1.484	39.9	84.3	0.50	0.38
11	126.6	114.1	8.650	10.05	1.242	1.472	38.5	68.5	0.59	0.41
12	90.4	93.2	10.100	8.65	1.376	1.339	62.0	84.45	0.42	0.32
F	3.681*	9.677**	20.042**	6.481**	19.948**	ns	79.606**	11.505**	28.355**	6.385**
LSD (5%)	17.875	9.339	0.454	1.088	0.088		9.201	8.558	0.095	0.160

Munns, 1980; Munns and Termaat, 1986; Pasternak, 1987). Yakushiji (1996) in his study of water stress on satsuma mandarins reports that in osmotic regulation sugars namely monosaccharides play a significant role. From the significant correlations between sugar fractions and salinity parameters, the osmoregulatory role of fructose in Satsuma mandarins becomes apparent.

Crude fibre which can be accepted as negative attribute in a berry fruit, was found to be highly correlated with the salinity parameters of the irrigation water namely EC, Na and Cl both in 1996 and 1997. The correlation coefficients of Na, Cl, EC and K contents of the irrigation water were significant mostly at 1 % level. Leaf Cl was positively related, as well.

Leaf potassium, on the other hand, had a positive effect in terms of overcoming the negative effect of salinity on crude fibre formation both in 1996 and 1997. The correlation coefficient of the significant relationship between the leaf K and fruit crude fibre content was determined as r = - 0.415* in 1996 and r = - 0.608 ** in 1997 (Table 5). It is reported that an increasing crude fibre with an increase in the EC value of the irrigation water had a consequent adverse effect on fruit quality due to thicker deposition in segment and vesicle walls (Aksoy et al., 1997). Holland et al. (1992) indicates the fibre fractions of satsuma mandarin fruits as 0.2 % cellulose and 0.6 % soluble and 0.1 % insoluble non-cellulosic polysaccharides.

Table 4. Sugar components of satsuma mandarin fruits sampled from the experimental site in 1997.

Orchards	Fructose (mg g[-1])	Galactose (mg g[-1])	α- Glucose (mg g[-1])	β- Glucose (mg g[-1])	Sucrose (mg g[-1])
1	117.69	13.50	57.05	62.05	240.05
2	127.30	16.46	59.15	63.99	238.19
3	201.10	11.40	71.64	88.58	302.10
4	155.36	10.11	57.89	66.68	292.95
5	177.19	14.65	71.27	83.52	262.90
6	136.15	11.01	50.24	60.63	270.41
7	200.03	8.01	62.78	77.66	184.82
8	227.59	19.90	95.22	104.15	474.00
9	185.56	11.44	64.05	72.12	363.45
10	177.28	10.01	61.20	64.02	261.98
11	176.53	11.18	63.73	60.52	173.71
12	99.93	9.17	42.32	46.71	236.68
F	ns	ns	ns	2.834*	ns
LSD (5%)				28.221	

Table 5. Correlation coefficients of some significant relationships.

	Fruit Weight		TSS		Crude Fiber		Fructose
1996	1996	1997	1996	1997	1996	1997	1997
Na (water)	-0.484*	-0.548**		0.470*	0.574**	0.596**	-0.473*
Na (leaf)							-0.538**
Cl (water)		-0.441*		0.468*	0.548**	0.577**	-0.482*
Cl (leaf)							-0.440*
K (water)	-0.490*	-0.498*		0.406*	0.488*	0.505*	
K (leaf)					-0.415*		-0.537**
Ca+Mg (water)	-0.407*				0.408*		-0.455*
EC (water)	-0.484*	-0.507*		0.474*	0.586**	0.539**	
1997							
Na (water)		-0.516**		0.516**		0.650**	-0.498*
Na (leaf)							
Cl (water)		-0.441*		0.477*		0.658**	-0.512*
Cl (leaf)		-0.409*		0.412*		-0.489*	-0.491*
K (water)		-0.469*		0.413*		0.521**	-0.573**
K (leaf)				-0.459*		-0.608**	
Ca + Mg (water)						0.578**	-0.582**
EC (water)				0.439*		0.646**	-0.549**

The Cl content of the citrus fruits is reported to range between 10 and 50 mg l^{-1} (Holland et al., 1992). In the fruit samples taken from the experimental orchards, Cl contents were found quite high in some of the orchards and varied between 38.5 and 124.6 mg l^{-1} in 1996 and 68.5 and 102.0 mg l^{-1} in 1997 (Table 3). Significant negative correlations were determined between the Cl content of the fruit juice and salinity parameters of water and leaf.

Among the analysed fruit quality parameters, salinity of the irrigation water within the tested range exerted a marked negative effect on the fruit attributes as the fruit weight, total soluble solids and chloride content of the juice and crude fibre of the pulp in both years. Increasing EC values (within the range of 0.6-2.4 dS m^{-1}) and Cl and Na contents of the irrigation water and Cl content of the leaves exerted a negative effect on leaf K. On the other hand, increasing potassium resulted in a decrease of Cl content (r = -0 .428 * in 1996 and r = - 0.826 ** in 1997) of the juice and the crude fibre content of the pulp and thus increasing fruit quality.

Acknowledgement

This researchwork is partially supported by the Avicenne Initiative of the Commission of the EU Contract No. 93 AVI 008.

4. References

Aksoy U, Anaç S, Anaç D and Can, H.Z, (1997) The effect of groundwater salinity on satsuma mandarins: preliminary results. Acta Horticulturae 449: 629-633.

Aksoy U, Hepaksoy S, Can H.Z, Anaç D, Okur B, Kılıç C.C, Anaç S, Ul M.A and Dorsan F, (1998) The effect rootstock on leaf characterisitcs and physiological response of satsuma mandarins under saline conditions. Proc. 25 th Int. Hort. Congress, Brussels (in press).

A.O.A.C. (1984) Official methods of analysis. Assoc. Official Analyt. Chemists, Virginia, USA.

Binzel M.L. and Reuveni M, (1994) Cellular mechanism of salt tolerance in plant cells. Horticultural Reviews 16:33-70.

Greenway H and Munns R, (1980) Mechanisms of salt tolerance in non-halophytes. Ann. Rev. Plant Physiol., 31:149-90.

Holland B, Unwin I.D. and Buss D.H, (1992) Fruits and Nuts: The composition of foods. The Royal Society of Chamistry, UK, p.70.

Martinez-Castro I, Calvo M.M and Olano A, (1987) Chromatographic determination of lactulose. Chromatographia 23 (2): 132-136.

Munns R and Termaat A, (1986) Whole plant responses to salinity. Aust. Jour. Plant Physiol. 13:143-160.

Olano A, Calvo M.M. and Reglero G, (1986) Analysis of free carbohydrates in milk using micropacked columns. Chromatographia Vol. 21, 538-540.

Pasternak D, 1987. Salt tolerance and crop production-A comprehensive approach. Ann. Rev. Phytopathology, 25:271-291.

Salinity Lab. Staff, (1954) Diagnosis improvement of saline and alkali soils. Agric. Handbook No. 6,US Gov. Print. Office, Washington DC.

Sugiyma N, Roerner K and Bünemann G, (1991) Gartenbauwissenschaft, 56 (3): 126-129.

Yakushiji H, Nonami H, Fukuyama T, Ono S, Takagi N and Hashimoto Y, (1996) Sugar accumulation enhanced by osmoregulation in satsuma mandarin fruit. Jour. Amer. Soc. Hort. Sci., 121 (3): 466-472.

Chapter 29

Effects of a temporary water and nitrogen deficit in the soil on tomato yield and quality

F. Montemurro, R. Colucci, V. Di Bari, D. Ferri.
Istituto Sperimentale Agronomico, Via C. Ulpiani n°5, 70125 Bari - Italy

Key words: water stress, nitrogen stress, total nitrogen, nitrate, tomato

Abstract: The effects of temporary water and nitrogen stress were investigated on tomato plant *(Lycopersicon esculentum* Mill) in a field experiment carried out in Southern Italy. The plants were stressed at different times during the cropping cycle; at 0, 50, and 100% flowering in the control treatment. Leaf area index, and the fresh and dry weights of tomato plants were determined at maximum development. In addition, total nitrogen and nitrate concentrations in stems, leaves and fruits were determined at the intermediate and final harvests. The results indicated that temporary stresses for water and nitrogen have significant influences on leaf area index, and the fresh and dry weights of tomato plants. Temporary stresses influenced nitrate accumulation by the intermediate harvest only , but did not influence nitrate content in the plant at the end of cropping cycle. On the other hand, total nitrogen and nitrate concentration behaved differently in the several different tissues (leaves, stems and fruits). This is evidence that water and nitrogen stresses influence plant development and nitrogen dynamics in tomato cropped in Southern Italy.

1. Introduction

Tomato *(Lycopersicon esculentum* Mill) is one of the most important horticultural crops in the world and particularly in the Southern Italy agronomic techniques for this crop have been widely studied.

The growth of the tomato plant has been investigated in relation to evapotranspiration (Vagliasindi *et al.,* 1993), but there is still an incomplete knowledge of the phenological behaviour and performance of the tomato under differing conditions of soil water.

The total N demand of a crop depends on the increase of dry matter, but is not linear during crop development and, in general, retranslocation of nitrogen occurs from older parts of the plant to newer ones (Schenk, 1996). On the other hand, nitrate accumulation in plant occurs naturally when the uptake of nitrate exceeds its reduction and assimilation into the plant. Then, the excess NO_3 is stored in vacuoles.

The presence of nitrate in tomato and in other vegetables can be a serious problem for human health, because vegetables are the major source of nitrate. For this reason the European Commission have recently proposed maximum acceptable levels for nitrate concentrations in vegetables (Gazzetta Ufficiale delle Comunita Europee, 1995; Santamaria, 1997).

The aim of the work reported here is to evaluate the influence of nitrogen and nitrate accumulation in tomato organs on growth, based on a field experiment with a several temporary deficits of water and nitrogen in the soil.

2. Materials and Methods

The experimental trial was carried out at Rutigliano - Bari in Southern Italy (41° 01' N, 17° 39' E, 122 m above sea level) in a "thermo-Mediterranean" climate. Four treatments were compared in a tomato crop of the hybrid variety "Elko". One treatment was well supplied with nitrogen and water during the whole crop cycle, and the other three were temporarily stressed at particular phenological stages: 34, 48 and 65 days after transplant, corresponding to 0, 50 and 100% of flowering of the control (indicated as F0, F50 and F100, respectively). Treatments were replicated three times in a randomized block design; each plot consisting of ten rows 18 m long. Three-week-old seedlings of tomato were transplanted on 14 May.1997 at a spacing of 33.3 x 100 cm within and between rows (plant density, 3 plant m^{-2}).

The irrigation system was drip irrigation. In the control treatment water was given to replace daily evapotranspiration. The other treatments were achieved by witholding irrigation at stages F0, F50 and Fl00. Applications of K_2O and P_2O_5 were at the rates normally used for tomato in Southern Italy, and were the same in all treatments. All treatments were given the same total application of nitrogen (120 kg N ha^{-1}). In the control teatment it was applied broadcast at 30 kg ha^{-1} on four occasions, viz.at 34, 48, 65, 70 days after transplanting, whereas the stressed treatments each received 40 kg N ha^{-1} applied on three occasions (F0 - 48, 65, 70 days after transplanting; F50 - 34, 65, 70 days; F100 - 34, 48, 70 days).

Table 1: Effect of temporary stress on leaf area index, fresh and dry weight on tomato plant.

Treatment	Fresh weight (t/ha)	Dry weight (t/ha)	LAI
Control	85.44a	6.91a	2.81a
F0	62.50b	5.11b	2.55a
F50	61.56b	5.48b	1.92b
F100*	55.59b	5.67b	1.84b

Values in a column, followed by different letters were statistically significantly different according to Duncan's multiple range test (DMRT) at P< 0.05.*F0, F50, F100 = beginning, middle and final flowering stages in the control treatment.

An intermediate harvest was made when about 30% of fruits were present on each plot (07 August 1997 - 76 days after transplanting). Plant samples were collected and analysed.

Leaf area index was determined by a laboratory area meter at the same time. The final harvest was made on 27 August 1997 - 85 days after transplanting.

Tomato plants sampled at the intermediate and final harvests were divided into leaves, stems and fruits for the determination of nitrate and total nitrogen concentrations. For nitrate determination each sample was extracted with hot pure water. NO_3 concentration, $[NO_3]$ was determined directly on aqueous extracts by reduction of cadmium and colorimetric determination. Total N concentration, $[N]$, was analysed using a Fison CHN elemental analyser (EA 1108).

Statistical analysis was performed using the General Linear Model (GLM) of the SAS software package (SAS Institute, 1990), considering time of stress and plant tissue as fixed effects. Differences between treatments were evaluated using Duncan's Multiple Range Test

3. Results and Discussion

The effects of temporary stress in the F_0 treatment, on leaf area index, and fresh and dry weights of tomato plants are presented at Table 1. At the earliest stage, stress did not show significant differences from the control, in leaf area index but there were statistically significant differences in fresh and dry weights. Differences between the F_0 treatment and the control in yield and number of flowers were not statistically significant (data not shown).

These results suggest that these temporary stress imposed at the earliest stage did not influence greatly the phenological development and yield of the crop.

The analysis of variance for total nitrogen and nitrate content in tomato plant is presented at Table 2. There were statistically significant differences in total $[N]$ and $[NO_3]$ in the different tomato tissues at both first and second harvests. This suggests possibly, a redistribution of N components during plant development. On the other hand, differences in $[NO_3]$ were statistically significant between the treatments, only in the tissue of the tomato fruit.

Table 2: Analysis of variance for nitrogen and nitrate concentration in the tomato plant.

Variable		Source of variation	d. f.	F	Pr>F
N content	First yield	Treatment (T)	3	1.31	n.s.
		Plant tissue (P)	2	14.33	***
		TxP	6	0.92	n.s.
		Replication	2	0.63	n.s.
	Second yield	Treatment (T)	3	1.84	n.s.
		Plant tissue (P)	3	11.52	***
		TxP	9	0.85	n.s.
		Replication	2	0.58	n.s.
$[NO_3]$	First yield	Treatment (T)	3	4.75	**
		Plant tissue (P)	2	17.74	***
		TxP	6	1.4	n.s.
		Replication	2	4.08	*
	Second yield	Treatment (T)	3	2.96	*
		Plant tissue (P)	3	6.74	***
		TxP	9	1.45	n.s.
		Replication	2	0.37	n.s.

*, **, *** = Statistically significant at P = 0.05, 0.01, and 0.001 levels, respectively; n.s. = non-significant

Total [N] was unaffected by treatment because the N fertilizer dose was the same in all of them.

The total [N], expressed as % of dry weight, and $[NO_3]$, expressed as mg kg^{-1} of fresh weight are presented at Table 3. The highest $[NO_3]$ in the tomato plant was observed in treatment F50 at the first harvest and in treatment F100 at the second harvest. These values were higher than those reported for tomato plant by other authors (Seitz, 1986; Santanaria, 1997), possibly because of the lateness of the last application of fertilizer N (30 kg ha^{-1} for the control and 40 kg ha^{-1} for the stressed treatments), given 70 days after transplanting. The first and the second harvests were made at 76 and 85 days after transplanting.

Concentrations of total nitrogen and nitrate in different tissues of the tomato plant are presented at Table 4. At the second harvest total [N] in the vegetative parts was less than in fruit, indicating translocation and of nitrogen to the reproductive parts. This had been evident at the first harvest, also. On the other hand, $[NO_3]$ at the second harvest as highest in the stem and leaves (3800 mg kg^{-1}), indicating that the nitrogen fertilizer last applied had not yet been re-distributed.

One year's data are not sufficient to draw firm conclusions, but the results of the present work suggest that temporary stress could be a way to increase the efficiency of nitrogen and water use of the tomato plant. Moreover, the data seem to indicate that the application of fertilizer nitrogen in the later stages of tomato plant development does not improve yield, but only increases the amount of nitrate in the plant.

Table 3: Total nitrogen (% of dry weight) and nitrate content (mg kg^{-1} of fresh weight) in tomato plant for the treatment tested.

Treatment	Total nitrogen		Nitrate	
	First yield	Second yield	First yield	Second yield
Control	1,7698	2,4022	1214.3c	2218.5b
F0	2,1615	2,6911	2577.3ab	2254.8b
F50	1,5881	2,4735	2810.7a	2652.2ab
F100*	1,9677	3,0817	1601.7bhc	3496.9a

Values in a column followed by different letters are significantly different at P< 0.05.F0, F50, F100 = beginning, middle and final flowering of the control treatment.

Table 4: Concentrations of total nitrogen (% of dry weight) and nitrate (mg kg^{-1} of fresh weight) in different tomato plant tissues.

Plant tissues	Total nitrogen		Nitrate	
	First yield	Second yield	First yield	Second yield
Stems	1.0712b	1.6209c	3427.Sa	3823.3a
Leaves	2.1233a	2.6542b	1830.4b	3026.la
Fruits	2.4209a	3.4337a	895.2c	1918.4b
Leaves + Stems		2.9376ab		1854.61
Red fruits				
Yellow fruits				
Green fruits				

Values in a column followed by different letters are statistically significantly different according to DMRT at P< 0.05.

4. References

Gazzetta Ufficiale delle Comunita Europee, (1995). C/ 39, 5/6/1995.

Santaria P. (1997). Contributo degli ortaggi all'assunzione giornaliera di nitrato, nitrito enitrosamine. Industrie altmentari XXXVI : 1329-1334.

SAS Institute (1990). SAS/Stat Software, SAS Institute.

Schenk M. K. (1996). Regulation of nitrogen uptake on the whole plant level. Plant and Soil 181:131-137.

Seitz P. (1986). La problematica dei nitrati in orticoltura. Colture protette 10:17-24.

Vagliasindi C., Tuttobene R. and Copani V.(1993). Analisi dell'accrescirnento e sue relazioni con il flusso traspirativo in pomodoro (*Lycopersicon esculentum*, Mill). Riv. di Agron. 27: 4, 323-327.

Chapter 30

Solute losses from various shoot parts of field-grown wheat by leakage in the rain

N. Debrunner, F. Von Lerber, and U. Feller

Institute of Plant Physiology, University of Bern, Altenbergrain 21, CH-3013 Bern, Switzerland (urs.feller@pfp.unibe.ch)

Key words: leakage, nutrient balance, potassium, rain, senescence, *Triticum aestivum* L., wheat

Abstract: Leakage from field-grown wheat was investigated during two seasons differing considerably in their rainfall patterns. For all solutes analyzed, these losses were low from non-senescing plant parts, increased after the onset of senescence and became maximal in fully senesced (dry, brown) organs. The cumulative losses of potassium by leakage in the rain were 65% of the content at anthesis for the flag leaf and 95% for the third leaf from the top, while these relative values were lower for magnesium (50 to 80%) and calcium (around 55%) and extremely low for sodium (<10%). The differences between potassium and sodium may be due to a different compartmentation on the tissue level or on the subcellular level. It became evident that for certain nutrients (e.g. potassium or magnesium) leakage in the rain may represent a major loss from senescing leaves and can be a relevant flux in maturing wheat.

1. Introduction

Nutrients losses from senescing leaves of field-grown cereals may be caused by the export of solutes via the phloem (Schenk and Feller, 1990; Stieger and Feller, 1994a), by the release of volatile compound into the atmosphere (Farquhar et al., 1979; Francis et al., 1993) or by leakage in the rain (Schenk and Feller, 1990; Tukey, 1970). Leakage was mainly investigated in plant populations (Tukey, 1970; Velhorst and van Breemen, 1989) and in detached leaves or leaf segments (Debrunner and Feller, 1995; Pennazio et al., 1982; Shyr and Kao, 1985). Highest solute losses by leakage have been reported for leaves under biotic or abiotic stress (Leopold et al., 1981; Pennazio and Sapetti, 1982; Vasquez-Tello et al., 1990), after injury of the foliage (Tukey and Morgan, 1963) and during senescence (Feller and Fischer, 1994; Pennazio et al, 1982; Shyr and Kao, 1985; Whitlow et al., 1992). The aim of the work presented here was to evaluate the importance of solute losses by leakage in the rain for the nutrient balance of maturing wheat in the field.

2. Materials and Methods

Winter wheat (*Triticum aestivum* L., cv. Arina) was grown in 1990 and in 1992 (in different fields) in Zollikofen near Bern. Polypropylene tubes (10 ml) were fixed with tape on the stem of main shoots shortly after ear emergence. The water was collected below the ear with a plastified wire fixed at the stem and endig in a polypropylene tube. This collar-like wire was necessary to direct the rain water into the tube. The water from the three uppermost leaf laminas, which were oriented downwards, was collected by fixing the tip of the leaf lamina in the opening of the tube. The water was collected after each major rainfall period from June 15 to July 31, the volume was determined and samples were stored frozen (-20°C) prior to analysis. Samples from 4 different plants were collected for each position and analyzed separately. Rain water collected above the canopy was also analyzed. The concentrations in these controls were subtracted from the concentrations in the water samples collected at the plants. The concentrations of all solutes analyzed were low as compared to the samples of water interacting with the plants.

The rain water samples were used directly to measure potassium, sodium, calcium, magnesium, carbohydrates and amino acids after appropriate dilution. For the analysis of the nutrient contents in plant parts, leaf laminas or ears were dried at 105°C, weighted and heated in glass tubes for 8 h at 550°C. After cooling, the ash was solubilized by adding 0.1 mL 10 N HCl and afterwards 2 mL deionized water. This solution was used to prepare the appropriate dilutions for the quantification of potassium, sodium, calcium and magnesium.

Potassium and sodium were detected by atomic absorption spectrophotometry after appropriately diluting the samples with 1000 ppm Cs as CsCl suprapur in 0.1 N HCl. Calcium and magnesium were also detected by atomic absorption spectrophotmetry, but for these elements the samples were diluted with 5000 ppm La as $LaCl_3 \cdot 7\ H_2O$ in 0.1 N HCl. Soluble carbohydrates were detected colorimetrically with an anthron reagent and amino acids with a ninhydrin reagent as reported previously (Stieger and Feller 1994b).

3. Results

The two seasons considered (1990 and 1992) differed considerably in the rainfall pattern during the grain-filling period (Fig. 1). The season 1990 was characterized by an extended rain-free period in July, when the uppermost wheat leaves senesced. In contrast, major rainfalls between July 10 and July 25 were relevant for the nutrient leakage in 1992.

The leakage of potassium from various shoot parts depended on the rainfall and on the senescence status (Table 1).

The differences between the two seasons may be explained by the meteorological situation on one hand and by the potassium status in the different fields on the other hand.

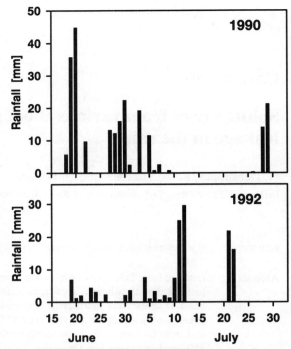

Figure 4. Daily rainfall during the wheat maturation phase in 1990 and 1992 (data for station Liebefeld-Bern from the Swiss Meteorological Institute).

In 1990, a considerable quantity of potassium was already detected in the rain water collected on July 1 below the third leaf from the top and to a minor extent also in the sample from the second leaf, while the leakage from the flag leaf and from the ear was very low for this date. The small water samples collected on July 4 and on July 9 confirmed the relative pattern observed for July. After the extended rain-free period in July 1990, major losses were also observed for the flag leaf lamina and for the ear on July 29 and July 30. In 1992 it became evident that the potassium leakage increased initially in the third leaf from the top (July 3), later in the flag leaf lamina (July 12) and finally also in the ear (July 23). Chlorophyll was degraded in lamina 3 after June 20 and was completely lost on July 14, 1992. In contrast, a major net chlorophyll degradation in the flag leaf lamina was observed after July 14, 1992. The leakage data for the two quite different seasons indicated consistently that the potassium leakage was low from non-senescing tissue and increased during senescence.

Table 1. Leakage of potassium from various shoot parts in the rain during maturation in 1990 and 1992. The leaves are numbered from the top. Means±SE of 4 replicates are shown for samples collected after major rainfalls. n.d.: not determined.

Date	Potassium leakage [µmol per plant part]			
	Ear	Lamina 1	Lamina 2	Lamina 3
Experiments in 1990:				
July 1	0.12±0.07	1.49±0.28	11.43±2.14	22.87±5.40
July 4	0.33±0.29	1.41±0.45	3.41±1.34	9.56±2.32
July 9	0.13±0.03	1.86±0.90	7.79±3.44	14.04±3.01
July 29	46.07±7.63	46.91±6.37	61.89±21.35	91.91±11.98
July 30	32.33±3.86	47.06±14.04	58.50±10.27	75.87±68.54
Experiments in 1992:				
June 20	0.95±0.20	0.68±0.08	n.d.	2.70±0.34
June 21	0.10±0.11	0.01±0.01	n.d.	0.74±0.13
June 26	0.06±0.04	0.15±0.04	n.d.	2.69±0.34
June 29	0.06±0.04	0.18±0.08	n.d.	2.05±0.26
July 3	0.17±0.04	0.42±0.02	n.d.	4.72±2.12
July 8	0.36±0.12	1.37±0.27	n.d.	15.32±1.25
July 12	<0.01	7.60±3.11	n.d.	21.46±4.53
July 23	5.76±1.32	16.07±2.82	n.d.	24.71±4.68

Table 2. Cumulative leakage of various solutes from June 15 to July 30, 1992. Water was collected below the ear, below the flag leaf lamina (leaf 1) and below the lamina of leaf 3 from the top. Water samples were analyzed after each major rainfall period (as listed in Table 1) and these results were used to calculate the cumulative leakage.

Solute	Leakage per plant part		
	Ear	Leaf 1	Leaf 3
Potassium [µmol]	7.1	26.5	74.5
Sodium [µmol]	0.2	0.3	0.3
Calcium [µmol]	1.3	26.7	23.9
Magneisum [µmol]	0.4	3.9	5.1
Amino acids [µmol]	<1.0	<1.0	2.8
Soluble carbohydrates [mg]	0.4	0.8	2.0

Highest losses by leakage in the rain were observed from senesced and at least partially dried plant parts.

The cumulative leakage during the period from June 15 to July 30, 1992 is shown for various compounds in Table 2. Only minor quantities of amino acids and carbohydrates were detected in the water samples, while large amounts of potassium and calcium were lost in the rain. The losses from the leaf laminas exceeded considerably the losses from the ear. A comparison of the cumulative leakage (Table 2) and of the contents in the plant parts (Table3) indicates that leakage in the rain from the upermost leaf laminas is highly relevant for potassium, calcium and magnesium. In contrast to potassium, only a minor percentage of sodium was lost in the rain.

4. Discussion

Losses by leakage in the rain are for some elements (e.g. potassium) high and must be considered for the overall nutrient balance in maturing wheat. Other macronutrients (e.g. nitrogen) can be better retained in the maturing plants. Therefore leakage losses are solute-specific and must be evaluated separately for each nutrient. The plant species and the variety may be important besides the climate (especially the rainfall pattern) for losses in the rain.

Leakage was low as long as the leaf surface was intact (before senescence an during early phases of senescence) and increased when leaf cells became accessible for rain water. It remains open, to which extent changes at the surface of the leaf (e.g. intactness of the cuticle) and changes in membrane properties of the leaf cells caused the increased losses in the rain during final stages of senescence.

Table 3. Dry matter, potassium, sodium, calcium and magnesium in the ear, the flag leaf lamina (lamina 1) and lamina 3 from the top. Means±SE of 4 replicates are shown for June 15, 1992 (anthesis) and July 31, 1992 (end of maturation).

| | Content in the plant | | | | | |
| | Ear | | Lamina 1 | | Lamina 3 | |
Leaf constituent	June 15	July 31	June 15	July 31	June 15	July 31
Potassium [µmol per plant part]	115±10	404±9	39±4	12±2	77±10	1±1
Sodium [µmol per plant part]	18±2	88±6	6±1	12±4	29±6	3±1
Calcium [µmol per plant part]	25±1	54±4	50±6	53±9	42±5	23±2
Magnesium [µmol per plant part]	20±1	154±8	8±1	7±2	6±1	2±1
Dry matter [mg per plant part]	375±19	1930±49	124±8	103±13	118±9	54±6

Major differences were observed between the monovalent cations potassium and sodium. This result is consistent with previous reports concerning excised plant parts (Debrunner and Feller, 1995; Pennazio et al., 1982.). The compartmentation of solutes on the tissue level (e.g. mesophyll cells versus cells in the bundles) or on the subcellular level (e.g. cytosol versus vacuole) may be relevant for such differences. The crucial steps on the cellular level (accessibility of solutes for leakage) and on the field level (e.g. importance of climate, genotype or nutrient status) for solute leakage from maturing wheat remain to be elucidated in the future

Acknowledgements

We thank the LBBZ Rütti in Zollikofen near Bern for growing the plants and Prof. J. Fuhrer for the rainfall data.

5. References

Debrunner N and Feller U 1995 Solute leakage from detached plant parts of winter wheat: influence of maturation stage and incubation temperature. J. Plant Physiol. 145, 257-260.

Farquhar G D, Wetselaar R and Firth P M 1979 Ammonia volatilization from senescing leaves of maize. Science 203, 1257-1258.

Feller U and Fischer A 1994 Nitrogen metabolism in senescing leaves. Crit. Rev. Plant Sci. 13, 241-273.

Francis D D, Schepers J S and Vigil M F 1993 Post-anthesis nitrogen loss from corn. Agron. J. 85, 659-663.

Leopold A C, Musgrave M E and Williams K M 1981 Solute leakage resulting from desiccation. Plant Physiol. 68, 1222-1225.

Pennazio S, D'Agostino G and Sapetti C 1982 Cation release from discs of tobacco leaves of different ages. Physiol. Vég. 20, 577-583.

Pennazio S and Sapetti C 1982 Electrolyte leakage in relation to viral and abiotic stresses inducing necrosis in cowpea leaves. Biol. Plant. 24, 218-225.

Schenk D and Feller U 1990 Rubidium export from individual leaves of maturing wheat. J. Plant Physiol. 137, 175-179.

Shyr Y-Y and Kao C-H 1985 Senescence of rice leaves. XV. Solute leakage and inorganic phosphate uptake. Bot. Bull. Academia Sinica 26, 171-178.

Stieger P A and Feller U 1994a Nutrient accumulation and translocation in maturing wheat plants grown on waterlogged soil. Plant and Soil 160, 87-95.

Stieger P A and Feller U 1994b Senecence and protein remobilisation in leaves of maturing wheat plants grown on waterlogged soil. Plant and Soil 166, 173-179.

Tukey H B 1970 The leaching of substances from plants. Annu. Rev. Plant Physiol. 21, 3305-324.

Tukey H B and Morgan J V 1963 Injury to foliage and its effect upon the leaching of nutrients from above-ground plant parts. Physiol. Plant. 16, 557-564.

Vasquez-Tello A, Zuily-Fodil Y, Pham Thi A T and Vieira da Silva J B 1990 Electrolyte and Pi leakages and soluble sugar content as phsyiological tests fro screening resistance to water stress in *Phaseolus* and *Vigna* species. J. Exp. Bot. 41, 827-832.

Velthorst E J and Van Breemen N 1989 Changes in the composition of rainwater upon passage through the canopies of trees and of ground vegetation in a Dutch oak-birch forest. Plant and Soil 119, 81-85.

Whitlow T H, Bassuk N L, Ranney T G and Reichert D L 1992 An improved method for using electrolyte leakage to assess membrane competence in plant tissues. Plant Physiol. 98, 198-205.

Chapter 31

The effect of drought in different growth stages on uptake, translocation and utilization of N in winter wheat

A.Öztürk and Ö. Çağlar

University of Atatürk, Faculty of Agriculture, Department of Agronomy, 25240 Erzurum, Turkey

Key words: drought, N uptake efficiency, N translocation efficiency, winter wheat

Abstract: The effect of drought on uptake, translocation and utilization of N by winter wheat has been investigated. Fully irrigated (FI), rainfed (R), early drought (ED), late drought (LD) and continuous drought (CD) treatments were tested. The treatments had significant effects on the characters studied. Minimum and maximum values were as follows in grain yield 1533 (CD)-4459 (FI) kg.ha^{-1}, protein content 10.47 % (FI)-12.37 % (CD), total N yield 55.5 (CD)-105.0 (FI) kg.ha^{-1}, N uptake efficiency 0.93 (CD)-1.75 (FI) kg N uptake/kg N applied, N translocation efficiency 59.4 % (CD)- 77.1 % (FI), agronomic efficiency 25.6 (CD)-74.3 (FI) kg grain/kg N applied and physiological efficiency 27.7 (CD)-42.5 (FI) kg grain/kg N uptake. The results showed that not only N uptake but also the conversion of N into grain were important for high yield. The negative effect of ED on uptake and utilization efficiency of N was more important than that of LD.

1. Introduction

Drought stress causes severe problems in wheat especially in dry regions. Nitrogen (N) requirement of winter wheat is provided by fertilizers which affect the physiological processes related to wheat growth. Soil water content during the growing season affects the response of plants to N fertilization (Clarke et al.,1990; Gauer et al., 1992). There is a close relationship between uptake and translocation of N and growth stage in which drought stress occurred. The rate of supply of nutrients decreases in a dry soil because the mobility of ions, water uptake, and rate of production of new roots are all affected adversely (Day et al.,1978). Robins and Domingo (1962) noted that drought during the booting- milky ripe period caused an important reduction in N uptake by wheat. When wheat suffers drought stress during grain filling, grain N % increases, which is associated with a reduced contribution of pre-heading accumulated stem carbohydrate reserves to grain weight (Morgan and Riggs, 1981). Therefore, knowledge of the response of the plant to drought at different growth stages helps to ensure optimum N application time and levels. To test this under Turkish conditions, a field experiment was conducted in 1995-96 and 1996-97, in Erzurum .

2. Materials and Methods

The field experiment was carried out in the 1995-96 and 1996-97 crop seasons in Erzurum at an altitude of 1869 m. Doğu-88 winter wheat variety (awned, red-grain and drought hardy) was grown. The effect of drought at different growth stages on uptake and translocation of N was investigated using nylon plot covers. Five treatments were tested. Based on the gravimetric water content of the surface 0.6 m of soil determined sampling of each plot from the beginning of stem elongation (Feekes 6.0) until maturity (Feekes11.4). **1.Fully irrigated (FI):** Plots were irrigated when 40 % of available soil water was used from the beginning of stem elongation until maturity (Giunta et al.,1995).

2. Rainfed (R): Plants were grown under natural conditions without irrigation and covering. **3. Early drought (ED):** Plots were covered starting from the second node visible stage (Feekes 7.0) to beginning of milky ripe (Feekes 11.1). Covers were set up 1.5 m above soil surface and extended 2.0 m beyond the plot sides to protect the soil from rainfall. Plants were irrigated from the beginning of milky ripe to maturity as in the irrigated treatment. **4. Late drought (LD):** These plots were irrigated until the beginning of milky ripe stage, and then covered till after maturity. **5. Continuous drought (CD):** Plots were covered between the second node visible stage and maturity.

A randomized complete block design with three replications was used. Doğu-88 was sown at a rate of 475 seeds m^{-2} using an 6 row planter with 0.2 m inter row spacing. Each plot was 6.0 m long and 1.2 m wide. Fertilizer was uniformly applied to plots as 60 kg ha^{-1} N (ammonium sulphate, 21 %) and 50 kg ha^{-1} P_2O_5. Half of the N and all the P_2O_5 were applied at sowing and the second half of the N was given at the beginning of stem elongation. The total annual precipitation was 279 and 385 mm in 1995-96 and 1996-97 respectively. The amount and distribution of precipitation and average temperatures were more favourable in 1996-97 than in 1995-96. The amount of intercepted rainfall to the CD, ED and LD was 32.7, 20.3, 12.4 mm in 1995-96 and 43.2, 33.6 and 10.0 mm in 1996-97 respectively. Available P_2O_5 and K_2O contents of the soil were 37 and 592 kg ha^{-1} respectively. The soil was a clay loam with pH 7.7 and organic matter content 1.8 %.

The mean water contents were 21.2 % at field capacity and 11.7 % at permanent wilting point. The minimum water content in the CD, ED, LD, R and FI was 11.0, 13.7, 14.8, 13.7 and 16.9 % in 1995-96 and 12.2, 14.4, 15.6, 14.9 and 17.2 % in 1996-97 respectively. At maturity, 3 m of the centre four rows of each plot was harvested. Plant samples taken were separated into grain and straw, oven-dried for 24 h at 80 ^{0} C before %N in the grain and straw were determined by the Kjeldahl method.

Grain protein content (GPC), grain N yield, straw N yield, N uptake efficiency (NUE), N translocation efficiency (NTE), agronomic efficiency (AE) and physiological efficiency (PE) were examined (Anderson, 1985; Grant et al., 1991; Gauer et al., 1992; Gonzales-Ponce et al., 1993)

3. Results and Discussion

Yields and nitrogen concentrations More favourable precipitation and temperatures in 1996-97 season increased grain yield, total N yield, NUE, NTE, AE and PE of N. The results indicated that treatments had a significant effect on yields and %N (Table 1). As an average of both years, the highest grain and straw yields were obtained with FI (4459 and 8007 kg.ha^{-1}) and the lowest with CD (1533 and 5644 kg.ha^{-1}) respectively. Drought treatments caused a marked reduction in grain yields resulting from fewer fertile spikes, less kernels per spike and lower kernel weight. Grain % N ranged between 1.82 (FI) and 2.15 % (CD) while straw % N ranged between 0.30 (FI) and 0.40% (CD). Significantly higher grain % N was obtained in the CD. This could be explained by reduced contribution of pre-heading accumulated stem reserves of carbon to grain weight (Morgan and Riggs, 1981). FI treatment had a negative effect on N percentage, but remarkably increased grain N yield through its positive effect on grain yield (Bole and Pittman, 1980). The higher straw N yield recorded in LD and a decrease in N translocation of N accumulated before anthesis under drought stress (Giunta et al., 1995). Total N yields per unit area was lower in CD, ED, LD, and R as a consequence of a reduced total dry matter. Drought may reduce either the rate of supply of N in the soil, thus reducing growth, or the rate of growth, thus causing a reduction in the total N uptake (Day et al., 1978; Giunta et al., 1995). GPC of the wheat was strongly influenced by the treatments (Table 2). GPC values were 10.47, 11.41, 11.01, 11.34 and 12.37 % in FI, R, ED, LD and CD respectively. CD greatly increased

Table 1. The effects of experiment treatments on yields and N percentages*

Treatment	Grain yield (kg.ha^{-1})	Straw yield (kg.ha^{-1})	Grain N (%)	Straw N (%)	Grain N yield (kg.ha^{-1})	Straw N yield (kg.ha^{-1})
FI	4459 a	8007 a	1.82 c	0.30 c	80.9 a	24.0 ab
R	3101 c	6612 b	1.98 b	0.34 b	61.4 c	22.5 b
ED	2648 d	6508 b	1.91 bc	0.34 b	50.7 d	22.4 b
LD	3391 b	7613 a	1.97 b	0.34 b	66.9 b	25.5 a
CD	1533 e	5644 c	2.15 a	0.40 a	33.0 e	22.5 b
LSD	121.3	453.4	0.09	0.002	3.24	2.23
1995-96	2874	6964	2.00	0.35	56.6	2.41
1996-97	3179	6700	1.94	0.34	60.5	2.27
Mean	3026	6877	1.97	0.35	58.6	2.34
Year (Y)	P<0.001	NS.	P<0.05	P<0.01	P<0.001	P<0.01
Treatment (T)	P<0.001	P<0.001	P<0.001	P<0.001	P<0.001	P<0.001
YxT	P<0.001	NS.	NS.	NS.	NS:	NS.

* The means with the same letter within each variable are not significantly different (NS = not significant).

Table 2. The effects of treatments on total N yield, grain protein content (GPC), N uptake efficiency (NUE), N translocation efficiency (NTE), agronomic efficiency (AE) and physiological efficiency (PE)*

Treatment	Total N yield (kg.ha^{-1})	GPC (% Nx5.75) (%)	NUE (kg N uptake / kg N applied)	NTE (%)	AE (kg grain / kg N applied)	PE (kg grain / kg N uptake)
FI	105.0 a	10.47 c	1.75 a	77.1 a	74.3 a	42.5 a
R	83.8 c	11.41 b	1.40 c	73.1 b	51.9 c	36.9 b
ED	73.0 d	11.01 bc	1.22 d	69.4 c	44.2 d	36.3 b
LD	92.4 b	11.34 b	1.54 b	72.4 b	56.5 b	36.9 b
CD	55.5 e	12.37 a	0.93 e	59.4 d	25.6 e	27.7 c
LSD	3.90	0.56	0.074	2.27	2.14	1.36
1995-96	80.7	11.49	1.35	68.9	48.0	34.7
1996-97	83.2	11.15	1.39	71.6	53.0	37.2
Mean	81.9	11.32	1.37	70.3	50.5	36.0
Year (Y)	P<0.01	P<0.05	P<0.05	P<0.001	P<0.001	P<0.001
Treatment	P<0.001	P<0.001	P<0.001	P<0.001	P<0.001	P<0.001
YxT	NS.	NS.	NS.	NS.	P<0.001	NS.

* The means with the same letter within each variable are not significantly different (NS = not significant).

GPC. This increase was associated with decreased synthesis and storage of carbohydrates, not with an increased protein storage (Löffler and Bush, 1982).

Uptake and translocation efficiency of N
Soil moisture was an important factor affecting NUE .Values were 1.75, 1.40, 1.22, 1.54 and 0.93 at FI, R, ED, LD and CD respectively (Table 2). Doyle and Holford (1993) also found that insufficient soil water resulted in a decrease in NUE.

The treatments had significant effect on NTE especially with the CD and ED treatments. With these two treatments the low NTE of CD and ED was perhaps due to a sink limitation, the death of tissue by dehydration occurred before translocation was complete, or restricted translocation of N to grain (Clarke et al., 1990).

Agronomic and physiological efficiency AE measured as grain yield per unit of applied nitrogen, is an indicator of N conversion into grain yield. AE was strongly influenced by the treatments (Table 2). The results showed that CD, ED and LD decreased AE by 65.5, 40.5 and 24.0 % respectively as compared to FI. The decrease in the AE in drought conditions was observed by Doyle and Holford (1993) and Gauer et al. (1992). The PE is known as the ratio of kg grain production to kg of N absorbed in the total dry matter at maturity. CD, ED, LD and R treatments had a negative effect on PE . PE decreased 34.8 % with CD as compared to FI. Clarke et al., (1990) also reported a decrease in PE depending on drought stress.

Nitrogen is an important variable that influences both grain yield and quality of wheat. This experiment indicated that conversion of N into grain yield was more important than N uptake for high yield. There were close relationships between this processes and soil water content during the growing season. The negative effect of ED on uptake and utilization efficiency of N was more important than that of LD. If the N application was made prior to the period of rapid uptake and growth, it could increase the potential for increased N uptake and N use efficiency under drought at after anthesis.

Acknowledgements

We thank Dr. Faik Kantar for critically reading the text.

4. References

Anderson WK (1985) Differences in response of winter cereal varieties to applied nitrogen in the field. II. Some factors associated with differences response. Field Crops Research 11: 369-385.

Bole JB and Pittman UJ (1980) Spring soil water precipitation, and nitrogen fertilizer: effect on barley grain protein content and nitrogen yield. Can. J. Soil Sci. 60:471-477.

Clarke JM Campbell CA Cutforth HW Depauw RM and Winkleman GE (1990) Nitrogen and phosphorus uptake, translocation, and utilization efficiency of wheat in relation to environment and cultivar yield and protein levels. Can. J. Plant Sci. 70: 965-977.

Day W Legg BJ French BK Johnston AE Lawlor DW and Jeffers DW (1978) A drought experiment using mobile shelters: the effect of drought on barley yield, water use and nitrogen uptake. J. Agric. Sci. Camb. 91:599-623.

Doyle AD and Holford ICR (1993) The uptake of nitrogen by wheat, its agronomic efficiency and their relationship to soil and fertilizer nitrogen. Aust. J. Agric. Res. 44: 1245-1258.

Gauer LE Grant CA Gehl DT and Bailey LD (1992) Effects of nitrogen fertilization on grain protein content, nitrogen uptake and nitrogen use efficiency of spring wheat cultivars, in relation to estimated moisture supply. Can. J. Plant Sci. 72: 235-241.

Giunta F Motzo R and Deiddo M (1995) Effects of drought on leaf area development, biomass production and nitrogen uptake of durum wheat grown in a Mediterranean environment. Aust. J. Agric. Res. 46: 99-111.

Gonzales Ponce R. Salas ML and Mason SC (1993) Nitrogen use efficiency by winter barley under different climatic conditions. Journal of Plant Nutrition 16: 1249-1261.

Grant CA Gauer LE Gehl DT and Bailey LD (1991) Protein production and nitrogen utilization by barley cultivars in response to nitrogen fertilizer under varying moisture conditions. Can. J. Plant. Sci. 71: 997-1009.

Löffler CM and Bush RH (1982) Selection for grain protein, grain yield and nitrogen partitioning efficiency in hard red spring wheat. Crop Sci. 22: 591-595.

Morgan AG and Riggs T (1981) Effects of drought on yield and malt characters in spring barley. J. Sci. Food Agric.32: 339-346.

Robins JS and Domingo CE (1962) Moisture and nitrogen effects on irrigated spring wheat. Agron. J. 54: 135-138.

Chapter 32

Nutrient balances in cropping systems in a soil at risk of salinity

G. Convertini, D. Ferri, N. Losavio, P. La Cava

Agronomical Research Institute, Via C. Ulpiani, 5 – 70125 Bari (Italy)

Key words: nutrient balance, salinity, irrigation

Abstract: Nutrient balances for N, P_2O_5 and K_2O have been determined for different rotations and cultural treatments on a soil near to the sea in southern Italy (Metaponto - Matera). The soil Typic Epiaquert had poor drainage with a superficial water table (from 0.4 to 0.8 m) which could cause Na accumulation in the soil profile and modify the chemical-physical soil characteristics. The experiment compare three rotations: continuous sorghum with three irrigation treatments; and traditional and minimum tillage on durum wheat with soybean and durum wheat with guar. At the harvest of each crop soil samples from 0-20, 20-40 and 40-60 cm depth were taken to determine $N-NO_3$, $N-NH_4$ available P_2O_5 and exchangeable K_2O; and crop samples (grain and residues) to determine N, P_2O_5 and K_2O . The results showed a large uptake of N, P_2O_5 and K_2O in all rotations and the balance (input-output) for K_2O was always negative. There were small differences among experimental treatments.

1.   Introduction

In the Mediterranean region fertilizers rarely produce the same increase in yield as in northern areas because water is the main limiting factor for crop yield. Due to the uneven rainfall distribution and the relatively high winter temperatures, conditions are frequently favourable for the leaching of nitrate; perhaps as a result of increased nitrifying activity in the autumn of ammonium nitrogen, accumulated in the soil. In terms of differences in the mineral nutrition to crops, water stress decreases the capacity of the crops to absorb nutrients because of the lesser root activity and the shorter vegetative period. In the Mediterranean areas the use of mineral fertilization has increased well beyond the amounts which would be consistent with the average yield obtained (Bonari et al., 1994; Caliandro et al., 1992).

In some Mediterranean areas the salinity of soil solution increases in some period of the year as does the osmotic potential of plants, whereas the availability of water for the crop decreases (Shalhevet, 1994). The water content of the plant progressively declines and eventually cellular turgor is lost, with the immediate stoppage of growth (Hsiao, 1993). In arid and semiarid areas salinity affects about 27% of irrigated land (Hamdy and Lacirignola, 1997). So that the application of mixtures of saline and sweet water and the supply of excess water to cause salt leaching (at the end of each irrigated treatment) can be considered optimal solutions to reduce the risk of salinity. However the salinity can modify the mineral nutrition of plants and in areas of southern Italy it is very important to study nutrient balance in typical cropping systems.

2.   Materials and Methods

The research was conducted at Metaponto (Matera) in southern Italy (40° 24' N; 16° 48' E) on a Typic Epiaquerts soil, with poor drainage and a superficial water table (from 0.4 to 0.8 m).

140

The experiment compared three rotations: two with durum wheat with conventional -TT and minimum tillage-MT and with either soybean or cluster bean as catch-crops (split-plot design with three replications).The third rotation was continuous grain sorghum with three irrigation treatments (randomized block design with three replications).

Soil samples from 0-20, 21-40 and 41-60 cm depth were collected each year from each plot to determine $N-NO_3$, $N-NH_4$, available P_2O_5 and exchangeable K_2O; and crop samples (grain and residues) to determine N, P_2O_5 and K_2O; the data were used to calculate nutrient uptake and then nutrient balance (input-output).

3. Results and Discussion

Figures 1 and 2 show the changes in N uptake and balance each year on plots with either minimum or traditional tillage. Comparing these treatments, two different trends appear: N uptake on MT plots decrease in the 2nd year and remained constant in 3rd year; while on TT plots, after a decrease (similar to MT plots) in the 2nd year, in the third year there was a remarkable increase. The N balance shows the same trends, indicating that climatic conditions (in the period July - September 1995 the rainfall was 199.4 mm vs. 53.4 as poliennal mean) interacted differently on crop N uptake when the soil was conventionally tilled (in the 3rd year N uptake was higher than other years and treatments).

This finding can be interpreted as deriving from a combination of a bigger root system (favored by deeper soil tillage) and better water availability.

The three year average values for N uptake and N balance on MT and TT plots were significantly different, showing that on this soil with 58 % of clay, conventional tillage improved nitrogen absorption by the crops tested compared to "minimum tillage".

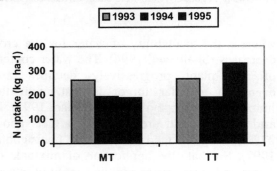

Figure 1. Effects of minimum tillage (MT) and conventional tillage (TT) on N uptake of DW + catch crops during the years.

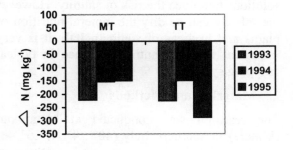

Figure 2. Effect of minimum (MT) and conventional tillage (TT) on N balance (input-output).

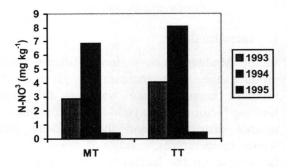

Figure 3. Effects minimum (MT) and conventional (TT) tillage on soil $N-NO_3$, and exchangeable $N-NH_4$

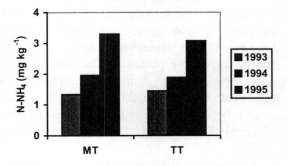

Figure 4. Effect of minimum (MT) and conventional (TT) tillages on soil N-NH4 exchangeable during the three years

However, only an economic assessment would be necessary to determine the advantages of a reduction of soil tillage, because with this treatment there was an average yield decrease of 18%. On the other hand, the average N balance is less negative in MT plots and this finding is important to conserve soil N fertility. The levels of nitrate and ammonium-N in Figures 3 and 4 show that: i) the average content of soil N-NO$_3$ was higher in TT than in MT plots; ii) differences between years in soil N-NO$_3$ and N-NH$_4$; iii) an increase of soil N-NH$_4$ over the three years probably due to reducing conditions related to the seasonal fluctuations of the saline water table and also the higher rainfall in the last year of the experiment.

Table 1 shows the variations in N uptake and balance when continuous grain sorghum was grown with three different irrigation treatments during five years .

When irrigated to establish 90% ETm there was the largest N uptake (as mean of 5 years) while the minimum value was with the T3 treatment. Although the C.Vs. were very high for all the treatments, the results suggest that it's not appropriate to water to 120% ETm but that there was not enough irrigation when watered to 60% ETm.

The larger N uptake was with the T2 treatment which suggests that supplying water volumes as 90% ETm was best. There were insignificant differences between the irrigation treatments and years on soil N-NO$_3$ and N-NH$_4$ in continuous grain sorghum (Table 2). At the end of the cropping cycle of sorghum (late autumn of each year) the processes of N transformation and transport (including mineralization, immobilization, leaching, etc.) were finished and the three irrigation treatments didn't determine significant variations in the soil N pools but only in the N uptake by crops.

Table 1. Effects of irrigation treatments and years on N uptake and balance of grain sorghum (T1=60% ETm; T2=90% ETm; T3=120% ETm).

Year/Treat.	N (kg ha^{-1})			ΔN (kg ha^{-1})		
	T1	T2	T3	T1	T2	T3
1993	97.37	113.06	97.05	2.63	-13.06	2.95
1994	69.32	76.50	74.47	30.68	23.5	25.53
1995	100.19	94.76	93.11	-0.19	5.24	6.89
1996	58.04	48.85	40.93	41.96	51.15	59.07
1997	48.93	52.29	48.73	51.07	47.71	51.27
Mean	74.77	77.09	70.86	25.23	22.91	29.14
C.V. (%)	45.8	55.7	48.3	//	//	//

Table 2. Effects of irrigation treatments and years on soil N-NO$_3$ and exchangeable N-NH$_4$ in continuous grain sorghum (T1=60% ETm; T2=90% ETm; T3=120% ETm).

Year/Treat.	N-NO$_3$ (mg kg^{-1})			N-NH$_4$ (mg kg^{-1})		
	T1	T2	T3	T1	T2	T3
1993	6.50	7.04	7.49	1.79	1.76	1.99
1994	5.00	5.55	3.56	3.03	2.24	3.49
1995	12.30	9.65	17.36	5.15	5.38	4.78
1996	6.96	6.51	7.48	2.07	2.79	2.69
1997	5.88	7.11	5.61	2.73	2.53	2.61
Mean	6.19	7.08	6.55	2.26	2.15	2.30
C.V. (%)	7.1	0.7	20.3	29.4	25.4	19.1

142

3. References

Bonari E., Mazzoncini M., Caliandro A. (1994) Cropping and farming systems in Mediterranean areas. Proc. 3rd ESA Congress, Abano-Padova, 636-644.

Caliandro A., Rizzo V., Mosca G., Stefanelli G., Manzini S., Bonari E., Mazzoncini M., Bonciarelli F., Archetti R., De Giorgio D., Basso F., Postiglione L., Rubino P., Lo Cascio B., Venezia G. (1992) Risultati sperimentali ottenuti su cereali a paglia con diverse tecniche dilavorazione al terreno. Riv. di Agron., 26, 3: 215-222.

Hamdy A. and Lacirignola C., (1997) Environment and water resources: major challenges. Intern. Conf. "water management, salinity and pollution control towards sustainable irrigation in the mediterranean region", 22-26 September, IAM, Valenzano (Bari), Vol. I, 33-84.

Hsiao T.C. (1993) Growth and productivity of crops in relatio to water status. Acta Horticolturae 335: 137-148.

Shalhevet J. (1994) Using water of marginal quality for crop production: major issue. Agric. Water Manag., 23: 233-269.

Chapter 33

Electrogenic acid extrusion by nutrient deficient barley seedlings

C. E Lee and M L Reilly

Environmental Resource Management Department, University College Dublin, Belfield, Dublin 4, Ireland

Key words: electrogenic proton pump, rhizosphere acidification, nutrient deficiency, phosphorus, potassium, magnesium.

Abstract: Investigations are reported on the capacity of seedling roots of barley (*H. sativum*, cv. Optic) to excrete acidity in response to selected inorganic nutrient deficiencies. Seedlings were grown hydroponically in half-strength Hoagland's No. 1 solution to provide a control, with modifications to separately exclude each of the major nutrients, phosphorus, potassium and magnesium, as accompanying treatments. Assays were conducted weekly at ages 1 to 4 weeks old on the basis of time taken for immersed root systems of intact, whole plants to induce colour change in half-strength Hoagland's, containing brome-cresol-purple (pH 6.8) as indicator. Acid excretion was generally enhanced by individual deficiency treatments and, with the exception of a drop between weeks 2 and 3, by plant age also. At one week, respective deficiencies of P, K and Mg induced responses some 60, 15 and 40% above control. Corresponding values at week 2 were about 50, 5 and 80%; at week 3 they were about 66, 20 and 60% up and at week 4 some 133, 100 and 200% higher. Mean excretion rates for week 4 were some 40% higher than those for week 1. The results are discussed in the context of limitations of the methodology and implications for fertiliser practice in the resolution of any possible conflict between quality food production and environmental quality.

1. Introduction

The conditions of the rhizosphere and their modification by the root system play an important role in mineral nutrient uptake (Marschner *et al.*, 1987). In particular, roots can establish a rhizosphere pH some two units below that of the bulk soil, sometimes in response to either individual or general nutrient stress. Solute transport through biomembranes like the plasmalemma may be either active or passive. Nutrients concentrated on one side with a higher free energy may diffuse across to a milieu of lower concentration or chemical potential (Clarkson, 1977). Transport against a potential gradient must be linked directly or indirectly to an energy-consuming mechanism, i.e. a pump in the membrane. An ATP-driven "proton pump", which counter-transports H^+ ions is the classical example of such a mechanism (Poole, 1978).

In the plant cell, such pumps are located in both plasmalemma and tonoplast to transport protons from the cytosol into the apoplast and the vacuole respectively (Kurdjian and Guern, 1989). They work to maintain electro and pH stat conditions in the cytoplasm at about -150 mV and pH 7.5, the pH of the apoplast/rhizosphere and of the vacuole being dropped some two units as a consequence. Cytosolic pH stabilisation involves two main components: the *biophysical* extrusion of H^+ ions and the *biochemical* formation, or else decarboxylation, of organic acids, notably malic and citric (Davies, 1986; Raven, 1986). Exchange (anion) extrusion of hydroxyl ions, particularly where overall anionic uptake rates exceed cationic rates, also contributes to cytosolic stasis. This latter process is somewhat less prominent and less well studied than that of the proton pump (Haynes, 1990). Plants typically absorb unequal quantities of nutritive cations and anions, with charge balance being maintained by excreting amounts of either H^+ or ^-OH stoichiometrically equal to the respective excess cationic or anionic uptake.

A classical example of the pH/charge balancing processes occurs in the differing response of plants to either ammoniacal or nitrate sources of nitrogen. About 70% of all the cations and anions taken up by plants are either NH_4^+ or NO_3^-, ammonium-fed plants being characterised by a high cation/anion uptake ratio and inversely for nitrate-fed ones (van Beusichem et al., 1988b). The feeding of ammonium engenders external acidification of the growth medium, a feature of fertiliser use long recognised in agronomic practice. This response is accentuated by the fact that each NH_4^+ assimilated to protein contributes one cytosolic proton, while each NO_3^- assimilated produces one cytosolic hydroxyl.

Localised rhizosphere acidification in response to nutrient deficiency has been observed in the case of Fe (Bienfait, 1988; Rhomheld and Marschner, 1986) and P (Gardner et al., 1983; Hoffland et al., 1998a, 1988b). Such responses seem to have been largely restricted to dicots and non grass monocots. The current investigation is an initial attempt to assess more widely the scope for such occurrences with other nutrients in graminaceous species, taking account of current world-wide developments in fertiliser usage.

2. Materials and Methods

Seeds of spring barley (H. sativum cv. Optic) were pregerminated at room temperature for 48 h and then transplanted into half-strength Hoagland's No. 1 solution as a control, along with modifications to exclude separately P, K and Mg as three other treatments. Solutions were replenished at tillering when the seedlings were three weeks old. The plants were grown on bench trolleys in the laboratory with supplementary fluorescent lighting on a 16 h photoperiod.

The root systems of intact plants were assayed for capacity to excrete acid at age one week and weekly thereafter until four weeks. The assay medium was half-strength Hoagland's No.1 to which brome-cresol-purple had been added as indicator (purple to colourless at pH 6.8). It was placed in shallow evaporating dishes in quantities sufficient to submerge all the roots in conditions approaching NFT coverage. Five ml per assay were used at week 1, and 10 ml thereafter.

Plant numbers per assay varied, depending on root bulk, from five at weeks 1 and 2 to four at week 3 and two at week 4. Assay times (to a colourless end point) varied between 10 and 60 minutes, after which the roots were blotted dry, separated from the shoots and weighed. During the course of the assay the plants were incubated in strong, mixed warm white and cold white, fluorescent lighting. Dishes were occasionally shaken to ensure uniform, complete root submersion and even solution distribution. Acidification capacity was recorded as $\mu mole\ H^+ \bullet gFW^{-1} \bullet h^{-1}$ by reference to a coincidental blank titration of 25 ml assay medium with $0.01\ mole \bullet l^{-1}$ HCl.

The experiment was established and analysed statistically as a split-plot design with four replicates, having plots allocated to treatments and weekly samples as splits.

3. Results

During the course of the experiment visual deficiency symptoms varied from mild initially to moderately severe at the fourth week. There was noticeably severe growth rate reduction in the absence of P and K but not for Mg. The principal results from the experiment are outlined in Table 1.

Table 1 Acid extrusion rate of spring barley seedling roots under nutrient deficiency. Values are micromoles hydrogen ions excreted per gram fresh weight per hour.

SEDs: Tmts (9 d.f.), .042; Wks (36 d.f.), .031; Tmts x Wks (>20 d.f.), .068; Wks x Tmts (36 d.f.), .062.

Tmt	Week	Rep 1	Rep 2	Rep 3	Rep 4	Means	
Ctrl	1	.136	.128	.320	.254	.209	
	2	.295	.224	.302	.220	.260	
	3	.195	.121	.112	.183	.153	
	4	.134	.154	.160	.125	.143	.191
-P	1	.227	.193	.521	.375	.329	
	2	.239	.310	.514	.550	.403	
	3	.354	.228	.170	.266	.255	
	4	.375	.300	.354	.240	.317	.326
-K	1	.162	.198	.162	.195	.179	
	2	.309	.306	.213	.280	.277	
	3	.209	.165	.167	.205	.186	
	4	.355	.312	.269	.212	.287	.232
-Mg	1	.144	.298	.487	.265	.299	
	2	.507	.698	.360	.337	.475	
	3	.224	.240	.278	.232	.244	
	4	.761	.557	.404	.633	.588	.402
Weekly Means		(1) .254	(2) .353	(3) .210	(4) .334		

One week old plants demonstrated an excretion capacity enhanced significantly by about 60% in the absence of P, by about 40% but not significantly in the absence of Mg and by about 15% in the absence of K. Between week 1 and week 2 there was a significant overall increase of about 40% in the rate of acid excretion averaged across treatments. By this time K was about 5% above control while P and Mg were both significantly higher at 50% and 80%, respectively.

At three weeks old a significant overall drop of about 40% was recorded. Relative performances and significances were largely unchanged with K up about 20%, P up about 66% and Mg up about 60% on the control. Nutrient replenishment of the growing solutions was conducted after the third weekly sampling, when there was a limited expression of tillering (the plants were grown at about 2 cm spacing).

There was also the first significant appearance of the long, white, unbranched, adventitious roots (Troughton, 1962) which characterises cereal, especially barley, development at this stage. Either solution depletion, or more likely the very rapid bulking of the roots may have contributed to reduced overall activity at this sampling.

Between week 3 and week 4 there was a significant overall increase of about 60%, although the control was down by about 5%. Deficiencies of P, K and Mg engendered significant increases of about 133%, 100% and 200% respectively over the control at the week 4 sampling.

Treatment means averaged across the four week duration of the experiment gave increases over the control of about 20% (not significant) in the absence of K, 80% for P and over 100% for Mg.

146

4. Discussion

This initial experiment with barley provides useful information on acidification potential and perhaps a reasonable basis for extrapolation to crops growing either hydroponically or in solid substrate. The simple technique employed has provided clear-cut data for the growing conditions employed. However, it only allows detection of net outflow of hydrogen ions, without any distinction possible between gross outflow and any simultaneous utilisation, activation or return processes. Neither can it distinguish counteraction/neutralisation by coincidentally excreted hydroxyl ions, nor any potential influence of pre-excreted ions already resident or concentrated in the rhizosphere. Against this background, however, it does seem to detect clearly identifiable influences of individual nutrient deficiencies.

The three nutrients examined were selected for being major nutrients, two of which, P and Mg, have a direct influence on ATP(ase) metabolism, the one being an anion and the other a cation. Potassium was a logical other choice, being the most prominent nutrient cation to exploit the electrogenic uptake system as well as a known activator of some ATPases. With hindsight, there is need to mention that the two most influential deficiencies in the experiment both had substantial additions of K_2SO_4 to modify their growing solutions: 0.01 moles$\bullet$l^{-1} for P and 0.005 moles$\bullet$l^{-1} for Mg. There is long-standing empirical evidence (Hiatt and Hendricks, 1967) that barley roots incubated in K_2SO_4 alone, even at 10^{-4} or 10^{-3}, exhibited a strong imbalance of cation uptake. This then prompted enhanced proton extrusion with its logical tendency to alkalisation/negativisation of the cytosol, in turn countered by enhanced organic acid formation. Such a phenomenon may have had at least a partial bearing on the main results of this experiment. However, further detailed studies seem warranted, encompassing a wider range of nutrients and crops in both laboratory and agronomic investigations. These might prove particularly relevant to current world-wide environmental concerns with more restricted use of fertiliser inputs, especially phosphate and nitrate. The metabolic responses reported here generally surfaced well in advance of any visual growth reduction or symptom expression.

5. References

Bienfait H F 1988 Mechanisms in Fe-efficiency reactions of higher plants. J Plant Nutr 11 : 605-629.

Clarkson D T 1977 Membrane structure and transport. *In* The Molecular Biology of Plant Cells, Ed H Smith. pp 24-63. Blackwell, Oxford, UK.

Davies D D 1986 The fine control of cytosolic pH. Physiol Plant 67 : 702-706.

Gardner W K, Barber D A and Parbery D G 1983 The acquisition of phosphorus by *Lupinus albus* L. III. The probable mechanism by which phosphorus movement in the soil/root interface is enhanced. Plant and Soil 70 : 107-124.

Haynes R J 1990 Active ion uptake and maintenance of cation-anion balance: a critical examination of their role in regulating rhizosphere pH. Plant and Soil 126 : 247-264.

Hiatt A J and Hendricks S B 1967 The role of CO_2-fixation in accumulation of ions by barley roots. Z Pflanzenphysiol 56 : 220-232.

Hoffland E, Findenegg G R and Nelmans J A 1989b Solubilisation of rock phosphate by rape. II. Local root exudation of organic acids as a response to P-starvation. Plant and Soil 113 : 161- 165.

Kurdjian A and Guern J 1989 Intracellular pH: measurement and importance in cell activity. Annu Rev Plant Physiol Plant Mol Biol 40 : 271-303.

Marschner H, Romheld V and Cakmak I 1987 Root-induced changes of nutrient availability in the rhizosphere. J Plant Nutr 10 : 1175-1184.

Poole R J 1978 Energy coupling for membrane transport. Annu Rev Plant Physiol 29 : 437-460.

Raven J A 1986 Biochemical disposal of excess H$^+$ in growing plants. New Phytol 104 : 175-206.

Romheld V and Marschner H 1986 Mobilisation of iron in the rhizosphere of different plant species. *In* Advances in Plant Nutrition, Vol 2. Eds, P B H Tinker and A Lauchli. pp 155-204. Praeger Publishers, New York.

Troughton A 1962 The roots of temperate cereals (wheat, barley, oats and rye). Mimeo 2/1962, CAB, Bucks, UK.

van Beusichem M L, Kirkby E A and Baas R 1988b Influence of nitrate and ammonium on the uptake, assimilation and distribution of nutrients in *Ricinus communis*. Plant Physiol 86 : 914-921.

4- Crop quality – nutrient management by the diagnosis of crop nutrition

Chapter 34

Relationships between nutrition of Ayvalık olive variety and its oil quality

S. Seferoğlu[1] and H. Hakerlerler[2]
[1] Adnan Menderes University, Faculty of Agriculture, Soil Science Department - Aydın –Turkey
[2] Ege University, Faculty of Agriculture, Soil Science Department - Izmir -Turkey

Key words: olive oil, quality, nutrition, leaves

Abstract: The research material is composed of olive leaf and oil samples collected from Ayvalık and Edremit provinces of Körfez region. Leaf N, P, K, Ca, Mg, Na, Fe, Mn, Zn, Cu and B contents were measured. At the same time, oil percentage, free acid, iodine number and some oil acids were analyzed.
The results of analysis have shown that there were significant relationships between leaf nutrients and quality parameters of olive oil.

1. Introduction

There are 898 million olive trees in the world and approximately in 9 million hectare of land olive is grown. Of these olive trees 98 % is in the Mediterranean countries. Turkey, with its 11 % share, is considered to be the homeland of olive (Anonymous, 1995). Among olive growing countries Turkey is the third in terms of area. Aegean Region has 76 % of all our olive wealth and 25.3 % of the olive trees in this region is Ayvalık olive variety (Anonymous, 1995). 25.7 % of the olive grown in Turkey is produced for table while 73.4% is used for oil production (Anonymous, 1989).

Ayvalık olive variety is grown especially in the region which is known as the gulf region (Ayvalık, Edremit etc.). Regarding its variety characteristics and the appropriate agronomic conditions, the best quality olive oil in Turkey, even in the world, is claimed to be produced from this variety (Dokuzoğuz and Mendilcioğlu, 1978; Karakır, 1979; Canözer, 1991). In order to increase productivity and to improve olive oil quality, precautions such as trimming, irrigation, fertilization, protection and cultivation should be taken in time.

In this present research, the relationships between the leaf nutrients of Ayvalık olive variety, which is extensively grown around Edremit and Ayvalık regions, and the quality parameters of olive oil were determined.

2. Materials and Methods

The research material consists of forty (40) leaf and forty oil samples taken from (40) forty olive groves where Ayvalık olive variety is extensively grown. Leaf and fruit (to produce olive oil) samples were collected in November and December which are the recommended times for sampling. For the assessment of leaf N, destilation method was used. For other leaf analysis, samples were extracted. Potassium, Ca and Na were determined flame photometrically; P by colorimeter; and Mg, Fe, Mn, Zn, Cu by AAS (Kacar, 1972). Boron was determined following Riehm, 1957. For the olive oil samples; oil percentage (oil %), peroxide number, acidity, iodine number (Anonymous, 1978) and oil acids content were analyzed using gas chromatography (Thies, 1971).The data were subjected to statistical analyses by MSTAT-C package program.

Table 1. The results of leaf mineral composition.

Nutrients	N	P	K	Ca	Mg	Na	Fe	Mn	Zn	Cu	B
	%						mg kg^{-1}				
Min	1.11	0.08	0.39	1.00	0.09	0.01	55	18	9	1.0	9
Max	1.99	0.14	0.88	2.10	0.31	0.03	150	57	22	10	26
Average	1.64	0.11	0.67	1.41	0.19	0.01	92	36	15	3.6	14

Table 2. Results of olive oil samples .

oil quality parameter	oil %	perox num.	acid %	iodine num.	palmitic acid C16:0	stearic acid C18:0	oleic acid C18:1	linoleic acid C18:2	linoleic acid C18:3
Min	31.73	0.50	0.50	75.76	13.68	0.12	69.07	6.24	0.64
Max	67.34	3.79	1.36	90.91	18.71	0.92	75.76	13.21	1.40
Opt	48.76	1.29	0.94	83.95	16.62	0.54	72.40	9.47	0.96

3. Results and Discussion

The results of leaf analyses were examined according to the limit values which have been set for olive by researchers as Bouat (1961), Hartman (1962), Eryüce (1980) and Reuter and Robinson (1986). N and P contents of the leaves were found sufficient while K contents of 85 % of the groves were measured under the sufficiency level. Ca content of all groves was above 0.8 % which has been set as the sufficient value by Hartmann (1958).

Likewise, Mg content, just like Ca content, was above the sufficiency level in all groves. When the groves were studied according to micro nutrients, results revealed that Fe, Mn and Zn contents of all the leaves were almost sufficient in all groves. When the Cu ranges were compared with respect to the limit values claimed by different researchers, 80% of the groves was under the sufficiency level. Boron which is an important micro nutrient for olive was examined according to the threshold concentration (11.7-24.5 mg kg^{-1}), which was set by Bouat (1961) and found that 70 % of the groves was sufficient while 30 % being under (Table 1).

Results of oil samples showed that oil percentage (oil %) of all groves was in accordance with the oil percentage Canozer (1991) had determined for Ayvalık olive variety. Peroxide number, acidity, iodine number of the oil samples were under the limit values which had been determined by Ersoy (1985) in Codex standards for natural olive oil. This shows that this region's olive oil is of good quality (Table 2).

Saturated (Palmitic and stearic acid) and unsaturated oil acid (oleic acid, linoleic acid) values of the oil samples of all groves were found within the acceptable limits according to the Codex standard values given by Ersoy (1985).

The correlation coefficients which give the relationships between the nutrient status of the leaves and of the oil quality parameters were determined. According to the statistical results, among the oil quality parameters there was a positive relationship (r= 0.365*) only between palmitic acid and leaf N. Cimato (1990) found out that nitrogenous fertilization increases the level of palmitic and linoleic acid in oil.

Between leaf P and the peroxide number in oil, there was a negative relationship (r= -0.381*) while a positive one was determined between P and palmitic acid (r= 0.275*). Likewise, Cimato (1990) notes that fertilization with P increases the amount of palmitic acid.

No statistically significant relationship was found between K content of the leaves and the quality parameters of oil. This case could be explained by the fact that K range of the leaves was considerably under the sufficiency level as it had been pointed out earlier. Between leaf Ca and the linoleic acid, which is desired to be present in small amounts, there was a positive relationship (r= 0.320*). A similar relationship was also found between Mg and peroxide number (r= 0.354*). A positive correlation significant at 1 % level existed between Na and Fe contents of the leaf samples and the oil % contents of the groves.

Between the peroxide number of the oil samples and Mn content of the leaves there was a positive (r= 0.513**) and between the considered parameter and Zn contents a negative correlation was existing (r= -0.384**). Also, Mn content of the leaves and palmitic acid in oil were negatively correlated significant at 5 % level.
Leaf Cu-linoleic acid (r=-0,326*) and leaf B – iodine number (r= 0.431**) were related.

4. Conclusion

Since, this study is not a fertilization experiment and has been carried out as a survey, not many significant relationships were found between acidity, peroxide number, oleic acid, linoleic acid, which determine the olive oil quality and leaf nutrients, especially N, P, K. However, results showed that they are within the acceptable limits when the quality parameters of the oil of the groves have been considered. Thus, we could conclude that not only the nutrition of olive trees but also the climate and ecological factors are effective on olive oil quality.

5. References

Anonymous, 1978., ISO, International standard, 1978,5508 and 5509.

Anonymous, 1989., Uluslararası Zeytin yağı Konseyi, Yemeklik Zeytin Sektörünün Dünya Durumu Hakkında Araştırma Cilt 2-B.

Anonymous, 1995., COI Report on Economic Matters CE/R. 42/Doc. No: 24 April.

Bouat, A., 1961.; Variable de I'alimentation minerale chez I'oliver. Inf. OI Int. No:16, s. 19-31

Canözer, Ö., 1991. Standart Zeytin çeşitleri Katalogu (Ed. Gökçe, N.H.). Tarım ve Köyişleri Bakanlığı Mesleki Yayınlar Serisi No:334, Ankara, s.44-46.

Cimato,A., 1990.; Effect of Agronomic Factors on virgin Olive Oil Quality Olivae, No:31 April, 1990. Pp.20-31.

Dokuzoğuz, M., ve Mendilcioğlu, K.; 1978. Zeytin. E.Ü. Ziraat Fak. Bahçe Bitkileri Bölümü Ders Notu. (Teksir)

Ersoy, B.; 1985. Zeytin Yağlarının Bileşim Özellikleri Çeviri. (International Olive Oil Council T. 15/Doc. No:, 1981). Zeytincilik Araştırma Enstitüsü Yayınları No: 34. Bornova- Izmir.

Eryüce, N. 1980. Ayvalık Bölgesi Yağlık Zeytin Çeşiti Yapraklarında Bazı Besin Elementlerinde Bir Vejetasyon Dönemi İçerisindeki Değişimler. E.Ü. Zir. Fak. Dergisi. 17/2 (209-2221).

Hartmann, H. T., 1958. Some Responses of the Olive to Nitrogen Fertilizer Proc. Amer. Soc. Hort. Sci. 72, 257-266.

Hartmann, H. T.,1962. Olive Growing in Australia. Econ. Bot. 16(1). 31-44.

Kacar, B., 1972. Bitki ve Toprağın Kimyasal Analizleri , II. Bitki analizleri, A.Ü. Zir. Fak. Yayınları, 453.

Karakır, N., 1979. Zeytinde Meyve Gelişmesi ve Meyvelerin Bileşimi Üzerinde Karşılaştırmalı Araştırmalar . E.Ü. Zir. Fak. Meyve ve Bağ Yetiştirme Islahı Bölümü . Doktora Tezi.

Reuter , D. J.and Robinson, J.B., 1986. Plant Analysis in Interpretation Manual. Inkata Press, Melbourne, Sydney. S.137.

Rhiem, H.,1957. Untersuchugen über die in Augustenberg ausgearbeitete Methode zur Bestimmung des heiss wasserjöslichen Bors in Boden nach Berger und Troug Agrochomica 1 (2) 91-106.

Thies, W., 1971. Schnozzle und einfache Analyses der Fettsaure zusammensetzung in einzelnen ropes Katy Ledonen I Gaschromatographische und Paper Chromatographische Methadone, Z. Pflanzenzücht 65181-202.

Chapter 35

Influence of magnesium deficiency on chestnut (*castanea sativa* mill.) yield and nut quality

E. Portela, J. Ferreira-Cardoso, M. Roboredo and M. Pimentel-Pereira
Universidade de Trás-os-Montes e Alto-Douro, Ap 202, 5000 Vila Real, Portugal

Key words: sweet chestnut, *Castanea sativa*, magnesium, Mg-deficiency, fruit quality

Abstract: Under conditions of poor management, severe Mg-deficiency is often observed in chestnut groves located on base-poor granites and schists, in NE Portugal. Many of the severe cases of Mg-deficiency are due to nutritional imbalances between Mg and a growth-stimulating nutrient, particularly nitrogen. In areas affected by Mg-deficiency sharp contrasts have been observed between chlorotic and symptom-free trees, particularly in young groves. A 20-years-old chestnut grove of a traditional Portuguese variety, Longal, was selected in order to study the influence of Mg-deficiency on several tree growth parameters, yield and chestnut quality. Trees were classified according to the intercostal yellowing chlorosis of their leaves into three categories: symptom-free, slight chlorosis and acute chlorosis. There is a strong negative correlation between severity of chlorosis and foliar Mg concentration, tree growth parameters, chestnut production and fruit quality. The mean Mg concentration in leaves of symptom-free trees was 1.2 g kg^{-1} and the lowest value in green leaves was 0.85 g kg^{-1}. Severe intercostal chlorosis was observed when the Mg concentration in leaves was lower than 0.55 g kg^{-1}. The N/Mg ratios increase with the severity of the deficiency, reaching values higher than 40 in the leaves of trees with more acute deficiency. Shortage of Mg causes poor growth of chestnut trees, severe yield reduction and lower nut calibre; nuts have lower levels of dry matter content and crude fat, but higher crude protein content.

1. Introduction

Most sweet chestnut (*Castanea sativa* Mill.) stands in northern Portugal are planted for nut production. Under conditions of poor management, severe Mg-deficiency is often observed in these groves. In areas affected by Mg-deficiency sharp contrasts have been observed between chlorotic and symptom-free groves mainly in young plantations. In more acute foliar discoloration, leaves often fall prematurely and trees have poor growth, low nut production and reduced nut size. In the last decade farmers have intensified management practices, including a greater use of mineral fertilisers in order to increase nut production. However, fertilisation is often imbalanced. Until recently, 50% of the fertilised groves were only dressed with ammonium nitrate and Mg is rarely applied as fertiliser (Portela and Portela, 1996).

Literature on nutrient disorders in chestnuts and nutrient threshold values in leaves is very scarce. Burg (1985) indicates normal values of leaf Mg concentrations for chestnuts to be in the range of 0.22-0.29 %. The same author refers to the value < 0.09% as being low in the fagacea family and that the normal values should be in the range of 0.16-0.28%. Bonneau (1992) indicates that normal values of Mg concentrations in broadleaf forest species are in the range of 0.12-0.23%. Ende and Evers (1996) report optimum values for *Fagus sylvatica* L. at higher than 1.5 mg g^{-1}. These authors draw the attention to the N/Mg ratios, which should be regarded and consider values between 30 and 35 as the threshold values for symptom expression; higher ratios indicate acute Mg-deficiency at absolute Mg contents lower than 0.6 mg g⁻¹.

Nutrient disorders not only affect nut production, but may also affect fruit quality and its technological characteristics and commercial acceptability. This study attempts to diagnose a recent nutrient deficiency and determine its effect on tree growth parameters and fruit quality as compared with some reference values for the same variety. The results may supply useful data to highlight current chestnut fertilisation trials and for the next chestnut fertiliser program.

2. Materials and Methods

A 20-years-old chestnut grove was selected to study the effect of Mg-deficiency on several tree growth parameters, and on nut yield and quality. Trees were classified according to the interveinal yellowing chlorosis of their leaves in tree categories, and foliar analysis was used to quantify Mg-deficiency. Five trees within each degree of Mg-deficiency were selected at random. Fully developed leaves (fourth to seventh from the terminal shoots), light exposed, from the middle third crown, were collected during the last week of August and analysed for macro and micronutrients. Leaf samples were dried at 65°C for 48 h and ground to pass through a 1-mm screen. They were digested as described by van Schouwenburg and Walinga (1978). Calcium, Mg, Fe, Cu, Zn and Mn were determined by atomic absorption spectrophotometry; K by flame photometry. For N and P analysis, the digestion was with sulphuric acid and their concentrations were determined on an autoanalyzer.

Chestnut fruit was harvested from the ground under each selected tree. The shells and pellicle were removed with a knife without damaging the kernel surface. The ease of peeling was evaluated by the time taken to peel each fruit by hand. At the same time, the rate of polispermic (divided nuts) and the groves occurrence were registered. Shelled chestnut kernel samples were dried at 65°C for 48 h to determine dry matter. Dried samples were analysed for neutral detergent fibre (NDF) by

the method of Robertson and van Soest (1981); crude protein by the micro-kjeldahl method (N x 6.25); crude fat by extraction with petroleum ether in a Soxhlet apparatus; starch by enzymatic hydrolysis (Salomonsson et al., 1984). Mineral concentrations in the fruit were determined as described before for foliar analysis.

Soil samples were taken from two depths beneath the crown of selected trees and chemical analyse were carried out (Table 3). Only the 0-20 cm depth is indicated in Table 3, because the subsurface horizon follows the same trend.

Several tree growth parameters and nutrient concentration in leaves, nutrient fruit concentration and fruit quality parameters were organised in two matrices, and Pearson's correlation coefficients were determined.

3. Results and Discussion

Leaves of Mg-deficient trees exhibit a discoloration between the veins, starting in the centre of the leaf. In the more acute stages the whole leaf becomes yellow with golden-brown spots near the main vein. A close negative correlation between the estimated leaf chlorosis and foliar Mg concentration (Table 2), tree growth parameters and chestnut production was observed (Table 1). Tree height and breast height diameter (BHD) are greatly reduced; the leaves are smaller, and older leaves fall prematurely during July and August. Leaf area index (LAI) and fruit production are reduced by more than 50%, with fruit size being particularly affected.

Individual Pearson's correlation coefficients reveal that tree parameters such as tree height, BHD and LAI are positively correlated with Mg content in leaves and negatively correlated with nut calibre.

The mean Mg concentration in leaves of symptom-free trees was 1.2 g kg⁻¹ (Sd= 0.3), the lowest value being in green leaves of 0.85 g kg⁻¹, and the maximum value, where chlorosis was already observed, being 1.7 g kg⁻¹. Severe intercostal chlorosis is observed when Mg concentration in leaves is lower than 0.55 g kg⁻¹. Leaf Mg concentration in trees with acute deficiency is reduced to one third in comparison with trees with no symptoms. According to Bonneau (1992), Kopinga and Burg (1995) and Ende and Evers (1996) the value of 1.2 g kg^{-1} is very low for some deciduous trees of the fagacea family, and they consider values of 1.6 g kg^{-1} and 1.5 g kg^{-1} as threshold values.

Nitrogen varies in opposition to Mg; the N/Mg ratios in leaves are negatively correlated with all tree growth parameters, and two fruit quality parameters : namely calibre and fruit dry matter (DM). The average N/Mg ratio of 16 in symptom free trees reaches 80 in trees with acute Mg-deficiency. Pronounced Mg-deficiency symptom expression was observed at a N/Mg ratio above 40. Comparison with the same ratios for beech and oak indicated by Ende and Evers (1996) shows that the N/Mg ratios obtained in the present study are in the same order of magnitude.

The Mg concentration in the fruit (Table 2) follows an opposite pattern to the leaves: in that it increases when the symptoms are aggravated. Although the Mg content in the soil (Table 3) beneath the more deficient trees is lower than under the symptom free trees, the Mg concentration in fruits is higher in the former. According to Slovik (1996), fruits are potent sinks for Mg; it is believed that even under conditions of low soil Mg supply, this nutrient is channelled to the fruit. However, a dilution effect in the trees with higher nut production may have also occurred. Because of the increase in Mg concentration in the fruit of deficient trees the N/Mg ratio is lower than in the leaves, and this ratio varies from 22 in symptom free trees to 13 in trees with acute deficiency.

As shown in Table 3 the values of exchangeable Mg in the soil are very low, even under the symptom free trees, and the range values of 0.08 to 0.15 cmolc kg^{-1} are far below the threshold values that Ende and Evers (1996) collected from several authors: ie. 0.24 to 0.45 cmolc kg^{-1}.

Chemical composition of the fruit (Table 4) shows that dry matter and crude fat have decreased significantly in Mg-deficient trees, while crude protein has increased. Crude fat and crude protein are negatively correlated, as is commonly observed (Weber, 1985). The data obtained by Ferreira-Cardoso et al. (1993) for 10 Portuguese cultivars shows the same tendency, lower Mg content in the fruit is associated with higher fat values and lower protein content.

Some technological characteristics such as peeling aptness, number of penetrations and percentage of polispermic fruit showed no change (Table 4). The standard quality parameters indicated by Ferreira-Cardoso et al. (1993) for several regional varieties show, in general, lower starch and crude protein contents and higher crude fat and NDF.

The percentage of dry matter in the fruit also decreased because of the Mg shortage. High dry matter content in the fruit is a relevant characteristic for fruit conservation. The fruit water content in the symptom free trees is more favourable to storage. According to Breisch (1993), values higher than 50% are detrimental, because promoting water condensation and the growth of mold. An important quality parameter that was most affected was nut size. While a calibre of 66 fruits kg^{-1} can be used for fresh consumption or 'marron glacé', calibres higher than 90 have less commercial value (Breisch, 1993).

Acid schists soils, which are usually base-poor are prone to Mg-deficiency, and it might be expected, in these locations, that orchard trees with high fruit production will develop Mg-deficiency symptoms more often than vegetative trees (Slovik, 1996). In addition, imbalanced N fertilisation may contribute to enhanced Mg-deficiency, since N is a growth stimulating nutrient and increases the Mg demand.

Acknowledgements

This study was supported by grants from the Trás-os-Montes Rural Development Project. Field assistance and laboratory work were provided by J. Pinheiro, R. Madeira and N. Vieira. Chemical analyse was carried out in the Laboratório de Nutrição Animal and Laboratório de Solos e Plantas of the UTAD.

156

4. References

Bonneau, M. 1992. Le magnésium chez les arbres forestiers: connaissances générales et problèmes identifiés en France. In: Le Magnésium en Agriculture. Coord. C.Huget et M. Coppenet. pp. 174-185. INRA, Paris.

Breisch, H. 1993. Harvesting, storage and processing of chestnuts in France and Italy. In: Proceedings of the International Congress on Chestnut. Org. ed. E. Antognozzi. pp. 429-436. Spoleto, Italy.

Burg, J. van den 1985. Foliar analysis for determination of tree nutrient status- A compilation of literature of data. Rijksinstituut voor Onderzoek in Bos-en Landschapsbouw "De dorschKamp", Wageningen.

Ende, H.P. and Evers, F.H. 1997. Visual magnesium deficiency symptoms (coniferous, deciduous trees) and threshold values (foliar, soil). In: Magnesium Deficiency in Forest Ecosystems. Eds. R.F. Hüttl and W.Schaaf. pp. 3-22. Kluvers Academic Publishers, London.

Ferreira-Cardoso, J.V., Fontaínhas-Fernandes, A.A. and Torres-Pereira, J. 1993. Nutritive value and technological characteristics of Castanea sativa Mill. fruits. Comparative study of some northeastern Portugal cultivars. In: Proceedings of the International Congress on Chestnut. Org. ed. E. Antognozzi. pp. 445-449. Spoleto, Italy.

Kopinga, J. and Burg, J.van den. 1995. Using soil and foliar analysis to diagnose the nutritional status of urban trees. J. Arboric. 21: 17-23.

Portela, E. e Portela, J. 1996. Práticas culturais em Soutos da Padrela. Universidade de Trás-os-Montes e Alto-Douro, Vila Real.

Robertson, J.B. and Van Soest, P.J. 1981. The detergent system of analysis and its application to human foods. In The Analysis of Dietary Fiber in Food. Ed.W.P.T. James and O. Theander. pp. 123-158. Marcell Dekker, New York.

Salomonsson, A., Theander, A. and Westerlund, E. 1984. Chemical characterisation of some cereal whole meal and bran fractions. Swedish J. Agric. Res. 14: 111-117.

Schouwenburg, J.C. van and Walinga, I. 1978. Methods of analysis for plant material. Agricultural University of Wageningen, Netherlands.

Slovik, S. 1997. Tree Physiology. In: Magnesium Deficiency in Forest Ecosystems. Eds. R.F. Hüttl and W.Schaaf. pp. 101-214. Kluvers Academic Publishers, London.

Weber, E.J. 1985. Role of potassium in oil metabolism. In: Potassium in Agriculture. Ed. R.D. Munson. pp. 426-442. Am. Soc. of Agron., Madison.

Table 1. Chestnut tree parameters under different symptoms of Mg-deficiency

	Tree height m	BHD m	Leaf length m	Leaf width m	LAI m^2m^{-2}	Yield kg tree^{-1}	Calibre fruits kg^{-1}
Green	5.8a	0.26a	0.152a	0.051a	2.04a	8.3a	66a
Slight yellowing	4.4b	0.20b	0.146b	0.051a	1.53ab	4.9b	78ab
Pronounced yellowing	3.1c	0.16c	0.129c	0.047b	0.97b	3.6b	99c

BHD - breast height diameter; LAI - leaf area index. Means in the same column with a letter in common are not significantly different (P<0.05) by Duncan's multiple range test

Table 2. Average mineral concentration in leaves and peeled fruit under different symptoms of Mg-deficiency

	N g kg^{-1}	P g kg^{-1}	K g kg^{-1}	Ca g kg^{-1}	Mg g kg^{-1}	Fe mg kg^{-1}	Mn mg kg^{-1}	Zn mg kg^{-1}	Cu mg kg^{-1}
Leaves									
Green	18.7a	2.0a	9.9a	3.9a	1.2a	153a	1099a	36a	7a
Slight yellowing	26.9b	1.9a	9.5a	4.3a	1.1a	216a	972a	49a	7a
Pronounced yellowing	25.0b	1.8a	10.4a	2.6b	0.4b	180a	707a	49a	9a
Fruit									
Green	9.5c	1.5c	8.3c	0.4c	0.4c	17c	97c	22c	18c
Slight yellowing	14.0d	1.4c	8.0c	0.3c	0.5c	25c	61c	26c	20c
Pronounced yellowing	17.8e	1.5c	8.7c	0.3c	0.7d	28c	70c	40d	32c

Means in the same column with a letter in common are not significantly different (P<0.05) by Duncan's multiple range test

Table 3. Chemical analysis of soils of the upper layer (0-20 cm) under the trees with different symptoms of Mg-deficiency

	OM %	pH KCl	P ext. mg kg^{-1}	Ca cmol$_c$ kg^{-1}	Mg cmol$_c$ kg^{-1}	K cmol$_c$ kg^{-1}	Na cmol$_c$ kg^{-1}	H+Al cmol$_c$ kg^{-1}	V %
Green	1.85a	4.0a	25a	0.28a	0.15a	0.18a	0.07a	1.22ab	36a
Slight yellowing	1.87a	3.9a	20a	0.28a	0.14a	0.10ab	0.08a	0.86a	42a
Pronounced yellowing	1.83a	4.0a	39b	0.16a	0.08a	0.07b	0.10a	1.35b	23b

OM- Organic mater by the modified Walkley-Black method; pH- with a 1:2.5 soil solution ratio; Pext.- extractable P by Egner-Riehm method; exchangeable cations by 1N ammonium acetate pH 7; H+Al by KCl method; V- base saturation. Means in the same column with a letter in common are not significantly different (P<0.05) by Duncan's multiple range test

Table 4. Chemical composition of fruit and some technological characteristics under different symptoms of Mg-deficiency

	DM g kg^{-1}	Starch g kg^{-1}	Crude protein g kg^{-1}	Crude fat g kg^{-1}	NDF g kg^{-1}	Peeling aptness min kg^{-1}	Grooves	Polispermie %
Green	494a	580a	59.6a	9.92a	223a	112a	few	<1
Slight yellowing	479ab	612b	85.6b	7.52ab	224a	124a	few	<1
Pronounced yellowing	457c	581a	106.6c	6.34b	219a	120a	few	<1

Means in the same column with a letter in common are not significantly different (P<0.05) by Duncan's multiple range test

Chapter 36

Soil fertility and plant nutritional status of strawberry in the Basilicata Region, Southern Italy

G. Lacertosa[1], V. Lateana[2], N. Montemurro[1], D. Palazzo[1] , S. Vanadia[1]

[1]Metapontum Agrobios, SS. 106, Km. 448.2, I-75010 Metaponto (MT), Italy
[2]A.A.S.D. Pantanello, SS. 106, Km. 448.2, I-75010 Metaponto (MT), Italy

Key words: strawberry, nutritional status, nutrient availability, soil analysis, leaf analysis, fruit quality

Abstract::   The use of leaf and soil analysis is a valuable tool for monitoring plant nutritional needs and planning the fertilisation of strawberries to improve yield and quality, to maintain soil fertility and to reduce environmental pollution. With this aim a survey of leaf mineral composition, soil characteristics and fruit quality was carried on commercial fields in the Basilicata region. Results showed that the availability of plant nutrients was strongly influenced by soil texture, organic matter and pH indicating that fertiliser made recommendations should be on the basis of the soil characteristics. Leaf nutrient content was influenced more by sampling date (most of the elements increased their concentration from winter to spring) than by strawberry variety. With the exception of manganese, a poor correlation was found between fruit quality and leaf mineral composition in the spring. Fruit quality parameters were within appropriate standards in most cases, but acid and sugar concentrations were inversely correlated to the content of nitrogen in the fruits, indicating that a proper control of N fertilisers could be effective in improving fruit quality.

1. Introduction

The importance of strawberry in the economy of Basilicata region of Italy has been increasing in the last two decades. Cultivation is widespread with about 700 ha in the Metaponto coastal plains where some productive Californian varieties have been successfully introduced thanks to the favourable climatic conditions. Most (86%) of the strawberries are cultivated under large plastic tunnels while the remainder are produced in open fields. About 20.000 tons of fruit are produced annually. Harvest time is between March and June. In order to ensure rapid growth and high yield, heavy fertiliser dressings are often applied independent of soil characteristics and crop nutritional status (Baruzzi, 1997). Therefore environmental pollution is likely to occur and, as consequence of nutritional imbalances, fruits of variable quality are often produced (Baruzzi,

1995). A survey of soil characteristics, leaf mineral composition, and fruit quality was carried out on commercial fields in the Basilicata region to obtain fertilisation guidelines based on the knowledge nutrient availability in different soil types and the nutritional needs of the strawberry at different growth stages.

2. Materials and Methods

Soil characteristics (122 samples), leaf mineral content (226 samples) and fruit quality (52 samples) were monitored on commercial strawberry fields of the Metaponto area in 1997/98.

Soils were sampled in summer before planting and analysed according to official guidelines (D.M. 11/5/92).

Leaf and fruit samples were collected from four strawberry varieties: Tudla, Pajaro, Miranda and Tethis.

160

Leaf samples were collected in October (15), November (43), December 1997 (104) and April 1998 (64). Leaves were washed, oven-dried and ground to powder. Total nitrogen (N) was determined with an automatic Kjeldahl method. Phosphorus(P), sulphur(S) and boron(B) were determined with an inductively coupled plasma. Potassium(K), magnesium(Mg), calcium(Ca) and micronutrients (Fe, Mn, Zn, Cu) were determined by atomic absorption spectrophotometry.

Fruit samples were collected in three different periods (15 April, 30 April, 25 May).Weight and hardness (penetrometer with 6 mm plunger) were measured on samples of 20 fruits. Titratable acidity (TA), pH, ash and sugars were measured on each juice sample. Total soluble solids (TSS) were determined with a refractometer and expressed as °Brix while quantitative determination of sugars (fructose, glucose and sucrose) were determined by refractometry (HPLC). Mineral nutrients were measured on freeze-dried samples with the same methodology used for leaf analysis.

Statistical analyses were performed with the aid of the SAS software. An ANOVA procedure was applied to test: 1) the effect of sampling period and variety on leaf mineral composition and 2) the effect of harvest time on quality parameters and mineral composition of strawberry fruits. A correlation analysis was applied to study: 1) the relationships between nutrient availability and soil properties, b) the relationships between leaf mineral composition and fruit quality and 3) the relationships between quality parameters and mineral composition of fruits.

3. Results and Discussion

Most of the soils analysed were sandy clay loams and clay loams (USDA textural classes). Soil pH ranged between 7.3 and 8.5 in 82% of the samples. In order to avoid deficiencies of P and some micronutrients (Fe, Zn, Mn), pH corrections and foliar fertilisation should be recommended. Organic matter content was strongly influenced by soil texture, with average values of 1.22%, 2.02% and 2.27% in

light, loamy and heavy soils respectively. These values are generally higher than found those (Lacertosa et al., 1997) in similar soils cultivated with other crops (orange, olive), indicating a large use of organic amendments for strawberry cultivation.

Composition of the cation exchange capacity (CEC) was not affected by soil texture. Exchangeable Na was, on average, 1.9% and well under the alkalinisation threshold of 15% of CEC in all the samples. In Figure 1 the frequency distribution of exchangeable K, Ca and Mg according the classification proposed by SISS, 1985 is shown. Exchangeable K was sufficient (between 2% and 5% of CEC) in 73% of the samples and low (less than 2% of CEC) in only 4% of the samples. Exchangeable Mg was abundant (higher than 10% of CEC) in most cases, and therefore Mg to K ratio was far above the threshold value of 1 according SISS (1985) at which Mg deficiency can occur.

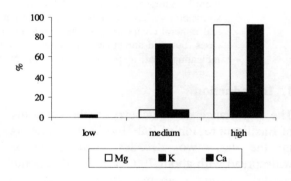

Figure 1. Frequency distribution of values of exchangeable cations (% of CEC) in three availability classes.

Concentration of available P was influenced by soil texture and was inversely correlated with pH (r = - 0.37) and free CaCO$_3$ (r =- 0.37). Low available P levels (less than 10 ppm, according to Cotienne, 1980) were found in many calcareous loamy and clayey soils, while in light textured neutral soils values were in many cases far above the sufficiency threshold. Micronutrient availability was clearly influenced by soil properties. Available B was lower in the lighter textured soil (r = - 0.63 with sand content) and increased with increasing soil salinity (r = 0.58 with the electrical conductivity of 1:2 soil water extract), organic matter content (r = 0.70) and free $CaCO_3$ (r = 0.65). Available Fe and available Mn were negatively correlated with soil reaction (r=-0.58 and r=-0.30 respectively) and free $CaCO_3$ (r = - 0.36 and r = - 0.45 respectively). Available Zn increased with P availability (r = 0.45). Nevertheless levels of available micronutrients (figure 2) can be considered sufficient in most samples according to the classification proposed by Marzadori (1992). Only 12% of soils had low levels of Zn (less than 1 mg kg^{-1}) and only 2% had less than 0.4 mg kg^{-1} B.

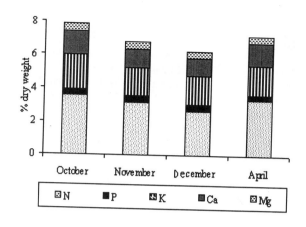

Figure 3. Variation of macronutrient leaf concentration during the growing cycle

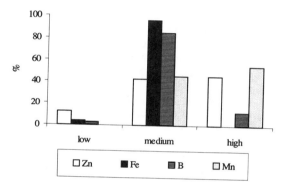

Figure 2. Frequency distribution of values of micronutrients in three availability classes .

Leaf concentrations of many nutrients varied during the growing season (Figure 3). However, nutrient concentrations in November and December samples were very close and not significantly different (data not shown), agreeing with the findings of other authors (John et al., 1976) who proposed this period as the most suitable for defining leaf analysis standards for fertiliser recommendations. In November and December N, K, Ca reached the lowest levels, while for P the variation from winter to spring was opposite (Table 1). Mg concentration was fairly constant in all the sampling periods. For micronutrients, with the exception of Cu, significantly higher levels were measured in leaf samples of April. Similar patterns of nutrient concentration in leaf blades were observed by May et al. (1994).

Table 1. Mean values of leaf nutrient concentration in two sampling time (different letters indicate that values are significantly different at p=0.99 according DMRT).

No. samples	Nov.-Dec. 97	Apr. 98
	147	64
N (%)	2.78 b	3.32 a
P (%)	0.40 a	0.34 b
K (%)	1.68 b	1.34 a
Ca (%)	1.12 b	1.34 a
Mg (%)	0.39 a	0.39 a
B (mg kg^{-1})	42 b	56 a
Fe (mg kg^{-1})	123 a	136 a
Mn (mg kg^{-1})	224 a	253 a
Zn (mg kg^{-1})	23 b	31 a
Cu (mg kg^{-1})	61 a	9 b

With few exceptions there were no significant differences of leaf composition among the four varieties, which are extremely grown in the area, suggesting that the same leaf analysis standards could be used throughout the this area of Italy.

In Table 2 some analytical characteristics of fruits at commercial ripening in three harvest periods are reported.

Table 2. Mean values of analytical characteristics of fruits at commercial ripening in three periods (different letters indicate that values are significantly different at p=0.99 according DMRT).

No. Samples	Early	Normal	Late
	20	13	18
Weight (g)	23.2 a	19.8 ab	17.0 b
pH	3.78 a	3.67 b	3.61 b
Ash (%)	0.30 b	0.33 ab	0.37 a
Hardness (g)	411 a	394 ab	347 b
Conduct.(μS cm^{-1})	3990 b	4887 a	5089 a
Acidity meq100g^{-1}	9.6 b	12.0 a	13.0 a
TSS (Brix°)	6.5 b	7.1 b	8.5 a
Fructose (%)	1.6 c	2.0 b	2.4 a
Glucose (%)	1.5 b	1.7 b	2.1 a
Sucrose (%)	0.7 b	0.8 b	1.1 a
N (%)	1.25 a	1.08 b	0.94 c
P (%)	0.26 a	0.20 b	0.20 b
K (%)	1.64 a	1.54 ab	1.38 b
Ca (%)	0.23 a	0.24 a	0.21 a
S (%)	0.08 a	0.08 a	0.07 b
B (mg kg^{-1})	15 a	6 b	7 b
Mn (mg kg^{-1})	46 a	40 a	40 a
Fe (mg kg^{-1})	98 a	59 b	59 b
Zn (mg kg^{-1})	21 a	18 b	12 b
Cu (mg kg^{-1})	10 b	12 a	13 a

Fruit weight and fruit hardness decreased significantly from the early to the late harvest, reflecting the behaviour of all the mineral nutrients. Conversely total soluble solids (TSS), sugars and acidity increased dramatically in the later harvests. However, fruit quality parameters were within the standards for marketable fruits in most samples, indicating a good level of crop nutrition. Indeed leaf nutrient content in April was poorly correlated with fruit quality and mineral composition (with the exception of manganese with r = 0.74), suggesting that plant nutritional status was above the sufficiency level in most fields and thus no quality response could be expected. In fact excessive nutrition appears to have occurred in some cases. Sugar concentrations were inversely correlated to the content of nitrogen in fruit (r = - 0.61) in the latest harvest, indicating that a proper control of N fertilisers could be effective in improving fruit quality.

4. References

Baruzzi G, Faedi W, Mazzotti A e Mosconi F, (1995). Variabilità di comportamento di alcune cultivar di fragola. L'informatore Agrario 44, 47-50

Baruzzi G, Scudellari D, Mosconi F, Fiori R, Sbrighi M, (1997). Indagine sullo stato nutrizionale dei fragoleti nel Cesenate. Rivista di frutticoltura. 9, 35-42.

Cotienne A. (1980). FAO soils bullettin 38/2, Rome

D.M. 11/5/92 N.79. Approvazione dei metodi ufficiali di analisi chimica del suolo. Ministero dell'Agricoltura e delle Foreste.

Lacertosa G, Montemurro N, Martelli S, Santospirito G and Palazzo D (1997). Indagine sulle caratteristiche chimico fisiche dei suoli agrari nella regione Basilicata. L'Informatore Agrario 26:2-4.

John, M.K., Daubney H.A. and McElroy F.D, (1976). Influence of sampling time on elemental composition of strawberry leaves and petioles. J. Amer. Soc. Hort. Sci. 100:513-517.

May G.M, Pritts M.P and Kelly M.J., (1994). Seasonal patterns of growth and tissue nutrient content in strawberry. J. Plant Nutri. 17(7):1149-1162.

Marzadori C, Orsini D, Schippa M, Sequi P, Vittori A L, (1992). Microelementi in agricoltura. Edagricole 89

SISS, Società Italiana di Scienza del Suolo (1985) Metodi normalizzati di analisi del suolo. Edagricole Bologna 100.

Chapter 37

Carbonhydrate fractions and nutrient status of watermelon grown in the alluvial soils of Küçük Menderes Watershed , Turkey

H. Hakerlerler, B. Okur, E. İrget and N. Saatçı
Ege University Faculty of Agriculture Soil Dept. 35100 Bornova-İzmir/Turkey

Key words: watermelon, carbonhydrate fractions,nutrient status

Abstract: Watermelon is one of the highly consumed summer fruits in Ege Region of Turkey. Ödemiş Valley in this region is important for its widespread watermelon fields. The aim of this research was to determine the nutrients status of watermelon by examining the leaf and fruit compositions and measuring the soil fertility. Fructose, the dominant sugar in fruits, and subsequently β-glucose, sucrose and α-glucose contents were determined. Statistically significant correlations were found between leaf blade Ca content and fruit brix values and sucrose contents.

1. Introduction

Watermelon (*Citrullus Lanatus*) is one of the highly consumed fruits in summer months in Ege region. Although sensorial tests in determining the quality of fruits is widely used, the fact that sugar content and the fractions play more significant roles should be considered (Hakerlerler et al., 1994). Fruit species is classified according to fructose, glucose and sucrose percentages (Ninkovski, 1984).

The aim of this research was to determine the nutrient status of watermelon. For this purpose, soil, leaf and fruit samples were taken from the alluvial soils of Ödemiş Valley in Küçük Menderes Watershed where watermelon is grown widely. Since fruits are the generative organs where carbonhydrates accumulate in general, the relationships between nutrient contents of fruits and leaves were also determined. Obtained results will be of further benefit in making recommendations for the fertilisation of watermelon.

2. Materials and Methods

Thirteen soil, leaf and fruit samples (totally 39) were taken from the watermelon fields which are intensively widespread in Ödemiş Valley of Küçük Menderes Watershed. Samples were taken at fruit maturity. Physical and chemical properties of soils (Black, 1965), macro and microelement contents of leaves and fruits and sugar fractions of fruits (Kacar, 1972; Neubeller and Buchloh, 1975) were determined.

3. Results and Discussion

The pH of field soils changed between acidic to neutral, with a low lime, organic matter and N contents. No salinity was identified.The texture was light (Akalan, 1987). Available P of soils was sufficient (Bingham, 1949), on the other hand K (Pizer, 1967), Ca and Mg contents were low (Loue, 1968). Fe and Mn contents were adequate and Cu and Zn contents low (Lindsay and Norwell, 1978).

Macro and microelement contents of leaf blade results showed that total-N (%N) varied from 2.69 to 4.21%, phosphorus 0.21 to 0.50 %, potassium 0.88-1.73 %, calcium 1.60 to 4.75 %, magnesium 0.42 to 1.06 % and sodium 0.018 to 0.070 % (Table-1,2).

Table 1. Macro nutrients in leaf blade of watermelon

Samp No	N	P	% K	Ca	Mg	Na
1	4.12	0.50	1.50	2.70	0.70	0.04
2	2.78	0.29	1.13	4.35	0.96	0.07
3	3.85	0.46	1.30	4.60	0.92	0.05
4	3.20	0.35	1.37	4.25	1.06	0.05
5	3.14	0.27	1.63	4.20	0.62	0.02
6	3.28	0.24	1.13	4.75	0.92	0.04
7	2.84	0.21	0.88	4.50	0.80	0.06
8	2.69	0.34	1.73	4.30	0.74	0.03
9	2.96	0.22	1.27	3.90	0.82	0.05
10	3.22	0.27	1.49	3.95	0.56	0.05
11	4.21	0.43	1.46	1.60	0.42	0.02
12	3.95	0.41	1.57	2.50	0.68	0.03
13	3.48	0.36	1.49	1.90	0.65	0.02
Min	2.69	0.21	0.88	1.60	0.42	0.02
Max	4.21	0.50	1.73	4.75	1.06	0.07
Avr	3.36	0.33	1.38	3.65	0.76	0.04

Table 2.Micro nutrients in leaf blade of watermelon

Samp. No	Fe	mgkg^{-1} Cu	Zn	Mn
1	232	11	55	820
2	265	13	43	210
3	210	12	41	266
4	240	12	21	106
5	255	12	29	120
6	250	16	33	108
7	185	12	38	88
8	385	16	32	142
9	240	12	23	144
10	240	12	26	120
11	150	12	27	82
12	200	12	26	74
13	280	42	33	34
Min.	150	11	21	34
Max.	385	42	55	820
Avr	241	15	33	178

In the case of microelements, Fe in the leaf blade was found between 150 and 375 mgkg^{-1}, Zn 21 and 55 mgkg^{-1} Mn 34 and 820 mgkg^{-1}, Cu 11 and 42 mgkg^{-1} (Table-2).

According to the reference values, total -N, P, Ca and Mg contents of leaves were sufficient in general and K contents low.

Regarding the microelements; Fe, Zn, Mn and Cu contents were at sufficient levels and these findings were similar to the cited reference values (Reuter and Robinson, 1986; Bergmann, 1993).

The quality analysis of watermelon put forth that total soluble solids (brix) changed in a quite narrow range as 8.00-10.80 %. Fructose content varied from 17.45 to 43.00 %, α-glucose 2.63 to 8.14 %, β-glucose 3.39 to 17.37 % and sucrose 1.00 to 20.58 %. Total sugars showed quite a wide variation as 34.46-79.25 % (Table-3). The dominant sugar in watermelon fruit was fructose and was succeeded by β-glu-cose>sucrose>α-glucose. According to some research results, sugars that are found in fruits are fructose, glucose and sucrose (Buchloh and Neubeller, 1969; Saatçı et al., 1997).

Table 3 .Sugars fractions of watermelon (%)

Samp No	Fru.*	α-G.*	ß-G.*	Suc.*	Total Sugar	Brix
1	32.26	6.07	13.20	12.82	64.75	10.2
2	42.98	7.92	16.06	9.91	76.87	8.2
3	32.77	5.85	11.73	7.77	58.12	9.7
4	17.45	3.40	7.39	8.20	36.44	9.6
5	42.19	8.14	17.37	9.88	77.58	10.4
6	34.37	5.87	13.75	20.58	74.57	10.4
7	35.58	5.83	14.07	16.48	71.96	10.8
8	34.16	6.07	13.30	12.99	66.52	9.8
9	38.01	6.62	14.08	9.37	68.08	9.7
10	42.70	7.30	17.08	12.07	79.25	9.4
11	45.09	5.22	11.83	1.00	63.14	5.1
12	35.09	5.34	11.18	5.90	57.51	8.0
13	43.00	2.63	3.39	6.74	55.76	9.3
Min.	17.45	2.63	3.39	1.00	36.44	5.1
Max.	45.09	8.14	17.37	20.58	79.25	10.8
Avr	36.62	5.86	12.65	10.28	65.43	9.27

* Fru. (Fructose), α-G. (α-Glucose),
ß-G. (ß-Glucose), Suc. (Sucrose)

In this study, the statistical relationships between leaf nutrient contents and quality properties of watermelon fruit were also examined. According to the multiple regression and correlation analysis; positive relationships were found between brix and sucrose content of fruit and Ca content of leaf blade. In some researches; it was stated that Ca is directly effective on the fruit meat formation (Seçer and Ünal, 1990; Hanada, 1981). Roemmer (1989) reports the average brix value in 15 different strawberry varieties as 7.99 %.

4. References

Akalan,İ.,1987. Toprak Bilgisi. Ankara Uni. Ziraat Fakültesi Yayınları 1058 Ders Kitabı 309 Ankara.

Bergmann, W.,1993. Ernährungsstorüngen bei Kulturpflanzen. Entstehung, visuelle und analytische Diagnose. Gustav Fischer Verlag. Jena-Stuttgart.

Bingham, F.T., 1949. Soil Test for Phosphate. Calif. Agriculture 3(7): 11-14.

Black, C.A., 1965: Methods of Soil Analysis. Part 1-2 American Soc. Agro. Inc. Publisher. Madison-USA.

Buchloh, G., Neubeller, J., 1969. Zur qualitativen und quantitativen Bestimmung von Zuckern und Zuckeralkoholen in einigen obstfruchten mittels. Gaschromatographie. "Der Erwerbsobstbau" 11. Jahrgang, Heft 2, 22-27.

Hakerlerler,H.,N.Saatçi.,S.Hepaksoy.,U.Ak-soy.,L.Üçdemir.1994.The relationships between the fruit carbonhydrate fractions of some Apricot and Peache varieties and nutrient contents in leaf and fruits.E.Ü.Journal of Agric..Fac. Vol : 31 (1), No : 17-24

Hanada, K.,1981. Studies on Nitrogen Nutrition in Muskmelons.IV.,The Influence of the Form of Calcium in Melon growth. Science Bulletin of the Faculty of Agriculture, Kyushu University 35 (1/2), 45-54.

Kacar, B., 1972. Chemical Analysis of Plant and Soil III. Plant Analysis. AUFA, No. 453 Ankara-Turkey.

Lindsay,W.L.,Norvell.,W.A.,1978.Development of a DTPA Soil Test for Zinc,Iron,Manganese and Copper.Soil Sci.Soc.of America Joırnal.Vol.42,No.3 , May-June,USA.

Loue,A.,1968.Diagnostis petiolaire de Prospection.Etudes sur la Nutrition et la Fertilization Potassiques de la Vigne.Societe Commerciale des Potasses d'Alsace Services Agronomiques.31-41._

Mengel,K., 1991. Erhänrung und Staffwechsel der Pflanze. Gustav Fischer Verlag. Jena.

Neubeller, J., Silbereisen, R., 1968. Anbauverhal ten und Fruchtqualitat einiger Mutanten der Apfelsorte "Gravensteiner". Mitt: Klosterneuburg 18: 274-285.

Neubeller, J., Buchloh, G., 1975. Zuckerbestimmung in Gartenbauprodukten in Hinblick auf die Qualitaetsbildung. Mitt. Klosterneuburg 25, Jahrgang 423-432.

Nınkovski, I., 1984. Sugars, Forms of Sugars and Acids in Stone Fruits Grown in Belgrade Region.Nauha Piaksi 14(1): 49-62.

Pizer, N.H.,1967. Some Advisory Aspects of Potassium and Magnesium. Tech. Bull. No. 14:84.

Reuter, D. J., Robinson, J.B., 1986. Plant Analysis. Inkata Press, Sydney.

Roemer, K., 1989. Das Zuckermuster Verdücdener Obstarten.Erwerbobstbau 31 Jg. 211-216.

Seçer, M., Ünal, A.,1990. Nährstoffgehalte in Blattspreite und Blattstiel der Zuckermelone und deren Beziehung zu Ertrag und Qualitat. Gartenbauwissenschaft 55(1), 37-41.

Chapter 38

Relations between nutrient status and quality properties of Çeşme Muskmelon

H.Hakerlerler, N. Saatçı, B. Okur, E. İrget
Ege University Faculty of Agriculture Department of Soil Science 35100 Bornova-İzmir/Turkey

Key words: muskmelon, macro and micro nutrients, sugar fractions

Abstract: Muskmelon (cv. Çeşme) is an important crop in the vicinity of Urla and Çeşme, towns of İzmir province known for their touristic potential also. Due to its high fruit quality, a study was designed to gather more information on soil, leaf and fruit characteristics. The correlations between the available primary and secondary plant nutrients in the soil and contents in leaves and fruits were determined. The sugar fractions, significant quality attributes in melon fruits, were analysed and correlated with the macro and micro plant nutrients. Obtained results will be of benefit in making further recommendations for fertilisation.

1. Introduction

Organoleptic analysis are currently considered worthwhile in determining the fruit quality. Taste, acidity, colour size and other aromatic properties are of interest for scientists (Neubeller and Silbereisen, 1968). Sugars are the fundamental factors that influence the taste and are unique quality components. They are easily affected by many physical and chemical side factors, which are quite hard to modify relatively (Neubeller and Buchloch, 1975).

Research results show that, fruits are pretty high in fructose, glucose and sucrose, on the other hand, poor in xylose, galactose and ribose. Buchloch and Neubeller (1969) state the importance of sugar alcohols, sorbite and inosite which are also present in considerable levels. Muskmelon is one of the highly consumed summer fruits in Turkey. Urla and Çeşme districts of İzmir are specially famous for its muskmelons and quite a considerable amount of land is under muskmelon cultivation. The objective of this study is to determine the relations between nutrient status and the quality properties of Çeşme muskmelon by examining the mineral composition of leaf blades, petioles, fruits and soils.

2. Materials and Methods

Twelve fields were selected in Urla and Çeşme districts of İzmir. Soil, leaf blade, petiole and fruits were sampled at the maturation period of muskmelon. Physical and chemical properties of the soils were determined (Black, 1965) and macro and micro element status of the leaf blades, petioles and fruits were measured. Furthermore, fruit sugar fractions were studied (Kacar, 1972; Neubeller and Buchloch, 1975).

3. Results and Discussion

Research area soils were neutral to slightly alkaline in reaction (6.98-7.72), high in $CaCO_3$ (0.41-40.53 %), low in organic matter and had clayey-loamy texture. Soils had no salinity problem but were poor in total N, available P, Fe and Zn. On the other hand, K, Ca, Mg, Cu and Mn contents were sufficient. Leaf blade N, P, K, Ca, Mg and Na contents ranged between 1.82-2.48, 0.09-0.20, 0.60-1.32, 3.30-8.20, 0.65-2.60 and 0.01-0.39 % respectively (Table 1). For microelements, Fe was measured between 115-505, Zn between 12-30, Mn between 30-150 and Cu between 9-13 mg kg^{-1} (Table-2). Research results revealed that leaf blade N, Ca and Mg concentrations were sufficient and P, K, Zn low compared to the cited critical values for muskmelon (Reuter and Robinson, 1986).

Table 1. Macro nutrients in leaf blade and petiole of muskmelon (%)

Samp.	Org.	N	P	K	Ca	Mg	Na
1	A	2.26	0.15	0.74	4.30	1.30	0.39
1	B	1.14	0.11	1.34	3.20	1.31	0.87
2	A	1.96	0.10	0.80	4.45	2.30	0.01
2	B	1.04	0.08	1.75	1.14	1.50	0.08
3	A	2.08	0.12	1.06	4.20	1.18	0.04
3	B	1.16	0.12	1.87	3.91	1.04	0.04
4	A	2.26	0.14	1.01	3.30	2.60	0.04
4	B	1.20	0.11	1.99	2.15	2.35	0.07
5	A	2.17	0.18	0.77	4.75	2.55	0.09
5	B	1.38	0.13	1.16	3.13	1.66	1.27
6	A	2.36	0.20	1.08	4.75	2.20	0.03
6	B	1.48	0.18	1.75	4.02	1.34	0.04
7	A	2.33	0.15	1.18	3.40	2.40	0.05
7	B	1.44	0.12	2.63	2.66	1.25	0.06
8	A	2.39	0.15	1.25	8.00	0.85	0.02
8	B	1.94	0.14	2.62	4.90	0.60	0.05
9	A	1.95	0.19	1.25	7.85	0.96	0.01
9	B	0.88	0.12	2.68	4.26	0.72	0.02
10	A	2.48	0.17	1.32	5.65	0.65	0.03
10	B	1.38	0.16	2.72	3.54	0.55	0.04
11	A	1.82	0.09	0.60	6.80	1.04	0.05
11	B	1.10	0.08	8.98	5.30	0.95	0.06
12	A	1.88	0.14	1.22	8.20	0.70	0.01
12	B	0.84	0.09	2.38	4.91	0.67	0.02
Min.	A	1.82	0.09	0.60	3.30	0.65	0.01
Min.	B	0.84	0.08	1.16	1.14	0.55	0.02
Max.	A	2.48	0.20	1.32	8.20	2.60	0.39
Max.	B	1.94	0.18	8.98	5.30	2.35	1.27
Aver.	A	2.15	0.14	1.02	5.47	1.56	0.07
Aver.	B	1.24	0.12	2.65	3.59	1.16	0.22

A: Blade, B: Petiole

Results were parallel with the findings of Wojtaszek and Sady (1982). On the other hand, K, Na, Zn and Cu content of leaf blades were higher than that of the petioles which were in agreement with Seçer and Ünal (1990) who report similar findings.

Results related to the sugar fractions of muskmelon revealed that sucrose (37.2-58.9 %) to be the primary fraction which is succeeded by β Glucose (5.17-12.81 %) and α-Glucose (2.35-7.09 %) (Table-3).

Table 2. Micro nutrients in leaf blade and petiole of muskmelon (mg kg^{-1})

Samp.	Org.	Fe	Cu	Zn	Mn
1	A	175	9	12	48
1	B	107	11	16	26
2	A	230	13	14	34
2	B	130	15	18	16
3	A	365	10	16	30
3	B	176	8	17	13
4	A	365	10	13	34
4	B	124	10	17	15
5	A	400	12	13	62
5	B	202	18	18	40
6	A	505	12	14	60
6	B	274	14	18	27
7	A	345	12	16	44
7	B	157	5	13	19
8	A	220	12	17	96
8	B	102	12	12	33
9	A	250	13	19	68
9	B	103	12	19	19
10	A	125	12	24	150
10	B	79	9	24	72
11	A	150	12	15	56
11	B	127	11	18	17
12	A	115	10	30	48
12	B	78	9	25	19
Min.	A	115	9	12	30
Min.	B	78	5	12	13
Max.	A	505	13	30	150
Max.	B	274	18	25	72
Aver.	A	270	11	17	61
Aver.	B	138	11	18	26

A: Blade, B: Petiole

Statistical studies showed significant positive correlations between leaf blade and leaf petiole Mg concentrations and fruit α-Glucose. These relationships show the significant role of Mg in carbonhydrate metabolism which is also stated by Mengel (1991). Moreover, an other significant positive correlation (5 %) was found between leaf blade and leaf petiole Ca/K ratio and total fruit sugar which also notified the significance of K in carbonhydrate metabolism.

Table 3. Sugar fractions in fruit of muskmelon (%)

Samp.	1	2	3	4	5
1	16.4	6.30	11.3	46.0	80.2
2	13.7	5.90	10.8	44.7	75.1
3	12.6	5.05	10.0	58.9	86.7
4	15.4	6.38	12.3	42.2	76.3
5	14.7	7.09	12.0	44.8	78.6
6	12.3	5.40	10.1	37.2	65.1
7	19.3	6.07	12.8	47.1	85.3
8	12.0	2.35	5.17	43.1	62.7
9	16.4	4.60	10.6	56.0	87.7
10	15.5	4.32	9.80	48.3	78.0
11	10.0	3.38	6.09	37.7	57.3
12	16.3	4.87	11.3	47.1	79.6
Min.	10.0	2.35	5.17	37.2	62.7
Max.	19.3	7.09	12.8	58.9	87.7
Aver.	14.6	5.14	10.2	46.1	76.0

1: Fructose, 2: α-Glucose, 3: β-Glucose, 4: Sucrose,
5: Total

4. References

Buchloch, G., Neubeller, J., 1969: Zur qualitativen und quantitativen bestimmung von zuckern und zuckeralkoholen in einigen obstfrüchten mittels gaschromotographie "Der Erwerbobstbau" 11. Jahrgang Heft 2, 22-27.

Kacar, B., (1972): Chemical Analysis of Plant and Soil III. Plant Analysis. AUFA, no: 453. Ankara

Lindsay, W.L., Norvell, W.A., 1978: Development of a DTPA Soil Test for Zinc, Iron, Manganese and Copper. Soil Science Society of American Journal, 42: 421-428.

Loue, A., 1968: Diagnostic Petiolaire de Prospection. Etude sur la Nutrition Et la Fertilisation Potassiques de la Vigne. Societe Commerciale des Potasses d'Alsace Services Agronomiques 31-41.

Mengel, K., 1991: Ernährung und Stoffwechsel der Pflanze.Gustav Fischer Verlag.Jena.

Neubeller, J., Silbereisen, R., 1968: Anbauverhalten und fruchtqualitäteiniger mutanten der apfelsorte 'Gravensteiner' Mitt.Klosterneuburg 18:274-285.

Neubeller, J., Buchloch, G., 1975: Zuckerbestimmung in gartenbauprodukten in hinblick auf die qualitätsbildung. Mitt. Klosterneuburg 25 Jahrgang.423-432.

Reuter, D.J., Robinson, J.B., 1986: Plant Analysis. Inkata Press. Sydney.

Seçer, M., Unal, A., 1990: Nährstoffgehalte in blattspreite und blattstiel der zuckermelone und deren beziehung zu ertrag und qualität. Gartenbauwissenschaft 55(1), 37-41.

Wojtaszek, T., Sady, W., 1982: The content of N, P, K, Mg and Ca in muskmelon leaves during three stages of growth in relation to the physical and chemical characteristics of the rooting medium. Soil and Fertilizers. Vol: 45.

Chapter 39

Spinach and heavy metal relations

N. Saatçı and M. Yaşar

Ege University Faculty of Agriculture Department of Soil Science 35100 Bornova- Izmir/Turkey

Key words: spinach, heavy metals

Abstract::   Spinach fields are intensively widespread in Menemen, Bulgurca and Torbalı districts of the city of Izmir-Turkey. Twenty one fields were selected and soil and plant (blade+petiole) samples were taken to study the heavy metal (Pb, Cd, Cr, Co, Ni, Fe, Cu, Zn and Mn) concentrations. Generally, Ni contents of soil samples were found high. Pb, Cd, Cr, Co, Ni, and Fe were high in some of the leaf blade samples. But Ni concentrations of the petioles were found higher than leaf blades.

1. Introduction

Heavy metals can also be attributed as quality factors in vegetables due to their importance in human nutrition and health. Since spinach leaves are the unique plant parts consumed by humans, heavy metal accumulations gain significance.

When plants are grown in contaminated soils and are further consumed by humans, various illnesses appear. Prawda (1997) state that heavy metals cause allergy in humans, especially Cu, Zn, Pb and Hg ions are very dangerous and damage the immune system of humans. Heavy metals also affect the lungs resulting in cancer (Merian, 1991).

2. Materials and Methods

This study was conducted in Menemen, Bulgurca and Torbalı regions of Izmir where intensive spinach (Spinacia oleracea l.) cultivation is realised. Twenty one fields were selected and plant (leaf blade, petiole) and soil samples (depth 0-30 cm.) (Bergmann,1986) were taken. From Menemen, Torbalı and Bulgurca 12, 5 and 4 sample fields were chosen respectively. Soils were analysed for their texture , pH, $CaCO_3$ and organic matter.

Soil heavy metals were analysed in 3HCl-1HNO$_3$ acid extract (soil / acid; 1/2.5) (Kick et al., 1980). Plant samples were wet digested (400-450° C) in 2N HCl. Heavy metals contents were measured by AAS (Slawin, 1968; Isaac and Kerber, 1969).

3. Results and Discussion

Spinach soils were slightly alkaline in reaction, low in organic matter and changed from low to high in $CaCO_3$ content. There wasn't any salinity problem of the soils. The texture of Menemen district soils were loam and sandy-loam, of Bulgurca and Torbalı districts were sandy-loam and sandy-clayey-loam.

Pb content of the soil samples was found low compared with 100 mg kg^{-1} threshold value reported by Kloke (1980) (Table-1). Leaf Pb concentrations of Menemen and Bulgurca were higher than 0.1-6 mgkg^{-1} critical ranges (Scheffer and Schachtschabel, 1989)(Table-2). On the other hand, Pb content of spinach petioles was low (Table-3). Hakerlerler et al., (1984) and Saatçı et al., (1988) report no Pb pollution in the soils of Izmir. In the latter study only one leaf with a high Pb is stated.

Table 1. Heavy metal content of spinach soils

Element (mgkg^{-1})		Location		
		1	2	3
	Min	18.7	21.8	12.5
Pb	Max	26.2	34.3	31.5
	Mean	21.8	26.3	21.5
	Min	0.56	0.56	0.50
Cd	Max	0.94	0.94	0.87
	Mean	0.76	0.67	0.66
	Min	13.7	15.1	10.3
Co	Max	16.8	21.5	17.8
	Mean	15.1	17.6	14.1
	Min	42.7	42.3	31.8
Cr	Max	57.3	65.2	42.3
	Mean	50.4	50.9	38.1
	Min	52.5	45.0	28.1
Ni	Max	71.8	75.0	47.5
	Mean	60.8	59.3	38.4
	Min	0.78	0.40	0.40
Fe(%)	Max	1.71	1.34	1.72
	Mean	1.23	1.01	1.18
	Min	16.4	14.4	15.0
Cu	Max	37.0	24.4	28.1
	Mean	26.0	21.2	21.7
	Min	28.2	24.0	24.0
Zn	Max	54.7	57.5	73.7
	Mean	43.2	45.4	52.0
	Min	519	707	370
Mn	Max	720	1187	610
	Mean	584	961	490

1:Menemen, 2:Bulgurca, 3:Torbalı

According to the 3 mg kg^{-1} Cd critical level claimed by Kloke (1980), the present spinach soils were found low in this aspect (Table-1). On the other hand, in the leaf blade samples of Bulgurca and Torbalı districts, there were higher Cd concentrations than the limit values (0.04-0.5 mg kg^{-1}) reported by Scheffer and Schachtschabel (1989) (Table-2). However, in the petioles there were negligible amounts of Cd. Sposite and Pagel (1984) point out that Cd uptake of spinach is 0.57 g ha^{-1} all through the year. Some soil and leaf samples of İzmir and its districts had Cd pollution, on the contrary to that of mandarine orchards in Gümüldür district with no such problems (Hakerlerler et al., 1994;Saatçı et al., 1988).

Co content of spinach soils was found relatively lower than the 50 mg kg^{-1} critical value (Kloke, 1980) (Table-1). On the other hand, Co content of spinach leaves was found high according to Scheffer and Schachtschabel, (1989). Excluding the spinach petioles of Bulgurca district, all the studied petioles contained trace amounts of Co (Table-2,3). Hakerlerler et al., (1994) report that mandarine orchards of Gümüldür district have Co pollution and Saatçı et al., (1988) had obtained some Co in their study.

The studied soil samples had lower Cr than the permissible amounts, 100 mg kg^{-1}, (Kloke, 1980). In the case of leaves, Bulgurca was high according to Scheffer and Schachtschabel (1989). Cr concentration of petiole samples was not at dangerous levels. Berrow and Reaves state that they (1986) had obtained 62 mg kg^{-1} Cr in Scotland soils. Hakerlerler et al., (1994) and Saatçı et al., (1988) found that Izmir soils had no Cr pollution.

Ni content of Menemen district soils were higher than the 50 mg kg^{-1} permissible level (Kloke, 1980) (Table-1). Leaf Ni contents of Bulgurca district were also found high and the average obtained value was 4 mg kg^{-1}. All of leaf petiole Ni concentrations were high (Table-2, 3) and the leaf petiole Ni concentrations were examined to be higher than blades. Ni content of agricultural soils is reported to change between 2-1000 mg kg^{-1}. Gümüldür mandarine orchards and the agricultural soils of İzmir and its districts have high Ni concentrations according to the examinations by Hakerlerler et al. (1994) and Saatçı et al., (1988).

HCl-HNO$_3$ acid extractable total Fe concentrations of the soils were determined to change between 0.4-1.7 % (Table-1). Saatçı et al. (1988) found Fe content of the soils in İzmir and its district as 2.0-4.2 %. Total iron concentrations of Gümüldür and Balçova orchards have been stated to change between 2.0-4.4 % (Hakerlerler et al., 1994). In the case of plant tissue analysis results showed that 2 N HCl extractable blade and petiole samples contained high total iron (Marschner, 1995; Pendias and Pendias , 1984).Zn content of soil samples was high according to Kloke (1980) who states 300 mg kg^{-1} as the critical level (Table-1). In general Zn concentrations were low in leaf blade and petiole samples (Bergmann, 1986; Pendias and Pendias, 1984) (Table-2,3).Cu concentrations of the soil samples and leaf blade and petioles were low (Kloke, 1980; Bergmann, 1986; Pendias and Pendias, 1984). Soil Mn concentrations was found lower than the 1500-3000 mg kg^{-1} permissible range (Pendias and Pendias, 1984). According to the results there wasn't any Mn toxicity in the soils. Mn content of leaf blades was optimum and high on the contrary of the petioles (Bergmann, 1986).

Table 2. Heavy metal content of leaf blades .

Element		Location		
(mg kg^{-1})		1	2	3
Pb	Min	5.0	5.0	5.0
	Max	7.4	7.4	5.0
	Mean	6.0	6.2	5.0
Cd	Min	0.2	0.3	0.2
	Max	0.4	0.6	0.7
	Mean	0.3	0.4	0.4
Co	Min	1.0	1.0	1.0
	Max	2.5	2.0	1.5
	Mean	1.4	1.5	1.3
Cr	Min	0.8	1.2	0.7
	Max	0.9	1.7	1.0
	Mean	0.8	1.3	0.8
Ni	Min	1.9	2.9	2.3
	Max	3.0	4.0	2.8
	Mean	2.6	3.4	2.5
Fe	Min	124	84	145
	Max	217	390	185
	Mean	150	261	171
Cu	Min	3.0	5.0	3.0
	Max	23	9.0	7.0
	Mean	10	8.0	5.0
Zn	Min	38	38	49
	Max	98	56	85
	Mean	58	47	63
Mn	Min	47	48	70
	Max	94	89	93
	Mean	61	66	83

1:Menemen, 2:Bulgurca, 3:Torbalı

Only one significant relationship was found between soil Co and Co content of leaf petiole significant at 5% level. Since heavy metal uptake of plants is affected by pH, CaCO$_3$, redox potential and organic matter, significant correlations were found with these criteria.

Soil pH and leaf blade Cu content had a negative relationship significant at 1 % level, similar relationship was found with leaf petiole

Table 3. Heavy metal content of petioles .

Element		Location		
(mg kg^{-1})		1	2	3
Pb	Min	2.5	5.0	2.5
	Max	5.0	6.0	5.0
	Mean	4.3	5.2	4.4
Cd	Min	0.1	0.3	0.1
	Max	0.4	0.4	0.3
	Mean	0.2	0.3	0.1
Co	Min	-	-	-
	Max	-	0.5	-
	Mean	-	0.5	-
Cr	Min	-	-	-
	Max	0.6	0.6	-
	Mean	0.3	0.3	-
Ni	Min	2.5	4.7	3.7
	Max	4.0	6.2	4.7
	Mean	3.5	5.2	4.0
Fe	Min	50	205	69
	Max	225	238	244
	Mean	105	222	145
Cu	Min	3.0	4.0	6.0
	Max	10	6.0	9.0
	Mean	5.0	5.0	8.0
Zn	Min	18	34	68
	Max	70	61	96
	Mean	34	49	84
Mn	Min	17	24	24
	Max	34	38	35
	Mean	24	29	29

1:Menemen, 2:Bulgurca, 3:Torbalı

Cr concentrations with a significance of 5 %. Between CaCO$_3$ content of soils and Cr and Ni content of leaf blades, there were also significant relations (1%). Co and Ni content of petioles had negative correlations (5%) .Apart from these significant relationships, between organic matter and leaf petiole Fe concentrations and between CEC and Zn content of leaf blades negative relations were determined significant at 1% and 5% respectively. Between clay fraction and the Cr, Ni content of leaf blades and Co content of leaf petioles, there were significant positive correlations.

4. References

Bergmann,W.,1986:Farbatlas,Ernaehrungsstörungen bei Kulturpflanzen. Gustav Fischer Verlag.Jena.

Berrow,M.L.,Reeves,G.A.,1986:Total chromium and nickel contents of Scottish soils. Geoderma 37,15

174

Hakerlerler,H.,Anaç,D.,Okur,B.,Saatçı,N.,1994: Heavy metal contamination in satsuma mandarine plantations in Gümüldür and Balçova districts. Ege University Research Fund, Project Number: 92-ZRF-47 Bornova İzmir, Turkey.

Isaac,A.R.,Kerber,J.D.,1969: Instrumental methods for analysis of soil and plant tissue. Perkin Elmer Corp. Atomic Absorption Dept. Norwalk. Conn.

Kick,H.,Bürger,H.,Sommer,K.,1980:Gesamtgehalte an Pb, Zn, Sn, As, Cd, Hg, Ni, Cr und Co in landwirtschaftlich und gaertnerisch genutzten Böden Nordrein_Westfallens.Land.Forschung 33(1):12-22.

Kloke,A.,1980:Das 'Drei-Bereiche- System' für die Bewertung von Böden mit Schadstoffbelastung. VDLUFA-Schriftenreihe, Kongressband, VDLUFA Verlag.Darmstadt:1117-1127.

Marschner,H.,1995:Mineral Nutrition of Higher Plants. Second Edition. Academic Press. London.

Merian,E.,1991:Metal Their Compound in the Environment. VCH, Weinheim-NewYork.

Pendias,K.A.,Pendias,H.,1984: Trace Elements In Soil And Plants. CRC Press. Boca Raton.

Prawda,W.,1997:The Congress of General Practitioner 30 Baden - Württernberg Hürriyet neswspaper,12.4.1997.

Saatçı, F., Hakerlerler ,H., Tuncay, H., Okur, B., 1988: A study on the pollution of agricultural soils and irrigation waters around some important industrial plantations in Izmir and its environs. Ege University Research Fund. Project Number: 127.Bornova Izmir Turkey.

Scheffer,F.,Schachtschabel,P.,1989:Lehrbuch der Bodenkunde 12 neu bearbe.Aufl.unter Mitarb. von W.R.Fischer.Ferdinand Enke Verlag.Stuttgart.

Slawin,W.,1968:Atomic Absorption Spectroskopy . Interscience Publishers New York

Sposito,G.,Page,A.L.,1984:In Metal Ions In Biological Systems. Ed.Sigel.H.Marcel Dekker, New York.

Chapter 40

Plant analysis as a basis for nitrogen fertilization decisions in winter wheat production.

R. Byrne; T. McCabe; E.J. Gallagher :
Department of Crop Science, Horticulture and Forestry, National University of Ireland, Dublin, Belfield, Dublin 4, Ireland.

Key words: foliar urea, nitrogen, protein, tissue analysis, wheat, yield.

Abstract: International studies have established critical nitrogen levels, in various plant parts, at different growth stages. They have endeavoured to establish a relationship between these levels and subsequent grain yield in wheat. Few studies have been carried out to establish the relationship between plant nitrogen levels, at various growth stages, with subsequent grain protein content. Experiments were carried out in 1996/1997, and continued in 1997/1998, to determine the relationship between whole-plant tissue nitrogen levels at growth stage 39, leaf blade nitrogen levels at growth stage 59 and subsequent grain protein content. There were good correlations between protein content and whole-plant nitrogen concentration at growth stage 39 and also between protein content and leaf blade nitrogen concentration at growth stage 59. The correlation between whole-plant nitrogen concentration at growth stage 39 and grain nitrogen yield was excellent. Future studies on the development of nitrogen status tests for commercial use should include this grain nitrogen yield parameter.

1. Introduction:

Nitrogen fertilization trial work in Ireland, on winter wheat, has been mainly focused on yield improvement, with considerable success, since national average yields are over 8 t ha^{-1}. Research on fertilization for protein increase has followed a number of paths. Application of supplemental nitrogen at a late growth stage (post growth stage 39) improves protein content by upwards of 0.5% units. Foliar urea, applied, as late as growth stage 75, tends to produce a more reliable increase. However, a degree of variability in response has prompted studies on tests to determine the nitrogen status of plants at various stages. These tests have been incorporated into the nitrogen fertilization trials described in this paper.

Most of the literature on this topic has concentrated on the relationship between plant nitrogen levels and yield, rather than protein. Although there is general comment on the variability of the results from plant analysis many workers, such as Batey (1979), Engle and Zubriski (1982), Roth et al (1989) were of the opinion that plant testing could give an indication of the nitrogen status of the plant, which might be used as an indication of the necessity for supplemental nitrogen, if that was needed.

It has also been well-established that plant nitrogen levels decrease with sequential growth stages. A review of the literature on this topic (Barraclough, 1997) describes work carried out to evaluate both total nitrogen and nitrate levels in stems and leaves. Most of the work has concentrated on levels in stems, probably because of ease of working with the material. The data from different countries on the critical % nitrogen in leaf dry matter for maximum grain yield has been summarised by Barraclough (1997). Although these data are prognostic, and yield may be affected by circumstances subsequent to the test, they showed good agreement, lying in the range of 3.5 - 4.6 % nitrogen.

2. Materials and Methods

Data from two trials were used in this study. Trial 1 was sown in October 1996 in good conditions. Six nitrogen levels (0-260 kg ha^{-1}) were tested across two cultivars, the winter-type Soissons and spring-type Chablis, using a 6x2 factorial arrangement. Trial 2 was sown in early December in difficult ground conditions with the same six nitrogen levels on Soissons, Chablis and Alexandria, another spring-type wheat. The nitrogen applications ranged from single to four-split programmes at growth stages 23, 32, 39 and 41. The nitrogen was applied in the form of calcium ammonium nitrate (26.5 % N) and as foliar urea (10 % solution) for the fourth application. In both trials whole-plant tissue samples were taken at growth stages 32 and 39 and leaf blade (upper two leaves) tissue samples at growth stages 39 and 59. Thirty plants were sampled at each growth stage, for each plant part, and were analysed for nitrogen concentration.

3. Results

In these trials the whole-plant and leaf-blade nitrogen concentrations (in the range 2.3 – 4.4 % N) were at or below critical nitrogen levels previously established or sourced from the literature. In both trials yields reached a plateau at approximately 140 kg ha^{-1} nitrogen but protein content tended to follow the rise in fertilizer nitrogen rates with a significant response to late-applied foliar urea (Table 1). There were good correlations between protein content and whole-plant nitrogen concentration at growth stage 39 and leaf-blade nitrogen concentration at growth stage 59. The correlation between whole-plant nitrogen concentration at growth stage 39 and grain nitrogen yield was excellent with r^2 = +0.93. The relationship between grain nitrogen yield and leaf-blade nitrogen concentration was also good but not as significant.

Table 1. Grain yield (t/ha at 15% m.c.) and protein content (%) for the variety and nitrogen programme treatments in the two winter-sown trials in 1997.

	Trial 1. - October-sown		Trial 2. – December-sown	
	Yield (t ha^{-1})	Protein (%)	Yield (t ha^{-1})	Protein (%)
Factor A. Fertiliser Nitrogen Programme				
Nitrogen Rates (kg ha^{-1})				
0	8.34	8.35	9.96	8.72
100 (40/60)	12.04	9.32	11.81	9.74
140 (40/100)	12.70	9.67	12.33	10.20
180 (40/100/40)	12.89	10.23	12.26	10.55
220 (40/140/40)	13.08	10.63	12.70	10.69
260 (40/140/40/40)	13.24	11.22	12.63	11.31
L.S.D. (5% level)	0.56	0.33	0.37	0.21
Factor B. Variety				
Soissons	11.51 b	10.05 a	11.67	10.32
Chablis	12.58 a	9.76 b	12.47	10.07
Alexandria	--------		11.70	10.22
L.S.D. (5% level)	Significant	Significant	0.26	0.15
Co-efficient of Variation (%)	4.6	2.3	3.3	2.5

4. Conclusions

These trials are part of a series investigating ways of improving protein content, while, at least, maintaining yield. The data would indicate that there is a reasonable possibility of identifying wheat crops where the whole-plant, but particularly leaf-blade nitrogen concentration, would indicate situations where supplemental fertiliser nitrogen could be used to improve protein levels. However, when these data were complemented, in 1997, by investigations with commercial wheat crops a number of situations were identified where low leaf-blade nitrogen concentrations did not correspond with low grain protein levels post-harvest and where high leaf-blade nitrogen results were followed by low protein levels. These investigations have been extended, in 1998, to a large number of commercial wheat crops in order to validate the conclusions.

5. References

Barraclough, P.B. (1997) Nitrogen requirement of winter wheat and diagnosis of deficiency. Aspects of Applied Biology 50, 117-123.

Batey, T. (1977) Prediction by leaf analysis of nitrogen fertiliser required for winter wheat. Journal of Science Food and Agriculture 18; 275-277

Engle, R.F. and Zubriski, J.C. (1982) Nitrogen concentration in spring wheat at several growth stages. Communications in Soil Science and Plant Analysis 13 (7); 531-544

Roth, G.W. Fox, R.H. and Marshall, H.G. (1989) Plant tissue tests for predicting nitrogen fertiliser requirements of winter wheat. Agronomy Journal 81; 502-507

5- Crop quality – nutrient management
in soilless culture

Chapter 41

Evaluation of a sewage sludge based compost for the production of container tree seedlings

H. Ribeiro[1], D. Ribeiro[2], E. Vasconcelos[1], F. Cabral[1], V. Louro[2] and J. Q. dos Santos[1]

[1] *Department of Agriculture and Environmental Chemistry, Instituto Superior de Agronomia, Tapada da Ajuda, 1399 Lisboa Codex, Portugal.*

[2] *Direcção Geral das Florestas, Avª João Crisóstomo 28, 1050 Lisboa, Portugal.*

Key words: *Pinus pinaster* Ait, *Pinus pinea* L., substrate, sewage-sludge-compost, nutrition.

Abstract: The effect of a sewage sludge based compost as a substrate component on the growth and nutrient status of two containerised tree seedlings (*Pinus pinea* L. and *Pinus pinaster* Ait.) was evaluated. Compost was mixed with a sphagnum peat at rates of 25, 50, 75 and 100% by volume and no additional fertilisation was applied. A control treatment consisting of 100% sphagnum peat and 3.5 g.L^{-1} of a controlled release fertiliser was used. Seedling and harvesting were carried out in March and October 1997, respectively. Plant growth increased with increasing doses of compost up to 50%. In fact, an increase of the total dry weight, plant height and diameter of the colar was observed. On the contrary, additions of compost greater than 75% led to a growth reduction. Despite this fact, plant growth in all compost treatments was always lower than in control treatment with fertilisation. The growth reduction that occurred in the 75% and, mainly, in the 100% compost treatments is likely a consequence of water stress since the available water (v.v^{-1}) of these substrates (13% and 5%, respectively) was quite low. The 50% compost treatment seems to supply all P, Ca, Mg, Cu, Zn and Mn needed for satisfactory plant growth. In addition, compost shows a liming effect, that is important when the basic substrate is acid.

1. Introduction

In Portugal peat is widely used as the main substrate component for the production of tree seedlings in containers. However, peat is imported from Nordic countries and recently has become more expensive and its properties more variable. Since the substrate performs an important role for the total crop costs, it is important to look for good quality and local low cost substitutes of peat, reducing production costs without sacrificing quality. A number of potential alternatives have been identified (Prasad, 1979, Bunt, 1988; Gouin, 1993).

In Portugal, the Maia municipality is composting sewage sludge with pine sawdust producing a peat-like material. Thus, as far as other municipalities begin to adopt this method important amounts of an universal compost, will become available. Previous work showed that this compost can be used as a substrate component for potted pelargonium plants (Ribeiro *et al.*, 1997). The main purpose of this study was to optimise the use of this compost as a substrate component for the production of tree seedlings in containers.

2. Materials and Methods

A compost obtained by composting aerobically a mixture of a dewatered anaerobic digested sewage sludge and pine sawdust – 1:3 by volume, respectively - was tested. The compost was mixed with sphagnum peat at rates of 25, 50, 75 and 100% (by volume) and no additional fertilisation was made. A control treatment consisting of 100% sphagnum peat and 3.5 gL^{-1} of a controlled release fertiliser with: 16%N, 8% P_2O_5, 12% K_2O; 2% MgO and micronutrients, was used.

Table 1. Peat and compost composition (dry weight basis)

	Peat	Compost
Dry matter (%)	40	36
Organic. Matter (%)	98	87
Total N (%)	1.1	1.1
Total P (%)	0.03	0.51
Total K (%)	0.01	0.10
Total Ca (%)	0.30	0.97
Total Mg (%)	0.12	0.10
Total Na (%)	0.02	0.1
Total Cu (mg.kg^{-1})	10	144
Total Zn (mg.kg^{-1})	10	509
Total Mn (mg.kg^{-1})	55	88
C/N ratio	52	46

pH and electrical conductivity (EC) of the five mixtures were measured in the water extract obtained at pF 1 (10 cm suction) conditions, according to Verdonck and Gabriëls (1992). Physical properties of the mixtures: Easily Available Water (EAW), Water Buffering Capacity (WBC), Air Capacity (AC) and Bulk Density (BD) as defined by De Boot and Verdonck (1972), were measured according to Verdonck and Gabriëls (1992).

A completely randomised experiment, with 3 replications, and 40 plants per replication was carried out in an open air nursery. Plastic containers with 87 mm height and 115 cm^3 of volume were used.

Seedling was performed in March and plants were harvested seven months after. Shoot and root material was dried at 65 °C for 48 h, grounded through a 1 mm screen and samples were taken for chemical analysis.

Nitrogen by the Kjeldahl method. After a hydrochloric digestion of the ash, phosphorus was determined by the vanadomolybdo-phosphoric yellow colour and all the other nutrients were determined by atomic absorption spectrosphotometry.

Data were subject to ANOVA analysis. In every table, means followed by the same letters, in the same column, do not differ at p≤0.05 by Scheffe F-test.

3. Results and Discussion

Physical properties of the substrate Physical properties of the substrate were significantly affected by the compost percentage in the mixture (Table2).

Table 2. Some physical and chemical properties of the substrates

Treat.	pH	EC mS.cm^{-1}	BD g.cm^3	EAW	WBC %,v.v^{-1}	AC
Control	3.7a	0.7a	0.09a	27.1a	4.9a	32.7a
25%C	3.9b	2.3b	0.14b	34.6b	5.5a	15.7b
50%C	4.9c	2.3b	0.17c	24.4c	1.7b	23.6c
75%C	5.4d	3.4c	0.20d	12.5d	0.5c	31.1a
100%C	6.0e	3.8c	0.21d	4.7e	0.1c	33.3a
Accept. Range	4.0-6.0[*]	0.75-1.99[**]	< 0.4[*]	20-30[*]	4-10[*]	20-30[*]

[*]Abad *et al.*(1989)
[**]Suitable range for seedlings - saturated media extract - (Warncke and Krauskopf, 1983)

The compost increased substrate bulk density (BD) from 0.09 (control) to 0.21 g cm^{-3} (100% compost), but all values found were within the adequate range. Compost addition to peat reduced (except for 25% compost treatment) the substrate water availability (Table 2). While EAW of control, 25 and 50% compost treatments are within the acceptable range, treatments with 75 and 100% compost presented very low values of EAW and WBC. All substrates showed a very good aeration capacity except 25% compost that was slightly lower than the "ideal substrate", but, for these species, 16% AC does not seem to act as a growth limiting factor.

Substrate pH and electrical conductivity Electrical conductivity (EC) significantly increased with increasing compost percentage of the media (Table 2). This is a result of the high compost soluble salts content, mainly K, Ca, Na and Cl (Table 1). According to these values it seems that EC of the treatments with 75 and 100% compost is slightly higher than desirable for seedlings. Compost additions also significantly increased pH values of the substrate (Table 2), having a liming effect on the peat.

Plant growth Plants of the control treatment showed the higher growth, consequence of the fertiliser applied. Increasing the compost percentage in the substrate from 25% to 50% led to a significantly increase of the shoot dry weight, plant height and diameter of the collar of the *Pinus pinaster* Ait. plants (Table 3).

Table 3. Some growth parameters of *Pinus pinaster* Ait. plants

	Shoot dw (g)	Root dw (g)	Height (cm)	Diameter of colar (mm)
Control	1.11a	0.37a	22.95a	2.91a
25% C	0.27c	0.18b	9.43d	1.99c
50% C	0.71b	0.29ab	18.47b	2.73ab
75% C	0.69b	0.27b	17.67b	2.49b
100% C	0.37c	0.23b	12.67c	2.00c

Although no significant differences were detected between 50 and 75% compost treatments, data show a general trend to the reduction of the above parameters when compost percentage is higher than 50%. These results seem to show that the growth reduction found on the 75% and, mainly, on the 100% compost treatment was, probably, a consequence of water stress, since the water availability of those substrates was very low (Table 2). The slightly high value of EC could also have a negative effect, but in the 75 and 100% compost treatments neither the germination percentage nor the germination period (data not shown) was affected.

Concerning the P*inus pinea* L. plants, a general trend similar to the *Pinus pinaster* Ait. plants was found (table 4). However, no significantly differences were found between control and 50% compost treatments.

Table 4. Some growth parameters of *Pinus pinea* L. plants

	Shoot dw (g)	Root dw (g)	Height (cm)	Diameter of collar (mm)
Control	1.29a	0.31a	28.50a	3.11a
25% C	0.75c	0.28a	20.52b	2.73b
50% C	1.12ab	0.30a	27.13a	2.94ab
75% C	0.90bc	0.28a	26.18a	2.82b
100% C	0.87bc	0.35a	20.78b	2.80b

Elemental composition Results obtained to macronutrients' composition of the shoots (table 5 and 6) showed that, in general terms, plants of control treatment presented higher contents in N and K and lower content in P than compost treatments.

Table 5. Macronutrients composition of the shoots - *Pinus pinaster* Ait.

	N	P	K	Ca	Mg
	(%)				
Control	0.86a	0.19b	1.03a	0.46b	0.15a
25% C	0.41b	0.78a	0.53b	0.83ab	0.17a
50% C	0.50b	0.61a	0.60ab	0.83ab	0.16a
75% C	0.58b	0.57a	0.81ab	0.85a	0.15a
100% C	0.46b	0.77a	0.98ab	0.68ab	0.16a

These data seem to indicate that on compost treatments N and K supplied are not enough to fulfil plant needs. On the contrary, compost seems to supply enough amounts of P, Ca and Mg namely in the treatments with 50, 75 and 100%. In fact, there are no significant differences in the macronutrients composition between treatments with compost, however the growth decrease of the plants in the 25% treatment led to a nutrients' concentration effect.

Table 6. Macronutrients composition of the shoots - *Pinus pinea* L.

	N	P	K	Ca	Mg
	(%)				
Control	1.9a	0.25b	1.00a	0.57a	0.19a
25% C	1.3b	0.40a	0.55c	0.64a	0.18a
50% C	1.4b	0.34ab	0.71bc	0.57a	0.16a
75% C	1.3b	0.35ab	0.88ab	0.56a	0.16a
100% C	1.2b	0.29ab	0.75bc	0.56a	0.14a

In compost treatments shoot Mn concentration tends to decrease as the percentage of compost increases (Table 7). This behaviour seems to be related to the pH of the mixtures. The low pH of 25 and 50% mixtures was, probably, responsible for the solubilisation of the Mn in the compost and, consequently, the availability and absorption of Mn increased. In addition, the growth decrease of the 25% treatment led, again, to a concentration effect. No significant differences were observed for Cu, while high levels of Zn were found in the 50% compost treatment.

Table 7. Levels of copper, zinc and manganese in the shoots.

	Pinus pinea L.			Pinus pinaster Ait.		
	Cu	Zn	Mn	Cu	Zn	Mn
	mg.kg^{-1}			mg.kg^{-1}		
Control	16a	52b	183b	13a	74d	212b
25% C	28a	104a	244a	18a	171b	381a
50% C	19a	111a	100c	11a	236a	164b
75% C	17a	90ab	55c	17a	199ab	86c
100% C	24a	65ab	60c	16a	119c	73c

In general the contents of Zn, Cu and Mn in the shoots are statically equal or higher than the control (except for Mn in the *Pinus pinaster*). These results indicate that the compost, namely in the 50% treatment, can fulfil the needs in Zn, Cu and Mn of these plants.

4. Conclusions

Results obtained seem to show that the sewage sludge+sawdust compost can be used, up to 50% by volume, as a component of a peat-based substrate for forestry seedlings production. Considering growth and shoot nutrients concentration, 50% compost treatment seems to supply all P, Ca, Mg, Cu, Zn and Mn needed for satisfactory plant growth. In addition, compost has a liming effect, that is important when the basic substrate is acid, like in peat. Levels of 75% or higher tend to reduced plant growth likely as a result of water stress, since the water availability becomes quite low.

5. Acknowledgement

This work had financial support from the project PAMAF IED n° 4110 4080 02 50

6. References

Abad M., Noguera V, Martinez-Cortes J and Martinez-Herrero MD (1989) Physical and chemical properties of sedge peat-based media and their relation to plant growth. Acta Horticulturae 238:45-56.

Bunt AC (1988) Media and mixes for container grown plants. Unwin Hyman Ltd. London.

De Boodt M and Verdonck O (1972) The physical properties of the substrates in horticulture. Acta Horticulture 26: 37-44.

Gouin FR (1993) Utilization of sewage sludge compost in horticulture. HortTechnology 3(2): 161-163.

Prasad M (1979) Physical properties of media for container-grown crops. I. New Zealand peats and wood wastes. Scientia Horticulturae 10: 317-323.

Ribeiro HM, Vasconcelos EV, Cabral F and Santos JQ (1997) Use of sewage sludge and sawdust compost for production of potted pelargonium plants. 11 th World Fertilizer Congress. 7-13 Sept Gent, Belgium.

Verdonck O and Gabriëls R (1992) Reference method for the determination of physical and chemical properties of plant substrates. Acta Horticulturae 302: 169-179.

Warncke DD and Krauskopf DM (1983) Greenhouse growth media: testing & nutrition guidelines. MSU AG Facts. Extension bulletin E-1736. Cooperative Extension Service. Michigan State University.

Chapter 42

Study of the substrate and fertilisation effects on the production of essential oils of *Rosmarinus officinalis* L. cultivated in pots

[1] G. Miguel, C. Guerrero, H. Rodrigues and J. Brito
[2] F. Venâncio, R. Tavares, A. Martins and F. Duarte

[1] *Unidade de Ciências e Tecnologias Agrárias, Universidade do Algarve, Campus de Gambelas, 8000 Faro Portugal*

[2] *Ineti, Ibqta, Departamento de Tecnologia das Indústrias Químicas, Estrada do Paço do Lumiar, 2699 Lisboa Codex Portugal*

Key words: *Rosmarinus officinalis* L., rosemary, Lamiaceae, essential oil composition, fertigation, myrcene, camphor, verbenone

Abstract: *Rosmarinus officinalis* L. is a perennial sub-shrub growing wild in all of the coastal regions of the Mediterranean Sea, and cultivated in some Mediterranean countries. In Portugal, rosemary grows wild and is very appreciated in culinary preparations as a spice and in herbal teas. This work submits studies on the essential oil composition leaves of *R. officinalis* kept in pots using three different substrates: fertilised turf (RTUFF2), non-fertilised turf (RTUNFF2) and a sandy soil (RTF2). These treatments were fertigated every fifteen days. These essential oils show high percentages of terpene hydrocarbons and oxygenated monoterpenes independently of the substrate used. In all cases, the main component was myrcene. RTF2 samples present the lowest concentrations of myrcene and camphor and the highest percentage of verbenone.

1. Introduction

Rosmarinus officinalis L. is a perennial sub-shrub growing wild in all of the coastal regions of the Mediterranean Sea, and cultivated in some Mediterranean countries, Spain being the greatest producer (Analytical Methods Committee, 1993). In Portugal, rosemary grows wild and is very appreciated in culinary preparations as a spice and in herbal teas.

Due to its aromatic and medicinal properties, Rosemary oil has been used for centuries as an ingredient in rubefacient liniments, inhalents, cosmetics, soaps, perfumes, room sprays and deodorants, flavoring and food conservants (Arnold *et al.* 1997). *R. officinalis* also possesses antioxidant properties (Domokos *et al.*, 1997) that can be used in food and pharmaceutical industries as a natural antioxidant more favourably accepted by the populations.

Maffei *et al.* (1993) reported some environmental factors, mainly temperature and humidity, that would affect the lipidic metabolism of *R. officinalis*. High temperatures increase the essential oil yield and the percentages of monoterpenes. Some authors (Li, 1996; Milia *et al.* 1996) reported several agronomic aspects, namely macronutrients (N, P, K) effects on the rosemary growth. Moretti *et al.* (1998) reported the effects of iron on yield and composition of rosemary essential oil proving that iron absorption either in irrigated or non-irrigated plants did not produce any significant increase in oil yield. However an increase in verbenone concentration was observed in the oil of irrigated plants.

This work submits studies on the essential oil composition of leaves of *R. officinalis* kept in pots using three different substrates: fertilised turf (RTUFF2), non-fertilised turf (RTUNFF2), and a sandy soil (RTF2). These treatments were fertigated every fifteen days.

2. Materials and Methods

Plant Material: The plant material was collected at Nave Barão, Algarve (Portugal), in December 1996 and October 1997.

In order to carry out the experiments in pots, *Rosmarinus officinalis* L. was collected at Nave Barão in December 1996 and the rootstocks were obtained in order to take roots in the greenhouse later, then, the plants were placed in pots (three plants in each pot).

The study of the various fertilisation types on the essential oil chemical composition was carried out using those pots. These were filled with a haplic arenosol (ARh), according to the FAO-UNESCO Soil Classification for soils composed of sand without calcareous material (RTF2), with a fertilised turf (RTUFF2), and a non-fertilised turf (RTUNFF2). These treatments were fertigated every fifteen days with a solution of 1:3:1 (N:P:K), 0,4 % of Mg ($5gL^{-1}$); the fertiliser used was the BASF Hakaphos.

Analysis of the essential oils: The oils were obtained from 30 g of leaves from fresh plant material by distillation, for 3 hours, using a Lickens-Nickerson-type apparatus.

The analytical gas chromatography was carried out on a Hewlett Packard 5890 Series II gas chromatograph equipped with a Permabond CW20M (50 m x 0.25 mm i. d., df = 0.25 μm) and a OV-101 (50 m x 0.25 mm i. d., df = 0.25 μm) columns. A flame ionisation detector (FID) was used for routine quantitative analysis. The oven temperature was held at 70 °C (5 min) and programmed to increase till 220 °C at $2.^{\circ}C\ min^{-1}$. Detector and injector temperatures were set at 260 °C and 250 °C, respectively. The carrier gas was helium and the working flow was $1\ ml.min^{-1}$. The percentage composition of the oils were computed from the GC peak areas without using correction factors. The data shown are mean values of two injections.

The gas chromatography – mass spectrometry analyses were performed using a Perkin Elmer 8320 gas chromatograph, equipped with a DB-5 (30 m x 0.25 mm i. d., df = 0.25 μm) column and interfaced with a Finnigan MAT 800 Ion Trap Detector (ITD; software version 4.1). Oven temperature was held at 70 °C and programmed to increase till 180 °C at $3\ ^{\circ}Cmin^{-1}$. Transfer line temperature, 250 °C; ion trap temperature, 220 °C; carrier gas helium adjusted to a linear velocity of 30 cm s^{-1}; splitting ratio, 1:100; ionisation energy, 70 eV; ionisation current, 60 μA; scan range, 30-400 μ; scan time, 1 s. The identity of the components was assigned by comparison of their retention times and mass spectra with corresponding data of reference oil components.

3. Results and Discussion

The composition of the essential oils of *R. officinalis* maintained on different substrates is given in Table 1.

Terpene hydrocarbons were the main group of constituents (52.5 %-58.5 %) immediately followed by oxygen-containing monoterpenes whose concentrations ranged from 34.9 % to 38.9 %, independently of the substrate used. β-caryophyllene, a sesquiterpene hydrocarbon, as well as caryophyllene oxide, an oxygen-containing sesquiterpene, and methyleugenol were also present although in small amounts (Table 1).

In all cases, the main component was myrcene followed by verbenone, α-pinene and camphor (Table 1). 1,8-Cineole limonene and were also present in relatively high proportions (Table 1).

Camphor (32.3 %) was the main component of the oil extracted from rosemary collected in the mountainous region of Andalucia in Spain (Lawrence, 1997). The main components of the essential oil composition of rosemary collected in the experimental fields of the Servicio de Investigación Agraria of Zaragoza, located in different geographical areas of Aragón, a region in the northeast of Spain with a drier continental climate, were camphor, α-pinene and 1,8-cineole (Guillén *et al.*, 1996).

Table 1. Constituents of essential oils of *R. officinalis* on different substrates and fertigated every 15 days

Components	Percentage		
	RTF2	RTUF F2	RTUNF F2
Monoterpene hydrocarbons	**52.5**	**52.8**	**58.5**
α-Pinene	10.8	10.0	13.5
Camphene	2.5	2.6	2.5
Sabinene	0.1	0.2	0.1
β-Pinene	3.2	2.0	3.1
Myrcene	21.0	26.9	26.8
α-Phellandrene	0.6	0.5	0.4
α-Terpinene	0.7	0.6	0.6
p-Cymene	0.8	0.8	0.6
Limonene	6.0	5.7	6.4
Cis-β-Ocimene	2.7	0.9	0.9
Trans-β-Ocimene	0.2	0.1	0.1
γ-Terpinene	2.7	1.5	2.3
Terpinolene	1.2	1.0	1.2
Oxygen-containing monoterpenes	**38.4**	**38.9**	**34.9**
1,8-Cineole	8.1	7.9	7.4
Trans-Sabinene hydrate	0.2	0.1	0.2
Cis-Sabinene hydrate	0.4	0.4	0.4
Linalool	1.4	2.3	1.3
Eucarvone	0.6	0.8	0.5
Camphor	7.2	13.2	8.9
δ-Terpineol	1.3	0.4	0.5
Borneol	0.5	0.4	0.3
Terpinen-4-ol	1.5	1.0	1.2
α-Terpineol	2.1	1.7	1.8
Verbenone	14.6	10.4	12.0
Linalyl acetate	0.2	0.2	0.2
Bornyl acetate	0.3	0.1	0.2
Sesquiterpene hydrocarbons	**2.0**	**0.7**	**0.8**
β-Caryophyllene	2.0	0.7	0.8
Oxygen-containing sesquiterpenes	**0.6**	**0.6**	**0.3**
Caryophyllene oxide	0.6	0.6	0.3
Others	**0.7**	**0.7**	**0.6**
Methyleugenol	0.7	0.7	0.6

Independently of the substrate used, the essential oils showed higher percentages of α-pinene, β-pinene, myrcene and limonene and lower concentrations of α-terpinene, p-cymene, 1,8-cineole, borneol and terpinen-4-ol when compared with those of Turkish origin (Lawrence, 1997).

In all samples, the rosemary essential oils presented lower percentages of camphene, 1,8-cineole and borneol and higher concentrations of myrcene, limonene and verbenone than the Spanish essential oil (Committee Draft ISO).

The percentage of monoterpene hydrocarbons was slightly higher in plants maintained in non-fertilised turf (RTUNFF2) when compared with those growing in RTF2 or RTUFF2 substrates (Table 1). On the other hand, the oxygenated monoterpenes concentration was lower in RTUNFF2 samples (Table 1).

The essential oils of RTUNFF2 samples were slightly richer in α-pinene (13.5 %) than the remaining samples while RTUFF2 essential oils were slightly poorer in β-pinene (2.0 %), γ-terpinene (1.5 %) and verbenone (10.4 %) and slightly richer in linalool (2.3 %) and camphor (13.2 %) (Table 1).

The essential oil of RTF2 samples showed lower percentages of myrcene (21.0 %), camphor (7.2 %) and higher percentages of cis-β-ocimene (2.7 %), δ-terpineol (1.3 %), verbenone (14.6 %) and β-caryophyllene (2.0 %) than RTUFF2 and RTUNFF2 samples (Table 1).

These preliminary results do not indicate yet that significant modification in the essential oil composition have occured through the effect of N, P, K macronutrients. However it was possible to observe a higher concentration of verbenone in some of the samples, which, in accordance with Moretti *et al.* (1998), could increase the value of *R. officinalis* L. essential oil for the perfume industry.

Acknowledgements

The authors are grateful to the "Fundação para a Ciência e Tecnologia-Praxis XXI Program, Portugal" and "Fundação Calouste Gulbenkian, Portugal" for financial support.

4. References

Analytical Methods Committee (1993) *Analyst*, 118: 1089-1098.

Arnold, N; Valentini, G; Bellomaria, B (1997) *J. Essent. Oil Res.*, 9: 167-175.

Committee Draft ISO; ISO/TC 54/SC; ISO/CD 1342.

Domokos, J; Héthelyi, E; Pálinkás, J; Szirmai, S; Tulok, MH (1997) Essential oil of rosemary (*Rosmarinus officinalis* L.) of Hungarian origin. *J. Essent. Oil Res.*, 9: 41-45.

Guillén, MD; Cabo, N; Burillo, J (1996) Characterisation of the essential oils of some cultivated aromatic plants of industrial interest. *J. Sci. Food Agric.*, 70: 359-363.

Lawrence, BM (1997) Progress in essential oils. *Perfumer & Flavorist*, 22: 71-73.

Li, TSC (1996) Nutrient weeds as soil amendments for organically grown herbs. *Journal of Herbs, Spices & Medicinal Plants*, 4: 3-8.

Maffei, M; Mucciarelli, M; Scannerini, S (1993) *Biochem. System. Ecol.*, 21: 765-784.

Milia, M; Pinna, ME; Scarpa, GM (1996) Preliminary examinations of agronomic aspects in *Rosmarinus officinalis* L. *Rivista Italiana EPPOS*, 19: 125-130.

Moretti, MDL; Peana, AT; Passino, GS; Bazzoni, A; Solinas, V (1998) Effects of iron on yield and composition of *Rosmarinus officinalis* L. essential oil. *J. Essent. Oil Res.*, 10: 43-49.

Chapter 43

Effect of different Fe-chelates on yield and nutrient uptake of butterhead lettuce in hydroponic vegetable growing

A.Demeyer[1], N. Ceustermans[2], F. Benoit[2] & M. Verloo[1]
[1]*Department of Applied Analytical and Physical Chemistry, University Gent, Belgium*
[2] *European Vegetable R&D Centre, Sint-Katelijne-Waver, Belgium*

Key words: lettuce, Fe, NFT, chelates, EDDHA, polyphosphate, citric acid

Abstract: The use of synthetic chelates such as EDDHA or HEDTA in hydroponic vegetable growing to sustain sufficient Fe concentrations for optimal plant nutrition is nowadays a common practice. The use of such chelates, however, is difficult to combine with the growing interest for UV and ozone disinfection of the nutrient solution. Both types of disinfection break down the chelates resulting in precipitation of Fe as $Fe(PO_4).2H_2O$ (strengite). As a consequence the necessary Fe-supply is much higher than the actual uptake. In this work citric acid and polyphosphate were compared to EDDHA and DTPA for the growth of butterhead lettuce (*Lactuca sativa var. capitata*). A first experiment was conducted on a laboratory scale with the plants in separate dishes and a second experiment on an industrial scale in a greenhouse. Analysis of the nutrient solutions showed that synthetic chelates were most efficient in maintaining adequate Fe in solution (30-40 μmol L^{-1}). Analysis of the plants revealed that Fe uptake was generally sufficient and was only to a lesser degree affected by various concentrations of Fe in the nutrient solution. Both experiments showed that the presence of synthetic Fe chelates reduced the uptake of Mn and Zn.

1. Introduction

Since the beginning of hydroponic vegetable growing the solubility of micronutrients, especially Fe, was always one of the major concerns. In earlier nutrient solutions without chelates, Fe was almost completely precipitated. Addition of citrate or EDTA kept a larger part of the added Fe in solution. Chelates such as EDDHA and DTPA more selective for Fe were later synthesised and applied.

Since nutrient solutions in NFT (nutrient film technique) are recirculated, disinfection is necessary to avoid spreading of diseases through the system. For the growth of butterhead lettuce, UV-disinfection was successfully applied from the nineties to avoid spreading of especially *Pythium* and *Olpidium* (Benoit & Ceustermans,

1993*).* This however induced a new problem; photolysis of the chelates and precipitation of Fe throughout the system, leading to blocking of tubes and pumps, bad functioning of sensors and excessive supply of Fe.

In this work citric acid and polyphosphate were compared to EDDHA and DTPA as chelates. The chelating capacity of citric acid is well known. Polyphosphates ($H_{n+2}P_nO_{3n+1}$) are strong chelates for Fe, but data on stability constants are limited.

2. Material and Methods

As a support for the plants pressed peat cubic pots (V: ±125 cm^3) were used. The peat was pre-treated with lime and nutrients to obtain optimal conditions for germination and growth of the seedlings. The experiments were set up for the growth of butterhead lettuce (*Lactuca sativa var. capitata*) and were started with ten days old seedlings.

For experiment 1 plants were placed in dishes with a diameter of 15 cm. Each dish contained 300 ml nutrient solution. The composition of the nutrient solution was adapted from Benoit et al. (1989).

Iron was supplied in four different forms:

EDDHA(ethylenediamine-di-O-hydroxyphenyl-acetic acid): a commercial mixture was used

DTPA (diethylenetriamine-pentaacetic acid): a commercial mixture was used

Citric Acid: is also subjected to UV-photolysis but can be a cheap alternative, was supplied in a molar ratio of 3:1 citric acid/Fe

Polyphosphates: are not subjected to UV-photolysis, but slowly hydrolysed to ortho-phosphate. Na-polyphosphate was supplied in a molar ratio 9:1 P/Fe

The plants were placed in a growth chamber with controlled humidity ($\pm 80\%$) and temperature (25-27°C) and artificial lighting with a day/night cycle of 12 hrs.

Plants were harvested after 25 days. During the growth period the nutrient solution was replaced 5 times. Demineralised water was added to compensate for evapotranspiration losses. To reduce the growth of algae to a minimum the dishes were covered with black plastic film. Each treatment was replicated three times. Results were statistically evaluated using ANOVA and Fisher's LSD test at the 0.05 level.

Experiment 2 was carried out on an industrial scale in NFT. Growth parameters such as geometry of the gullies, plant interspacing, management of the nutrient solution (concentration of nutrients, EC, pH,...), light intensity, etc... were established according to Benoit et al. (1989). The nutrient solution was pumped at a rate of 2 L/min through a UVc-disinfection unit (Benoit et al., 1993). Three independent circuits were used, each containing 340 plants and only differing in the chemical form of Fe-supply; chelated with EDDHA (circuit 1), citric acid (circuit 2) or polyphosphate (circuit 3). The same Fe/chelate ratios were used as for experiment 1.

Chemical Analysis

Plants were dried at 60°C for determination of dry weight. For elemental analysis dry plants were ashed for 4 hrs at 550°C and ashes were dissolved in boiling 6 M HNO_3. Plant extracts and nutrient solutions were analysed with atomic absorption spectrophotometry (Varian SpectrAA-200 SIPS).

3. Result and Discussion

Experiment 1

Some selected properties of the plants from experiment 1 are presented in *Table 1*. Concentrations of Cu, Mg, K and Ca (not presented in the Table) showed no significant difference among the treatments and the averages for all treatments (mean±S.D.) were 4.67±1.09, 2460±144, 82496±2323 and 6561±330 mg kg^{-1} DM, respectively.

Table 1. Selected properties of plants from experiment 1

	dry mat	Fe	Mn	Zn
	gram		mg kg^{-1} DM	
EDDHA	3.92 a	93.70 ab	106.5 d	43.95 cd
citric ac.	3.98 a	82.47 bcd	194.1 ab	56.03 ab
poly-P	4.17 a	79.12 bcd	207.4 ab	54.78 ab
DTPA	4.07 a	71.97 bcd	136.7 c	41.90 cd
LSD$_{0.05}$	0.35	12.36	19.11	8.61

The dry matter yield was not significantly different among the treatments. The Fe content was significantly higher in the treatment with Fe-EDDHA, and was lowest with Fe-DTPA. Uptake of Mn was strongly suppressed in the presence of EDDHA and DTPA. This effect was highest for EDDHA, and the Mn content in this treatment was significantly lower than in all other treatments.

This effect of Fe-EDDHA was noticed and described for many plants in soil-culture by several authors. The effect is strongest on calcareous soils with Mn-accumulating plants such as Flax and Soybean (Moraghan, 78 and 93). The uptake of Zn was, although less severely than of Mn, also significantly depressed by EDDHA and DTPA.

Experiment 2

Results of two growth cycles will be discussed.

The first growth cycle with planting on 24/07/96 and harvest on 28/08/96; the second with planting on 06/09/96 and harvest on 23/10/96.

First growth cycle

The average Fe concentrations in the nutrient solution over the growth period were 59.4±21.5, 1.11±0.64 and 7.04±3.25 µmol Fe L^{-1} and the Fe content in the plants was 122.3, 89.4 and 89.2 mg kg^{-1} DM for circuit 1 (EDDHA), 2 (citric acid) and 3 (poly-P), respectively. The average yields per head were 284, 319 and 332 g fresh material for circuit 1, 2 and 3, respectively.

Citric acid obviously had the lowest capacity to keep Fe in solution. This could be expected from the stability constants (log K): Fe(III)-citric acid ~12 and Fe(III)-EDDHA ~33. Apparently the stability of Fe(III)-poly-phosphate is situated between the two other chelates, but no evidence was found in literature. However plants were able to absorb sufficient Fe for all treatments and crop yield and quality were not negatively affected by low Fe concentrations. Obviously Fe from weaker chelates is more easily absorbed by plants and high optimal levels of 40 µmol Fe L^{-1} (Benoit, 1989) are necessary for synthetic chelates since Fe is more slowly released from these compounds by the plant. In general, plants are also able to make good growth over a wide range of nutrient concentrations in hydroponics (Burrage, 93 and Adams, 93). Manganese contents in plants were similarly affected by the Fe-chelates as in experiment 1.

The fraction of Fe taken up was only 7 %, 4.3 % and 2.8 % for circuit 1, 2 and 3, respectively. This was lowest of all elements and therefore the supply of Fe was reduced for the second growth cycle.

Second growth cycle

Circuit 1, 2 and 3 were supplied with 79.3, 62.3 and 63.0 mmol Fe, respectively. After 20 days of growth 5.4 mmol Fe-EDDHA was supplied additionally to circuit 2. All the other nutrients were equally provided. The average Fe concentration in the nutrient solution (mean±S.D.) was 7.00±5.13, 0.77±0.59 and 1.52±1.21 µmol Fe L^{-1} and the yields per head were 303, 287 and 313 g fresh material for circuit 1, 2 and 3, respectively. The nutrient

concentration in the plants at the harvest are presented in *Table 2*.

Table 2. Nutrient contents in plants from experiment 2 (harvest 23/10/96), expressed in mg kg^{-1} DM

circuit	Fe	Mn	Zn	Cu
1	69.6	98.0	68.7	8.0
2	62.7	147.5	85.5	10.9
3	55.0	152.9	91.2	9.3

For this growth cycle the plants from circuit 2 and 3 were of minor quality: they were pale and heading was less optimal. The difference in Fe content was however very low. Since the quality of the plants from circuit 1 was satisfactory, it must be concluded that the critical concentration for market acceptable crop quality was approximately 70 mg Fe kg^{-1} DM. The concentration of Zn and Mn was affected in the same way by the treatments as noticed in experiment 1. The Cu content was also slightly lower in the presence of EDDHA.

During this growth cycle subsamples were taken after 18, 24 and 38 days of growth. Plants were harvested after 47 days. In *Figure 1* the evolution of the Fe, Zn and Mn content in the plants is depicted. For 18 days old plants the difference between the Fe contents was low. At this moment the plant roots were still mainly inside the peat pots and plants fed on Fe reserves from the peat. When the growth continued the roots stretched out in the gully and the plants took nutrients directly from the nutrient solution. As a result of this, the nutrient content of the plant reflected more directly the concentration in the nutrient solution.

The accumulation of Mn and Zn in circuits 2 and 3 was very high in young plants. In older plants the Mn and Zn content decreased, and at the harvest period Mn concentrations were only 50 % higher in circuit 2 and 3 than in circuit 1. In the early growth stage plants rapidly accumulated Zn and, especially, Mn from the peat. In the presence of Fe-EDDHA this accumulation was hindered. When the roots grew out of the peat pots plants were limited to the solution, where supply of Mn and Zn is restricted to avoid luxury uptake, and the concentrations quickly decreased to normal levels. In some cases this effect of Fe-EDDHA can be wishful, when high availability of Mn causes Mn toxicity or disturbs the heading process (Benoit, 89).

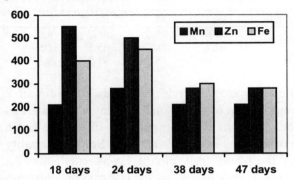

Figure 1. Evolution of Fe, Mn and Zn content in plants in circuit 1, 2 and 3 from experiment 2 (mg kg^{-1} DM)

4. Conclusion

Promising results were obtained using polyphosphate and citric acid as chelates for Fe. Compared to synthetic chelates such as EDDHA and DTPA, these compounds are less expensive and ecologically more acceptable. They also allow the plant to absorb sufficient Fe at lower Fe concentrations in the nutrient solution, but excessive supply of Fe is still necessary. Iron supply better adapted to the actual needs might reduce considerably loss of nutrient.

Analysis of the plants revealed that a minimum content of 70 mg Fe kg^{-1} DM is necessary for market acceptable butterhead lettuce. The type of Fe-chelate also has a considerable influence on the uptake of Mn and Zn; EDDHA and DTPA depressed the uptake of Mn and to a lesser degree of Zn. More experiments are necessary to study the consequences on yield and crop quality.

Acknowledgements

The research was subsidised by the Board of Research and Development of the Ministry of Middle Class and Agriculture.

5. References

Adams, P. 1992. Crop nutrition in hydroponics. Acta Horticulturae 323: 289-305.

Benoit, F. and Ceustermans, N., 1989. Recommendations for the commercial production of Butterhead Lettuce in NFT. Soilless Culture, V (1): 1-11.

Benoit, F. and Ceustermans, N., 1993. Lage-druk UVc-ontsmetting ook efficient voor NFT-kropsla (Low pressure UVc-disinfection also effective for NFT-lettuce). Tuinbouwvisie, 5 (196): 12-16.

Burrage, S.W., 1993. Nutrient film technique in protected cultivation. Acta Horticulturae 323: 23-38 .

Moraghan, J.T. and Freeman, T.J., 1978. Influence of FeEDDHA on Growth and Manganese Accumulation in Flax. Soil Sci. Soc. Am. J.,42: 455-460.

Moraghan, J.T., 1992. Iron-Manganese Relationships in White Lupine Grown on a Calciaquoll. Soil Sci. Soc. Am. J., 56: 471-475.

Sillen, L.G. and Martell, A.E., 1964. Stability constants of metal-ion complexes. London: The Chemical Society, Burlington House, W.1.

Chapter 44

Relationships between fruit quality characteristics and leaf nutrient contents of cucumber plants

Ö. Tuncay, M.E. Irget, A. Gül, N. Budak
Ege University, Faculty of Agriculture 35100 Izmir- Turkey

Key words: cucumber, quality, leaf analysis

Abstract: The relationships between leaf nutrient contents and quality characteristics were determined by using path analysis in cucumbers grown in perlite. There were significant correlations between leaf nutrient contents and fruit diameter, fruit length, TSS, titratable acidity, pH, dry matter, fruit firmness and colour as quality characteristics.

When the effects of leaf nutrient contents on quality characteristics were considered, K had significant positive direct effects on all of the quality traits with the exception of dry matter, which had been affected by P positively. Leaf Ca content had negative direct effects on all of the quality traits determined in the study.

1. Introduction

Turkey has nearly 15,000 ha greenhouse area, of which 95 % is used for vegetable growing (Tüzel and Eltez, 1997). Production is generally realized conventionally in soils. Soil-borne pathogers and nematodes lead to significant yield and quality losses. Therefore soil sterilization is a very common practice and chemical sterilization has been preferred due to the difficulties in steam sterilization.

Soilless cultivation has been spread in protected cultivation in many countries in the early eighties and it is reported that the change-over to soilless growing methods was induced mainly to rule out the problems in connection with the methyl bromine residues (Benoit and Ceustermans, 1995). Although commercial soilless cultivation area in Turkey is only around 6 ha, it is obvious that it will increase in the future due to the facts mentioned above.

Crop production in soilless culture, which is now considered as one of the superior growing techniques for a sustainable agriculture free from environmental pollution, requires an adequate supply of all the elements essential for plant growth be maintained in nutrient solution.

It is necessary to monitor the elemental content of plants in order to ensure that all of the essential elements are being supplied in sufficient quantity to satisfy the crop requirement. Leaf analysis is suggested to provide the information needed to establish the proper nutrient solution management in soilless growing systems (Jones, 1983). The objective of this study is to determine the relationships between leaf nutrient contents and fruit quality characteristics of cucumber plants grown in perlite.

2. Materials and Methods

Three different cucumber cultivars (Afrodit, Efes, and Rawa) were grown in substrate culture during the spring season of 1997. Perlite was used as substrate in an amount of 8 liters per plant. The nutrient solution was prepared according to Day (1991) and its chemical composition was as follows (mg kg^{-1}): N 180, P

40, K 250, Ca 150, Mg 50, Fe 2, Mn 0.75, Zn 0.50, B 0.40, Cu 0.10 and Mo 0.05.

For determining the fruit quality characteristics, 9 samples were taken each week from May 2nd to June 20th and fruit diameter, fruit length, fruit colour, fruit firmness, pH, total soluble solids (TSS), titratable acidity (TA) and dry matter were measured.

The upper fully expanded leaves (4th or 5th leaves below the growing tip) were taken from April 11th to July 9th, 1997 in order to determine the nutrient content of the leaves. After cleaning, the leaves were dried at 65-70°C and ground for analysis. Total nitrogen was measured by modified Kjeldahl method. In the wet digested leaf extracts, P was measured colorimetrically; K, Ca and Na flamephotometrically; and Mg, Fe, Zn, Mn and Cu by Atomic Absorption spectrophotometer (Kacar, 1972)

To determine the relationships between leaf nutrient contents and fruit characteristics, correlation and regression analysis were done by using mean values (Li, 1975).

3. Results

Leaf K content had the highest direct positive effect on all the fruit quality traits except the dry matter. P and Mg in the leaves were found to be the second effective elements on fruit characteristics, and their effects were positive except TA/pH and fruit length, respectively. Leaf N content affected fruit length, fruit diameter, TA, pH and TA/pH ratio in positive direction, whereas, it had negative effects on fruit firmness, TSS, TSS/TA, dry matter and fruit colour. Effects of leaf Ca content were negative on all quality traits excluding the dry matter. Among micronutrients, Mn, Zn and Fe had the similar correlation coefficients. Mn and Zn contents of the leaves were found to have positive direct effects on all quality traits. Leaf Fe content negatively affected the fruit characteristics excluding diameter and pH. Cu measured in the leaves had significant correlation coefficients at p<0.05 probability level while the rest of all coefficients were significant at p<0.01 probability level. Leaf Cu

content had only direct positive effects on TA and TA/pH. Leaf Na content giving the second lowest correlation coefficient following Cu had positive highest direct effects for TSS/TA and negative highest direct effect on TA while the other direct effects were considerably smaller (Table 1).

After the most positively and negatively effective elements on each quality characteristic were determined (Table 1), the effects of these elements were tried to be explained by taking into account the indirect effects of the other elements (Table 2).

It was determined that leaf Mg and K content had the highest negative and positive direct effects on fruit length, respectively. Adverse effect of leaf Mg on fruit length was mostly decreased by K (r:0.832) and P (r:0.251), whereas its effect was increased by Ca (r:-0.121).

Fruit diameter was affected mostly by leaf Ca and K contents negatively and positively, respectively. Ca had the highest positive indirect effects through K (r:0.624) and Mg (r:0.218).

The highest positive and negative direct effects on fruit firmness was obtained by K and Ca contents of leaves, respectively. Ca had the highest positive indirect effects through K (r:0.515), Mg (r:0.242) and P (r:0.236). The same conclusion could be made for TSS.

Leaf Ca content, which had the highest negative direct effect on TA, had the highest positive indirect effect through K (r:0.708). The highest direct effect on TA was obtained by K content of leaves. K had the highest positive effects through Mg (r:0.189) and N (r:0.120), and had the highest negative indirect effects through Ca (r:-0.197) and Na (r:-0.115).

Leaf K content had a positive direct effect on pH of fruit juice while leaf Ca content had negative.

On the TSS/TA ratio, leaf K content had the highest positive (r:0.448) and negative (r:0.121) indirect effects through P and Ca, respectively.

Leaf Ca content, which had the highest negative direct effect on TA/pH ratio, had the highest positive indirect effects through

K (r:0.800) and Mg (r:0.159) and the highest negative indirect effect through Na (r:-0.110).

Fruit dry matter was directly affected by leaf Fe and P contents in negative and positive directions, respectively. They had the highest negative and positive indirect effects through N and Mg, respectively.

Leaf Ca and K contents had the highest direct effects on fruit colour (blue/yellow). The adverse effect of leaf Ca content on fruit colour was mostly decreased by leaf K (r:0.702), Mg (r:0.295) and Mn (r:0.154) contents.

4. Discussion

It was determined that cucumber fruit characteristics were mostly affected by leaf K content followed by P or Mg in a positive direction, whereas leaf Ca content had negative direct effects on fruit quality parameters. All nutrients are transported from roots to the shoots via xylem and some are redistributed in the phloem. Since Ca is only xylem mobile, the accumulation of Ca by fruits is disproportionately low. In contrast, the intensive accumulation of K by fruit is mainly from the phloem sap, together with assimilates (Ho and Adams, 1995). It is generally believed that Ca uptake is by passive means and that its movement within the plant is by means of the transpiration stream (Jones, 1983; Marschner, 1995). It is reported that only 14.7 % of total Ca has been uptaken by fruits of cucumber plants, whereas leaves uptake 77.6 % of total Ca (Röber and Schaller, 1985). Therefore, increase in leaf Ca content does not necessarily lead to increases in fruit quality properties.

Results show that there are significant relationships between leaf nutrient contents and fruit quality of cucumbers. Leaf analysis, which is a common method for determining the nutrional status of plants may be used for fruit quality assessment, as well.

5. References

Benoit, F., N. Ceustermans, 1995. Horticultural aspects of ecological soilless growing methods. Acta Hort. 396: 11-24.

Day, D., 1991. Growing in Perlite. Grower Digest No. 12, Grower Pub. Ltd., London.

Ho, L.C., P. Adams, 1995. Nutrient uptake and distribution in relation to crop quality. Acta Hort. 396: 33-44.

Jones, J.B., 1983. A Guide for Hydroponic & Soilless Culture Grower. Timber Press, Portland, Oregon, USA.

Kacar, B., 1972. Bitki ve Toprağın Kimyasal Analizleri. II Bitki Analizleri, Ankara Üniv., Ziraat Fak. Yay. No: 453, Ankara.

Li, C.C., 1975. Path analysis-a primer. The Boxwood Press, USA.

Marschner, H., 1995. Mineral Nutrition of Higher Plants. Academic Press, Harcourt Brace and Comp. Pub., London, UK.

Röber, R., K. Schaller, 1985. Pflanzenernahrung im Gertenbau. Verlag Eugen Ulmer, Stuttgart, Germany.

Tüzel, Y. and R.Z. Eltez, 1997. Protected cultivation in Turkey. In: A Contribution Towards a Data Base for Protected Cultivation in the Mediterranean Region.Ed. Ayman F. Abou-Hadid.

Table 1. The path coefficients between the quality characteristics and nutrient elements.

	DIRECT EFFECTS									
	N	P	K	Ca	Mg	Cu	Fe	Mn	Na	Zn
Fruit Length (cm)	0.023	0.292	0.919	-0.132	-0.216	-0.029	-0.048	0.044	0.013	0.076
Fruit Diameter (cm)	0.070	0.024	0.706	-0.274	0.239	-0.027	0.102	0.099	-0.008	0.075
Fruit Firmness	-0.047	0.286	0.583	-0.291	0.265	-0.062	-0.046	0.204	0.013	0.084
TSS (%)	-0.029	0.384	0.451	-0.226	0.320	-0.016	-0.169	0.018	0.050	0.197
Titratable Acidity	0.133	0.051	0.801	-0.223	0.209	0.069	-0.072	0.039	-0.164	0.156
pH	0.073	0.093	0.651	-0.152	0.164	-0.010	0.002	0.118	0.027	0.067
TSS/TA	-0.097	0.371	0.448	-0.137	0.148	-0.103	-0.062	0.134	0.208	0.068
TA/pH	0.137	-0.038	0.904	-0.231	0.174	0.066	-0.046	0.059	-0.180	0.146
Dry Matter (%)	-0.162	0.595	-0.069	0.016	0.394	-0.003	-0.184	0.146	0.099	0.220
Blue/Yellow	-0.017	0.046	0.794	-0.311	0.322	-0.022	-0.059	0.196	0.049	0.005

Table 2. Direct and indirect effects of the nutrient elements on different quality characteristics.

		Direct effects	INDIRECT EFFECTS										Total
			N	P	K	Ca	Mg	Cu	Fe	Mn	Na	Zn	
Fruit length	Mg	-0.216	0.018	0.251	0.832	-0.121		-0.009	-0.034	0.029	0.008	0.058	0.816
	K	0.919	0.020	0.268		-0.116	-0.195	-0.011	-0.036	0.033	0.009	0.057	0.948
Fruit diameter	Ca	-0.273	0.053	0.020	0.624		0.218	-0.010	0.073	0.077	-0.005	0.050	0.827
	K	0.706	0.064	0.022		-0.242	0.216	-0.011	0.078	0.075	-0.005	0.056	0.959
Fruit firmness	Ca	-0.291	-0.035	0.236	0.515		0.242	-0.022	-0.033	0.161	0.008	0.056	0.837
	K	0.583	-0.042	0.262		-0.257	0.240	-0.024	-0.035	0.156	0.009	0.063	0.955
TSS	Ca	-0.226	-0.022	0.317	0.398		0.293	-0.006	-0.121	0.014	0.030	0.131	0.808
	K	0.451	-0.026	0.352		-0.200	0.289	-0.006	-0.129	0.013	0.034	0.148	0.926
TA	Ca	-0.223	0.101	0.042	0.708		0.191	0.024	-0.051	0.031	-0.100	0.104	0.827
	K	0.801	0.120	0.046		-0.197	0.189	0.027	-0.055	0.030	-0.115	0.118	0.964
pH	Ca	-0.152	0.055	0.077	0.576		0.150	-0.004	0.002	0.093	0.014	0.046	0.857
	K	0.651	0.066	0.085		-0.134	0.149	-0.004	0.002	0.090	0.017	0.052	0.974
TSS/TA	Ca	-0.137	-0.073	0.306	0.396		0.135	-0.037	-0.045	0.105	0.127	0.045	0.822
	K	0.448	-0.088	0.340		-0.121	0.134	-0.040	-0.048	0.102	0.145	0.051	0.923
TA/pH	Ca	-0.231	0.103	-0.031	0.800		0.159	0.023	-0.033	0.046	-0.110	0.097	0.823
	K	0.904	0.123	-0.034		-0.204	0.157	0.025	-0.035	0.045	-0.126	0.110	0.965
Dry matter	Fe	-0.184	-0.114	0.423	-0.052	0.011	0.282	-0.001		0.124	0.060	0.134	0.683
	P	0.595	-0.146		-0.063	0.013	0.338	-0.001	-0.131	0.107	0.052	0.136	0.900
Blue/yellow	Ca	-0.311	-0.013	0.038	0.702		0.295	-0.008	-0.042	0.154	0.030	0.003	0.848
	K	0.794	-0.015	0.042		-0.275	0.292	-0.009	-0.045	0.150	0.034	0.004	0.972

Chapter 45

Lettuce (*Lactuca sativa sp.*) responses to shared aquaculture media

P.A.. Chaves[1]; L.M. Laird[2]; A.V. Machado[3]; R.M. Sutherland[4]; J.G. Beltrão[5]

[1]*Zoology Department/ Aquaculture and Fisheries Group/ Unidade de Ciências e Tecnologias Agrárias, University of Aberdeen, Scottish Agricultural College, Universidade do Algarve;*
[2]*Zoology Department, University of Aberdeen*
[3]*Unidade de Ciências e Tecnologias Agrárias, Universidade do Algarve* [4]*Aquaculture and Fisheries Group, Scottish Agricultural College;* [5]*Unidade de Ciências e Tecnologias Agrárias, Universidade do Algarve*

Key words: aquaculture, effluents, integration, lettuce, recirculating systems, sustainable

Abstract: The aims of this work were to assess the potential to integrate lettuce production and ell (*Anguilla anguilla*) production sustainably in recirculating aquaculture units in greenhouses. It analysed the response of lettuce to fish culture effluents. Lettuce was hydroponically grown in a closed recirculating system at different levels of intensity ranging gradually from 10 to 30 heads per m². Head densities were linked to increased fish stocking densities and consequently increased food input to the system. Apart from iron (as iron chelate), which was introduced at a concentration of 2mg/l throughout the production cycles, all the other nutrients were from fish waste. The production cycle was 5 weeks with mean weights of 115-155 g per head of lettuce. The experiments included water samples taken on a regular basis for analysis of nutrient concentration in the system for comparison to commercial hydroponics nutrient solutions. During the production cycles no disease of fish occurred, and there were no pests, diseases or nutrient deficiencies on plants. Such features could likely be major technical limiting factors for the development of these systems on a commercial basis. For the highest stocking density (30 plants/ m²), the current density practised in commercial hydroponics systems, a taste panel was established and nutrient leaf content was analysed. Results from the panel reveal good acceptance of the produce when compared to industrially grown lettuce.

1. Introduction

The existence of arid regions or regions where water is becoming increasingly scarce, such as the Algarve region of the south of Portugal, has generated interest to study production systems which utilise water more efficiently. Wastes from aquaculture include all materials used in the process which are not removed from the system during harvesting. The quantity of the total waste produced, and which leaves the system to load the environment, is closely linked to the culture system used.

Environmental impacts of different aquaculture systems are described by Pullin (1989). Nitrate (NO_3) and phosphates (PO_4) are major pollutants for receiving water bodies. On the other hand the same nutrients are among the essential elements for good plant growth (Clarkson and Lane, 1991). Integrated fish/plant systems are designed and managed to meet the production requirements of both fish and vegetables. The primary attraction for integrated systems is efficient resource utilisation (water and nutrient) and potential system profit enhancement.

2. Materials and Methods

Three separate treatments were applied where European eels (*Anguilla anguilla*) were stocked at three different intensities, (5, 10, 15 kg m^{-3}) integrated with lettuce (*Lactuca sativa sp.)* at different linked stocking densities (10,20 and 30 plants m^{-2}). The control comprised a setup with no eels present. Prior to introduction in the system, eels were treated in a 100 mgl^{-1} formalin bath for 30 minutes and afterwards in a 1 mg l^{-1} malachite green bath for a further 30 minutes to eliminate parasites and disease organisms. They were allowed to acclimatise and one month later lettuce seedlings were introduced into the system. Iron as iron chelate (Fe 10%) was introduced in the system at a concentration of 2mg l^{-1}.

Table 1. System description: treatment and control replicated twice

	Treatment	Control
Hydroponically grown lettuce	+	-
pH	7-8	7-8
System volume (l)	169	100
Daily water renewal (l)	6	6
Production cycle (days)	35	35
Temperature ^{0}C	20-25	20-25
Rearing tank volume (l)	72	72
Dissolved oxygen (mg l^{-1})	6.6-7.2	6.6-7.2
Flow rate (l h^{-1})	120	120

Water samples were taken on days 1, 15, 25, 35 after seedlings were introduced into the system and their nutrient content compared to those of solutions utilised in commercial hydroponically grown lettuce. The analysis of nutrient uptake by lettuce was done through comparison of treatments with controls after standardisation. Statistical tests (t-tests) were carried out to assess effects on water quality improvement of incorporating lettuce into the culture system. Fresh lettuce head weights were used in all experiments and differences assessed statistically. For the last set of experiments, lettuce was integrated at a density of 30 plants m^{-2} with eels at an initial stocking density of 15 kgm^{-3}. After harvest, lettuce heads were weighed and oven dried at 105 ^{0}C. Dry weights were taken prior to leaf nutrient content analysis for P, K, Mg, Ca, Fe, Zn, Mn and Cu. A control (lettuce without fish) was conducted

to test effects of well water utilisation to fertigate lettuce. A taste panel was also established and organoleptic analysis was done where lettuce heads were analysed in relation to subjective features ranked from 1 (poor) to 5 (excellent).

3. Results

Nutrient pattern

Results from water sampling showed that increased fish stocking densities intensified differences between control and treatments. This happened in relation to two nutrients, nitrogen (as NO$_3$) and phosphorus (as PO$_4$), where treatment differences ranged from 13-14% to 14-19%, below control, respectively. However results were not significantly different. Other elements analysed, (Ca, K, Mg and Fe,) were all present in the well water (Table 2). Apart from nitrates, all other nutrients were present in the culture water at much lower concentrations. Whilst, the plants did not present any perceptible symptoms of nutrient deficiency, neither did the level of salt accumulation seem to cause any toxic effects to lettuce. Thus the maximum electrical conductivity values achieved in this study were 1.8.dS m^{-1} (1155 mg l^{-1} as total dissolved solids) which is low when compared to levels of 2200 mg l^{-1} described as causing lettuce yield reductions in such systems (Costa Pierce, B.A and Rakocy, J.E., 1997).

Table 2. Nutrient contents of well water, recirculating aquaculture effluent (experiment 3), control solution (experiment 3) and hydroponics nutrient solution (Resh, 1989).

Element	Well water (mg l^{-1})	Treat. 3 (day 35)	Contr 3 (day 35)	Nutrient solution (mg l^{-1})[1]
Ca	97.9	129.5	135.6	197
Mg	33.7	38.1	39	44
K	20	28.5	42	400
PO$_4$	0	17.81	20.82	65
NO$_3$	0	151.9	172.4	145
Fe[2]	0.0294	0.45	0	2
Conduc[3]	582	1155	1191	-

[1]Recommended nutrient solution (Resh, 1989); [2]Electrical conductivity as total dissolved solids; [3]Fe added to the system at a concentration of 2 mg l^{-1}.

Lettuce Productivity

Fresh weights of lettuce heads (Table 3) did not vary significantly between treatments 2 and 3, however, in all experiments lettuce have grown significantly bigger when compared to the system with no fish integrated, (p<0.01). Also, results from treatments 2 and 3 were significantly better results than in treatment 1, (p<0.02).

Table 3. Lettuce production in response to variation in both plant and fish density shared aquaculture media

Fish density (kgm^3)	Plant density (n m^{-2})	Head weight (g)	Dry weights (g)
5	10	115	-
10	20	151	-
15	30	155	7.36
0 control	30	34.5	-

Leaf nutrient content

The nutrient leaf analysis is presented in Table 4. Apart from K and Mn which were present at a lower concentration and Zn which was present at a higher concentration when compared to commercial hydroponics, -6%, 40-50 mgl^{-1} and 30-36 mgl^{-1}, respectively (Adler, P. et al, 1996; Wheeler, P.M., et al., 1994)- all the other nutrients were fairly acceptably provided.

Table 4. Nutrient analysis of hydroponically-grown lettuce in a recirculating system for fish culture.

P (%)	K (%)	Mg (%)	Ca (%)	Fe ppm	Zn ppm	Mn ppm	Cu ppm
0.44	0.83	0.24	0.49	98.6	98.9	7.8	6.1
0.02	0.02	.004	.025	5.1	2.43	0.15	0.82

Taste panel

Results from the organoleptic analysis reveal the produce to have good acceptance within the panel (Table 5).

Table 5. Sensory evaluation of hydroponically grown lettuce fertigated with fish culture media.

Features	Mean
Taste	4
Texture	4.38
Colour	4.44
Size	3.75
Consumption rates	4.75
Consumer appeal	4.8
Acceptability	4.19

Scores for all features varied from poor (1) to excellent (2)

4. Discussion

The results from this experimental study suggest that fish culture water with slight nutrient supplementation might be used to fertigate hydroponically-grown lettuce. Thus, under the conditions of the studies, lettuce was produced at three different densities and did not present any symptoms of nutrient deficiency. The lower plant density (10 heads m^{-2}) gave significantly poorer results than the higher densities of 20 and 30 heads m^{-2}, probably due to the lower nutrient concentrations in the culture water. However, for all densities productivity was significantly better when lettuce was integrated with fish than when the system did not have any fish. Nutrient removal recovery benefited from introducing lettuce in the fish system reducing phosphates and nitrates, by 14-19 and 13-14%, respectively, when compared to systems which did not have any plants. This is an important feature; firstly because there is potential to enhance system profit through lettuce production and secondly, because of the increasing public awareness and restrictive EU directives concerning effluent discharges into the environment (Rosenthal et al., 1993). According to our work, when lettuce is stocked at a density of 30 plants m^{-2} the integrated system is discharging 20g and 3g (NO_3 and PO_4, respectively) less per m^3 discharged than a fish recirculating system not having plants incorporated. The water provided for the hydroponics had a much lower nutrient concentration than commercial hydroponics nutrient solutions (Table 2). However, the plants did not present perceptible nutrient deficiency symptoms, which agrees with previous works where a wide range of nutrient concentrations is possible in hydroponics systems before either nutrient deficiency or phytotoxicity occurs (Copper, 1979; Resh, 1989). Moreover, results from the taste panel reveal the produce to have good acceptability, size being the lowest scored but still with a score above the average (Table 5).

Acknowledgements

This work was conducted under grants from Fundação para a Ciência e Tecnologia. The research team appreciates the support and collaboration of Dra. Maria Teresa Dinis, UCTRA; Eng. Armindo Rosa, Direcção Regional de Agricultura do0 Algarve; Members of the taste panel and staff of the horto, UCTA, and Eng. José Barracha, Direcção Geral do Ambiente.

5. References

Adler R, Summerfelt S, Glenn DM, Takeda F0 (1996) Evaluation of the effect of a conveyor production strategy on lettuce and basil productivity and phosphorus removal from waste water. Water Environment Research 68 (5): 836-840.

Clarkson R, and Lane D (1991) Use of a small scale nutrient film technique to reduce mineral accumulation in aquarium water. Aquaculture and Fisheries Management 22: 37-45.

Copper AJ (1979) The ABC of NFT. Gower books, London.

Costa-Pierce BA and Rakocy JE (1997) Tilapia aquaculture in the Americas, World Aquaculture Society.

Parker D, Anouti A and Dickerson G (1990) Experimental results: integrated fish/plant production systems. Report N. ERL 90-34. University of Arizona, Environmental research labor atory, Tuckson, A.Z.

Pullin RSV (1989) Third world aquaculture and the environment. NAGA, the ICLARM Quarterly 12: 10-13.

Resh HM (1989) Hydroponic food production: a definitive guide book of soilless growing methods. Woodbridge press publishing company, Santa Barbara, CA.

Rosenthal H, Hilge V and Kamstra A (1993) Workshop on fish farm effluents and their control in EC countries-report. Hamburg, Nov. 23-25, 1992. University of Kiel: Kiel, Germany, pp. 12-13.

Wheeler PM, Mackowiak CL, Sager JC Yorio NC and Knott WM (1994) Growth and gas exchange by lettuce stands in a closed, controlled environment. Journal of the American Society of Hort. Science 119 (3): 610-615.

Chapter 46

Study of the substrate and fertilization effects on the production of essentials oils by *Thymus mastichina* (L.) L. ssp. *mastichina* cultivated in pots

[1] G. Miguel, C. Guerrero, H. Rodrigues and J. Brito
[2] F. Venâncio, R. Tavares, A. Martins and F. Duarte
[1] *Unidade de Ciências e Tecnologias Agrárias, Universidade do Algarve, Campus de Gambelas, 8000 Faro Portugal*
[2] *Ineti, Ibqta, Departamento de Tecnologia das Indústrias Químicas, Estrada do Paço do Lumiar, 2699 Lisboa Codex Portugal*

Key words: *Thymus mastichina* subsp. *mastichina*; Lamiaceae; essential oil composition; fertigation; 1,8-cineole, camphor, intermedeol.

Abstract: *Thymus mastichina* (L.) L. subsp. *mastichina* is endemic in the Iberian Peninsula, largely spread in Portugal and growing in stony and arid places. Plants growing either wild in Algarve fields or maintained in pots, using three different substrates: fertilised turf (TTUFF2), non-fertilised turf (TTUNFF2) and a sandy soil (TTF2), were analysed for their essentiel oil composition. In all samples, oxygenated monoterpenes were the main components of *T. mastichina* essential oils, 1,8-cineole being the most representative component. TTF2 and TTUNFF2 showed the highest concentrations of sabinene, *trans*-sabinene hydrate and linalool and the lowest concentrations of *p*-cymene and camphor. The highest concentration of 1,8-cineole was found in the essential oils of TTUFF2 samples whilst the highest amount of borneol was registered in the essential oils of TTF2 samples. Essential oils of plants kept in pots seem to be characterised by higher intermedeol levels than those maintained wild in field.

1. Introduction

Thymus mastichina (L.) L. subsp. *mastichina* is endemic in the Iberian Peninsula, largely spread in Portugal and growing in stony and arid places. This plant is an aromatic 20 to 50 cm shrub, that may constitute large populations in almost all of the country, excepting the calcareous regions (Fernandes Costa, 1945; Amaral Franco, 1983).

In Portugal, this taxon has been largely studied by some authors (Proença da Cunha, 1987; Salgueiro, 1994, Salgueiro *et al.*, 1997; Martins, 1995; Venâncio *et al.*, 1996) in view of acquiring knowledge of its essential oil chemical composition. At least three main chemotypes of *Thymus mastichina* (L.) L.

subsp. *mastichina* namely the 1,8-cineole, the linalool and the linalool/1,8-cineole were reported. Geographically, the 1,8-cineole chemotype can be found in Trás-os-Montes, Beira Alta, Beira Baixa, Estremadura, Ribatejo, Alto Alentejo and Algarve provinces, while the two remaining chemotypes can be found in the seaside of Estremadura, mainly in Arrábida and Sesimbra.

Also in Spain the same chemotypes were detected (Garcia-Vallejo *et al.*, 1983). Despite these authors consider a genetic factor to be responsible for the diverse chemotypes found in Spain, Salgueiro *et al.* (1997) established some correlation between the amount of linalool in the essential oil of *Thymus mastichina* (L.) L. subsp. *mastichina* and the Atlantic humidity.

This work submits studies on the essential oil composition from leaves of *T. mastichina* kept in pots using three different substrates: fertilised turf (TTUFF2), non-fertilised turf (TTUNFF2) and a sandy soil (TTF2). These treatments were fertigated every fifteen days.

2. Materials and Methods

Plant Material: The plant material was collected at S. Brás de Alportel, Algarve (Portugal), in December 1996 and October 1997.

In order to carry out the experiments in pots, *Thymus mastichina* (L.) L. subsp. *mastichina* was collected at S. Brás de Alportel in December 1996 and the rootstocks were obtained in order to take roots in the greenhouse later, then, the plants were placed in pots (three plants in each pot).

The study of the effects of the various fertilisation types on the essential oil chemical composition was carried out using those pots. These were filled with a haplic arenosol (ARh), according to the FAO-UNESCO Soil Classification for soils composed of sand without calcareous material (TTF2), with a fertilised turf (TTUFF2), and a non-fertilised turf (TTUNFF2). These treatments were fertigated every fifteen days with a solution of 1:3:1 (N:P:K), 0,4 % of Mg (5 g.L^{-1}); the fertiliser used was the BASF Hakaphos.

Analysis of the essential oils: The oils were obtained from 30 g of leaves from fresh plant material by distillation, for 3 hours, using a Lickens-Nickerson-type apparatus.

The analytical gas chromatography was carried out on a Hewlett Packard 5890 Series II gas chromatograph equipped with a Permabond CW20M (50 m x 0.25 mm i. d., df = 0.25 µm) and a OV-101 (50 m x 0.25 mm i. d., df = 0.25 µm) columns. A flame ionisation detector (FID) was used for routine quantitative analysis. The oven temperature was held at 70 °C (5 min) and programmed to increase till 220 °C at 2 °Cmin^{-1}. Detector and injector temperatures were set at 260 °C and 250 °C, respectively. The carrier gas was helium and the working flow was 1 ml.min^{-1}. The percentage composition of the oils were computed from the GC peak areas without using correction factors. The data shown are mean values of two injections.

The gas chromatography- mass spectrometry analyses were performed using a Perkin Elmer 8320 gas chromatograph, equipped with a DB-5 (30 m x 0.25 mm i. d., df = 0.25 mm) column and interfaced with a Finnigan MAT 800 Ion Trap Detector (ITD; software version 4.1). Oven temperature was held at 70 °C and programmed to increase till 180 °C at 3 °C.min^{-1}. Transfer line temperature, 250 °C; ion trap temperature, 220 °C; carrier gas helium adjusted to a linear velocity of 30 cm s^{-1}; splitting ratio, 1:100; ionisation energy, 70 eV; ionisation current, 60 µA; scan range, 30-400 µm; scan time, 1 s. The identity of the components was assigned by comparison of their retention times and mass spectra with corresponding data of reference oil components.

3. Results and discussion

The composition of essential oils of *T. mastichina* (L.) L. subsp. *mastichina* collected wild at S. Brás de Alportel and those maintained on different substrates is given in Table 1.

In all samples, oxygenated monoterpenes were the main components of *T. mastichina* essential oils whose concentrations ranged from 70.8 % to 76.6 % immediately followed by monoterpene hydrocarbons, the 1,8-cineole being the major component with percentages varying from 46.6 % to 51.6 % (Table 1). As all samples were characterised by high amounts of this compound, these plants seem to belong to cluster I according to Salgueiro *et al*. classification (1997). Camphor was the second main component present in essential oils of all samples, in contrast to the results by Salgueiro *et al* (1997) who reported α-terpineol (4.2 %) as the second major constituent.

α-Pinene, camphene and β-pinene were the most representative components in the monoterpene hydrocarbons group and β-elemol was the principal oxygenated sesquiterpene, although the plants kept in pots in several

Table 1. Constituents of essential oils of *T. mastichina* subsp. *mastichina* wild in field and maintained on different substrates fertigated fortnightly

Components	Percentage			
	Wild plants	TTF2	TTUF F2	TTUNF F2
Monoterpene hydrocarbons	**21.4**	**19.1**	**15.3**	**22.6**
α-Pinene	4.6	4.4	3.0	5.0
Camphene	5.2	4.2	3.8	4.4
Sabinene	1.7	2.3	1.5	2.9
β-Pinene	3.5	3.4	2.7	4.1
Myrcene	1.1	1.0	0.7	1.4
α-Terpinene	0.4	0.4	0.3	0.4
p-Cymene	0.8	0.2	0.6	0.1
Limonene	2.1	1.4	1.2	1.7
Trans-β-Ocimene	0.9	0.8	0.7	1.6
γ-Terpinene	0.8	0.7	0.6	0.7
Terpinolene	0.3	0.3	0.2	0.3
Oxygenated monoterpenes	**71.7**	**75.2**	**76.6**	**70.8**
1,8-Cineole	46.6	48.3	51.6	46.1
trans-sabinene hydrate	0.4	0.9	0.3	1.2
Linalool	0.2	1.3	0.4	1.2
Camphor	10.4	8.0	10.9	7.7
δ-Terpineol	1.5	1.6	1.7	1.7
Borneol	5.6	7.6	4.7	5.4
Terpinen-4-ol	2.5	2.7	2.6	1.9
α-Terpineol	4.5	4.8	4.4	5.6
Sesquiterpene hydrocarbons	**0.3**	**0.2**	**0.1**	**0.2**
β-Caryophyllene	0.3	0.2	0.1	0.2
Oxygenated sesquiterpenes	**2.7**	**3.3**	**4.1**	**4.2**
β-Elemol	1.1	1.3	1.8	2.0
Caryophyllene oxide	0.1	0.1	0.1	Traces
γ-Eudesmol	0.3	0.2	0.2	0.2
β-Eudesmol	0.6	0.3	0.4	0.3
α-Eudesmol	0.4	0.2	0.2	0.3
Intermedeol	0.2	1.2	1.4	1.4

Traces ≤ 0.05 %

substrates also showed significant intermedeol amounts (1.2 %-1.4 %) (Table 1). The essential oil of *T. mastichina* collected wild in field showed slight differences when compared with those maintained in pots and fertigated fortnightly. Thus, in monoterpene hydrocarbons group, the percentages of camphene, *p*-cymene, limonene were slightly higher, whilst the concentration of linalool, an oxygenated monoterpene, was lower in the essential oil of wild plants, than in pot cultivated plants (Table1).

β-Elemol and intermedeol concentrations in the wild plants were lower comparatively to the remaining samples, with β-eudesmol in higher concentration (Table 1).

In the monoterpene hydrocarbons, TTF2 and TTUNFF2 showed the highest concentrations of sabinene and the lowest concentrations of *p*-cymene (Table 1).

In oxygenated monoterpenes, the highest concentration of 1,8-cineole was found in the essential oils of TTUFF2 samples (51.6 %) while the highest amount of borneol was registered in the essential oils of TTF2 samples (7.6 %). TTF2 and TTUNFF2 samples had higher concentrations of *trans*-sabinene hydrate and linalool, and lower concentrations of camphor (Table 1). Independently of the substrate used, essential oils of plants cultivated in pots had higher intermedeol amounts than the wild plant essential oil (Table 1).

Acknowledgements:

The authors are grateful to the "Fundação para a Ciência e Tecnologia-Praxis XXI Program, Portugal" and "Fundação Calouste Gulbenkian, Portugal" for financial support.

4. References

Amaral Franco, J. (1983) Botânica das labiadas portuguesas e suas potencialidades. In: 1[as.] Jornadas Nacionais de Plantas Aromáticas e Óleos Essenciais. Coimbra, pp. 6-13.

Fernandes Costa, A. (1945) Subsídios para o estudo das plantas aromáticas portuguesas. Algumas essências de Thymus L. (Th. zygis subsp. sylvestris, Th. zygis subsp. zygis, Th. mastichina, Th. caespititius, Th. capitellatus, Th. villosus). Coimbra: Faculdade de Farmácia. PhD Thesis.

Garcia-Vallejo, M. C.; Garcia Martin, D.; Muñoz, F.; Bustamante, L. (1983) Avance de un estudio sobre las esencias de Thymus mastichina (l.) L. español ("mejorana de Espanha"). In: 1[as.] Jornadas Nacionais de Plantas Aromáticas e Óleos Essenciais. Coimbra, pp. 241-266.

Martins , A. S. (1995) Tecnologias de produção do óleo essencial de manjerona brava (Thymus mastichina L.). Faculdade de Ciências e Tecnologia da Universidade de Lisboa. Master Thesis.

Proença da Cunha, A. (1987) O polimorfismo químico em espécies do género Thymus. 2[as.] Jornadas Nacionais de Plantas Aromáticas e Óleos Essenciais. Lisboa, pp. 8-25.

Salgueiro, L. (1994) Os tomilhos portugueses e os seus óleos essenciais. Coimbra: Faculdade de Farmácia. PhD Thesis.

Salgueiro, L.; Roque, O. R.; Proença da Cunha, A. (1992) Composition de l'huile essentielle de Thymus mastichina spontanée du Portugal. Riv. Ital. E.P.P.O.S.. nº Sp. Feb., pp 491-495.

Salgueiro, L.; Vila, R.; Tomàs, X.; Cañigueral, S.; Proença da Cunha, A.; Adzet, T. (1997) Composition and variability of the essential oils of Thymus species from Section mastichina from Portugal. Bioch. Syst. Ecol., 25, pp 659-672.

Venâncio, F.; Martins, , A. S.; Duarte, F:L: (1996) Etude comparative de l' Huile Essentielle de Thymus mastichina L. type linalol obtenue par différentes méthodes déxtraction. Riv. Ital. E.P.P.O.S.. nº Sp. Gennaio, pp 417-425.

Chapter 47

Sewage sludge as a horticultural substrate

J. M. C Brito, R. Lopes, A. M. V Machado, C.A.C Guerrero, L. Faleiro. and J.Beltrão
Unidade de Ciências e Tecnologias Agrárias, Universidade do Algarve - Campus de Gambelas Faro – Portugal

Key words: urban sludges, *Pelargonium* hybrid, growth substrate, chemical analysis ,microbiological analysis

Abstract: This work aims to justify of the use of urban sludges, mixed with either fertilized or unfertilized peat, through the study of their effects on the germination and growth of a *Pelargonium* hybrid as well as effects on substrata and plant pollution (pathogens). Several treatments of urban sludges mixed in varying proportions with fertilized or unfertilized peat were used. Germination rate in all instances was above 70% and reached 100% for 3 treatments. Plant growth (height) and leaf production were generally better when fertilised peat was used in mixtures but plant height also equalled the maximum when grown in pure urban sludge. Chemical and microbiological analysis of the different treatments showed that there was not any potential pollution hazard for the use of this urban sludge as a substrate.

1. Introduction

Seedbeds are used for plant production, in nursery-gardens. They provide plants with excellent characteristics for transplanting, with ready adaptation to local production to ensure fine quality crops. The factors that influence the final quality of plant produced in nursery-gardens are: characteristics, control of climate and management of the substrate, plant space and applied treatments.

For 20 years, different types of substrata have been used in nursery-gardens, such as fertilized or unfertilized peat, with or without sand or other inert products. The use of peat has some problems, namely, high price because of extraction and transportation costs (Abad *et al.*, 1993) and limited sources (Hoitink and Grebus, 1994). Peat deposits condition the local fixation of atmospheric carbon dioxide and their destruction allows an increase of that gas in the atmosphere (Reis, 1997). Diseases may appear faster in peat medium, when some pathogenic agents could be propagated when unsterilized peat is used, e.g. *Pythium spp.* (Chen and Hadar, 1987). Alternative substrata can be used

with the same or better results than peats (Chen and Hadar, 1987; Marchesi and Cattivelli, 1988), namely tree barks, wood shreds, either aged or composted (Rankov, *et al.*, 1987). Human activities produce a set of organic residues such as urban sludges, which must be dispensed in appropriate areas, or incinerated or applied in agriculture. The latter solution can reduce environmental impact and facilitates an increase in organic matter levels. Urban sludge is potentially a serious environmental problem, with production likely to be about 8 millions ton.year[-1] in AD 2000. This residue has high levels of organic matter, and considerable amounts of nutrients, such as nitrogen (N), potassium (K) and phosphorous (P) (Brito, 1986 ;Machado, 1996). The overall goal of this work was to study the viability of the use of sewage sludges as substrate. Urban sludges (US), unfertilized peat (UFP), fertilized peat (FP) and mixtures of these composts (0, 25, 50, 75 and 100%) were tested in *Pelargonium* hybrid production. Chemical and microbiological analyses of substrate mixtures, seed germination, plant growth and number of leaves, were studied.

2. Materials and Methods

Three different types of composts were tested as described in Table 1.

Table 1. Treatments of urban sludges (US) and peat (FP and UFP) applied in treatments

Treatment	Compounds (US, FP and UFP)
M1	100% FP
M2	75% FP + 25% US
M3	50% FP + 50% US
M4	25% FP + 75% US
M5	100% US
M6	100% NFP
M7	75% UFP + 25% US
M8	50% UFP + 50% US
M9	25% UFP + 75% US

These treatments were distributed on randomized seedbed plots with seven plants for each plot.

Urban sludges were collected in the wastewater plant of Tavira - Algarve (Portugal). The residues with peat admixtures were dried at an average temperature of 25°C for one week. They were then ground to an average particle size of 2 mm diameter. Afterwards, then were submitted to microbiological analyses (Table 2).

Table 2. Microbiological analyses of the urban sludges (MPN = most probable numbers)

Urban Sludge (g)	Total Colif. (MPN)	Fec. Colif. (MPN)	E. coli (MPN)	Clost Sulf. Red. (g^{-1})	Salmonella (MPN)
1	$15*10^{-4}$	0	absent	10^{-3}	absent

After mixing urban sludges and peats, one sample was collected from each treatment for microbiological analyses. Total and Fecal Coliforms, *Escherichia coli*, *Salmonella* and *Clostridium* Sulfite Reductors were determined according to Costa and Loureiro (1990). From each treatment, one sample was collected for physical and chemical analyses. The macro-Kjeldahl method (Morgan *et al.*, 1957) was used for the quantification of nitrogen. After calcination at 550±50°C and digestion with hydrochloric acid (3N), total phosphorous was determined by ascorbic acid method (Boltz, 1958) and total potassium was determined by flame photometry. Organic matter was analyzed by determination of weight differences, using the calcination method; pH and electric conductivity were determined by a potentiometric method in a saturation paste. Chemical and microbiological analyses, seed germination, plant growth and average number of leaves, were submitted to variance analysis (ANOVA) at a significancy level of 95%; the Tukey test being used to locate significant differences.

3. Results

According to Table 3, the enhanced urban sludge mixtures with fertilized peat (treatments M1 to M5) and unfertilized peat (M6, M7, M8, M9 and M5) increased the pH, EC, N, P, K and organic matter (OM), as referenced to M6, the 100% UFB treatment. It may be observed that the levels of pathogens of all treatments (Table 4) were under the microbiological standards according to Felgueiras (1983). However, it was verified that increasing the proportion of urban sludges applied enhanced the microbiological levels of the studied mixtures.

According to the Tukey test it can be concluded that the average seed germination was not significantly different between the treatments. Seed germination of *Pelargonium* hybrid varied between 71.43% to 100%. All seeds germinated in M1, M4 and M6 treatments (Figure 1). However, seeds germinated faster using 100% of fertilized peat (M1).

Increasing the amounts of urban sludges with unfertilized peat, decreased seed germination, but for fertilized peat treatments no systematic pattern was obvious as these admixture included both the highest and the lowest (M2) germination. The several mixtures of urban sludges with fertilized and unfertilized peat did not cause significant differences on the final plant height. However, when urban sludges were increased in mixtures with unfertilized peat the plant heights were greater (Figure 2).

Table 3. Physical and chemical parameters of the different treatments

Treat	pH	EC dS.m^{-1}	N %	P %	K %	OM %
M1	5.6	1.7	1.05	0.29	0.12	11.8
M2	6.0	2.8	1.52	1.37	0.27	22.2
M3	6.2	4.0	1.86	1.77	0.27	28.6
M4	6.4	4.2	2.05	1.71	0.35	33.1
M5	6.8	4.9	2.57	2.00	0.33	37.9
M6	4.8	0.1	1.04	0.11	0.07	11.5
M7	4.9	1.1	1.37	1.41	0.23	18.8
M8	5.3	1.8	1.98	1.76	0.32	29.8
M9	5.9	4.1	2.18	1.54	0.26	34.9

Table 4. Microbiologial analyses of the substrata.

Treat	Total Colif. (MPN)	Fec. Colif. (MPN)	E. coli (MPN)	Clost Sulf. Red. (g^{-1})	*Salmonella* (MPN)
M1	0.9*10^2	0	0	0	absent
M2	2.5*10^2	0	0	0	absent
M3	3.5*10^3	0	0	0	absent
M4	45*10^3	0	0	0.1	absent
M5	>140*10^2	0	0	0.01	absent
M6	9.5*10^5	0	0	0	absent
M7	>140*10^2	0	0	0	absent
M8	7.5*10^4	0	0	0	absent
M9	>140*10^2	0	0	0.1	absent

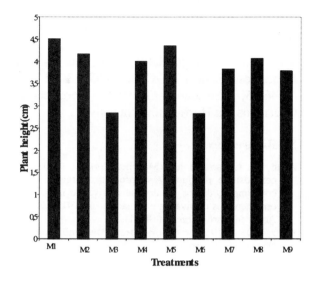

Figure 2. Plant growth

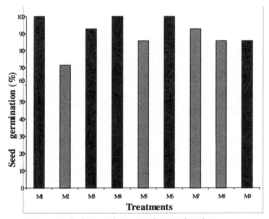

Figure 1. Pelargonium seed germination

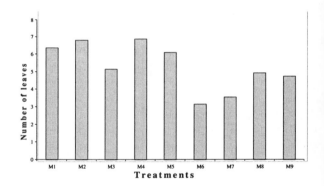

Figure 3. Number of leaves

4. Discussion

The use of urban sludges as substrate in plant production at nursery gardens should not be a total alternative to peat. Mixtures of 75% of urban sludges and 25% of fertilized peat have shown good results on *Pelargonium* hybrid production. These results showed a great interest for the use of these kind of experiments with other plant species.

Parameters such as electrical conductivity and pathogen levels of these kind of substrata

From Tukey test results the average number of leaves obtained were significantly different between M4 *vs.* M7, M4 *vs.* M6, M1 *vs.* M7, M1 *vs.* M6. In general, fertilized peat mixed with urban sludges produced plants with a higher number of leaves (Figure 3).

208

can restrict plant production and planting management.

Acknowledgements

This work was partially supported by the Portuguese Project PRAXIS XXI/3/3.2/Hort /2146/95, entitled "Novos" Fertilizantes Orgânicos em Horticultura Intensiva.

5. References

Abad, M, Cegarra, J and Martinez-Corts, J (1993). El compost de resíduos y subproductos organis como componente de los medios de cultivo de las plantas ornamentales cultivadas em maceta, p 1191-1196. In: Actas de Horticultura da SECH, 10.

Boltz, DF (1958). Colorimetric Determination of Nonmetals. Interscience, New York.

Carrasco de Brito, JM (1986). As lamas pretas como fertilizante: contribuição para o seu estudo. PhD thesis, Instituto Superior de Agronomia, Lisboa, Portugal.

Chen, Y and Hader, Y (1987). Composting and use of agricultural wastes in container media. p. 71-77. In M.de Bertoldi et al. (eds). Compost: Production, Quality and Use. Elsevier Applied Science Publishers Ltd. Essex, Reino Unido.

Costa, MLF and Loureiro, V (1990). Preparação de amostras para análise microbiológica. Universidade Técnica de Lisboa, Instituto Superior de Agronomia.

Felgueiras, MI (1983). National and International microbiological characteristics of foods. Laboratório Nacional de Engenharia e Tecnologia Industrial. DCEAI n.º 141, R-23. Lisboa.

Hoitink, HAJ and Grebus, ME (1994). Status of biological control of plant diseases with compost. Compost Science & Utilization:6-12.

Jackson, ML (1958). Soil Chemical Analysis. Prentice-Hall, Englewood Cliffs, New Jersey.

Machado, AMV (1996). Resposta da variedade de tomate PEX-1307, à aplicação de resíduos orgânicos biodegradáveis. Licentiate degree thesis.

Marchesi, G and Cattivelli, F (1988). A comparison of different substrates, with or without fertilizer, for the production of capsicum and aubergine seedlings. Colture Protette 17(1): 91-93.

Morgan, GB, Lackey, JB and Gilcreas, FW (1957). Quantitative determination of organic nitrogen in water, sewage and industrial wastes. Anal. Chem., 29: 833-840.

Page, AL, Logan, TJ and Ryan, JA (1982). Methods of Soil Analysis, Parts 1 and 2. Madison, U.S.A..

Rankov, V, Simidchiev, H, and Kanazirska,V (1987). Use of pine bark tree in container growing of glasshouse tomatoes. Plant Science XXIV(8): 119-123.

Reis, MMF (1997). Compostagem e caracterização de resíduos vegetais para utilização como substratos Horticolas. Ph.D thesis.

6- Crop quality – nutrient management by alternative sources

Chapter 48

Citrus nitrogen nutrition within organic fertilization in Corsica

P.Berghman [1], M. Crestey [2], G. de Monpezat [3], P. de Monpezat[3]
[1] Demeter, Moriani-Plage, F 20230 San Nicolao
[2] Civam Bio Corse, Pôle Agronomique, F 20230 San Giuliano
[3] Cedat, Cidex 120, F 06330 Roquefort Les Pins, France

Key words: citrus, organic farming, N

Abstract: Organic farming for citrus cultivation is expanding in Corsica since the latest ten years. But yields are generally much lower than under inorganic fertilization. For the last five years, we have elaborated tests to research the origin of that low yield and also to define means for cultural improvement. The study included comparison of leaf mineral composition under organic and inorganic fertilization for a characteristic soil, test of different organic produces, soil analysis changes in the different orchards.
Low yields are related to bad nitrogen assimilation by roots, and it is necessary to use two complementary organic produces :
- one, rich in nitrogen, providing a sufficient amount of nitrate,
-the other, rich in organic carbon, securing a steady nitrogen feeding during the vegetative period.

1. Introduction

Organic farming for citrus cultivation is expanding in Corsica since 1980. Thanks to works by INRA, IRFA, and then CIRAD and CEDAT, nutrient management methods were defined and adjusted to soils and climates of Corsica (Bathurst, 1944; Chapman, 1949, 1968; Martin-Prevel et al., 1965, 1966; de Monpezat, 1968; Marchal et al., 1978).

But yields of clementine orchards cultivated according to the regulations for organic farming are always 50% lower than under traditional inorganic farming.

Different studies showed difficulties for plants to absorb enough nitrogen (Cassin et al., 1977; Koo, 1988; Garcia et al., 1988; de Monpezat., 1997). The compost used in Corsica to provide organic nitrogen releases nitrate-nitrogen with a pattern spread along a great number of months. In most cases, the peak of nitrogen requirement in May-June is not satisfied.

2. Materials and Methods

Orchards selection This experiment was conducted in two orchards of clementine, cv. "SRA 92" on Poncirus trifoliata rootstock, 25 years old, with a drip irrigation system. One is cultivated under traditional farming (inorganic fertilization) and the other under organic farming (called "bio" in the following text). Both of them were controlled for 5 years with soil and leaf analysis. Last year, the "bio" orchard was subdivided into two sectors, one still receiving the plant compost, the other receiving a mixture of castor oil cake (COC) and seabird fertilizer (guano).

Soil and leaf analyses
-Complete soil (5-25 cm) and sub-soil (40-60 cm) analysis at the beginning of the experiment.
-Annual analysis of variable soil characteristics: pH, phosphorus (Olsen), exchangeable calcium, magnesium, potassium.

-Annual foliar analysis between 15th and 20th October, sampling on fruiting shoots.

-Measuring out of Nitrate nitrogen in situ (Nitracheck method) in May, July, August, September and November 1997. Measures effected at two depth (between 5 and 25 cm, then between 25 and 35 cm). Presented results are averages of 11 replications.

Fertilization

In 1993 : both orchards received 2000 kg ha^{-1} of calco-magnesium limestone with high carbonic solubility.

From 1993 to 1997 :

The traditional orchard received 100 to 130 kg ha^{-1} of N in three applications in spring and 0 to 80 kg ha^{-1} of P_2O_5 and K_2O according to foliar analysis results.

"The bio" orchard received in autumn / winter a plant compost and an organic "bio" fertilizer corresponding to approximately 100 kg ha^{-1} of N.

In 1997 : half "bio" orchard received instead of compost a COC + guano mixing corresponding to 130 kg ha^{-1} of N.

Inorganic fertilizers were applied on the whole surface, organic fertilizers were applied under the canopy and slightly buried.

3. Results

Soil The two orchards are situated on a brown leached soil from old alluvium of medium Quaternary era. Pebble rate exceeds 60%. Soil (Table1a) is not calcareous and not hydromorphous, but susceptible to acidification by calcium leaching.

Limestone application on the first year increased immediately pH and calcium rate, which then remained steady. Magnesium rate, because of the lower solubility of this element, increased during the three or four following years (Table 1b).

Plant Vegetative growth of "bio" trees was lower than traditional trees : narrower leaves, shorter shoots, lower fruit number.

The steady increase of potassium rate in leaves after liming was significant for both parcels (Table 2).

Calcium and magnesium variations depended on soil aeration in spring.

In traditional trees leaf nitrogen rate was stable and remained steady in the standard interval. In "bio" trees, leaf nitrogen, always insufficient, fluctuated between low and deficient rates.

Micro-nutrients rates reflected mineralogical soil origin : zinc and manganese deficiency, correct rates for boron, iron and copper in both orchards.

Yields (Table3) in the traditional orchard changed according to years in accordance with climatic conditions in spring, responsible of the worsening of physiological fruit fall. Those of "the bio" orchard were systematically lower in all years.

Organic matter experiment Application of COC and guano allowed leaf nitrogen (Table 2) to increase from deficiency (1.87 %DM) to normal rate (2.36 %DM), and doubled the yield (Table 3), primarily through the reduction of physiological fall in June.

Nitrate rates in soil (Table 4) were similar for both sectors. On the other hand, "bio" sub-soil reference was very poor in nitrates whilst the sector receiving guano and castor oil presented normal rates with a peak at 57 mgkg^{-1} in the end of July.

4. Discussion

In spite of the liming which created biological and physico-chemical satisfactory conditions for root activity, "bio" cultivation with compost applications did not ensure a sufficient nitrogen absorption. These results are in agreement with those obtained by Legaz et al. (1992) who noted a difference in leaf nitrogen rates between nitrogen applications as nitrate and ammonium or urea. The two latters induce a low cellular multiplication and a worsening of the physiological fruit fall.

Incorporation into the soil of organic amendment rich in carbon as castor oil cake on one hand , and rich in organic nitrogen easily mineralizable as seabird guano on the other hand, allows a good nitrogen assimilation and a reduction of physiological fruit fall. This root absorption is confirmed by the absence of nitrate accumulation in the soil during the mineralization period, which contributes to environment respect. This old concern (Embleton et al. 1978) is meanwhile also topical. Koo (1986) already showed the interest of controlled-release nitrogen to limit pollution risk together with an improvement of application efficiency (25% less nitrogen applied without yield effect).

Lower yields in "bio" citrus cultivation in Corsica are really caused by insufficient release of nitrate in the soil. Positive results obtained since the first year with the mixture of castor oil cake and seabird guano must be verified in the following years in order to set up a really controlled biological fertilization, well adjusted to this crop.

Table 1a. Physico-chemical characteristics of soil (S) and subsoil (SS) of orchards under traditional agriculture (Chemical) and organic fertilization (Organic)

Orchard	depth	clay %	silt 1 %	silt 2 %	sand 1 %	sand 2 %	O.M. %	Total lime %	CEC meq%	Total N ‰	C/N
Chemical	S	17	13	7	15	45	1.3	0	12.2	0.85	8.41
	SS	23	15	11	19	30		0	13.8		
Organic	S	17	24	16	19	21	1.9	0	12.6	0.93	11.24
	SS	24	17	13	18	26		0	14.7		

Table 1b. Exchangeable cations and P in soil from 1993 to 1997

Orchard		pH water	K meq %	Ca meq %	Mg meq %	P (Olsen) mg kg^{-1}
Chemical	1993	6.0	0.55	6.6	0.40	44
	1994	7.4	0.84	13.9	0.84	79
	1995	6.6	0.30	7.2	1.41	210
	1996	7.1	0.78	9.9	1.28	57
	1997	7.1	0.50	12.0	1.23	120
Organic	1993	6.6	0.30	5.0	0.60	66
	1994	7.2	0.33	8.6	0.97	74
	1995	7.0	0.43	10.5	1.51	119
	1996	7.6	0.71	11.1	1.69	108
	1997	7.3	0.25	6.9	2.29	54

Table 2. Mineral contents in leaves

Orchard	Year	N	P	K	Ca	Mg	B	Zn	Mn	Fe	Cu
		% DM					mg kg^{-1}DM				
Chemical	1993	2.61	0.16	1.05	4.21	0.20	31	21	44	322	18
	1994	2.60	0.07	0.91	4.75	0.09	38	17	11	154	29
	1995	2.51	0.13	1.04	4.19	0.26	55	16	28	136	43
	1996	2.49	0.17	1.40	3.31	0.23	56	17	19	193	19
	1997	2.53	0.17	1.92	3.77	0.34	57	16	19	84	38
	Mean	2.57	0.17	1.49	3.99	0.27	44	19	32	203	28
Organic	1993	1.82	0.12	0.90	3.71	0.33	42	22	30	192	12
	1994	2.08	0.23	1.48	5.25	0.42	70	23	21	261	30
	1995	1.84	0.18	1.01	4.35	0.37	38	17	19	277	10
	1996	1.98	0.17	1.38	3.55	0.27	50	27	25	173	59
	1997	1.87	0.22	2.01	3.28	0.27	55	17	27	90	44
	Mean	1.85	0.17	1.46	3.50	0.30	49	20	29	141	28
COC+ guano	1997	2.36	0.19	1.73	4.18	0.27	78	16	19	126	39
Standard values		2.3	0.13	0.70	2.50	0.27	40	20	25	150	12
		2.7	0.17	1.10	4.70	0.60	100	60	80	200	25

214

Table 3 . Yield in kg ha^{-1}

Orchard	1993	1994	1995	1996	1997
Chemical	25300	14800	25300	20100	18300
Organic	8400	6700	18300	10700	7100
COC+guano					14700

Table 4 . Nitrates in soil (S) and subsoil (SS). mg kg^{-1}, 1997

Orchard	depth (cm)		07/05	25/07	08/08	27/08	19/11
Organic	S	5-25	23	25	42	25	17
Reference	SS	25-35	11	16	19	9	17
Organic with	S	5-25	28	27	45	35	23
COC+guano	SS	25-35	32	57	26	32	23

5. References

Bathurst A.C. (1944) Method for sampling citrus leaves for diagnostic purposes. Farm. South Afric. 19, 329-330.

Cassin P.J., Favreau P., Marchal J., Lossois P., Martin-Prevel P. (1977) Influence of fertilization on growth, yield and leaf mineral composition of "Clementine" mandarin on three rootstocks in Corsica. Proc. Int. Soc. Citriculture, 1, 49-57.

Chapman H.D. (1949) Citrus leaf analysis. Nutrient deficiencies, excesses and fertilizer requirements of soil indicated by diagnostic aid. Calif. Agr., 3, 10-14.

Chapman H.D. (1949) Tentative leaf analysis standards. Calif. Citrograph., 34, 518.

Chapman H.D. (1968) The mineral nutrition of citrus, in: The Citrus Industry, 2, 127-289, Reuhther W., Bachelor L.D., Webber H.J., Ed. Univ. Calif.

Embleton T.W., Jones W.W. (1978) Nitrogen fertilizer management programs, nitrate-pollution potential and orange productivity. Nitrogen in the Environment. Vol. 1, Academic Press, New York, p275-296.

Garcia M., Chambonnière S., de Monpezat G., Cassagnes P. (1988) Incidence des facteurs pédo-climatiques sur la nutrition minérale des fruits à pépins. 7ème Coll. Int. AIONP. Nyborg. 552, 1-6.

Koo R.C.J. (1986) Controlled release sources of nitrogen for bearing citrus. Proc. Florida State Hortic. Soc., 99, 46-48.

Koo R.C.J. (1988) Use of controlled-release nitrogen for citrus in a humid region. Proc. 6th Int. Citrus Cong. Tel Aviv. March 6-11. 633-641.

Legaz F., Serna M.D., Primo-Millo E., Perez-Garcia M., Maquieira A., Puchades R. (1992) Effectiveness for the N form applied by a drip irrigation system to citrus. Proc. Int. Soc. Citriculture, 2, 590-592.

Marchal J., Cassin P.J., Favreau P., Lossois P., Martin-Prevel P. (1978) Diagnostic foliaire du clémentinier en Corse. Fruits, 33, 822-827.

Martin-Prevel P., Del Brassine J., Lossois P., Lacoeuilhe J.J. (1965) Echantillonnage des agrumes pour le diagnostic foliaire. II. Influence de la position des feuilles sur l'arbre. Fruits, 20, 595-603.

Martin-Prevel P., Lossois P., Lacoeuilhe J.J., Del Brassine J.(1966) Echantillonnage des agrumes pour le diagnostic foliaire. III. Influence du caractère fructifère ou non fructifère des rameaux, de leur hauteur et de l'ombrage. Fruits, 21, 577-587.

de Monpezat G. (1968) Les bases de la fertilisation des agrumes en Corse et son contrôle. Comité de liaison de l'agrumiculture méditerranéenne.

de Monpezat G. (1997) Effet des apports de matière organique sur la fertilité du sol. Rapport annuel d'activités du programme Européen Proterra : Réhabilitation des cultures en terrasses, 14p.

Chapter 49

Effects of some organic biostimulants on the quality of table grapes

C. Köse and M.Güleryüz

Department of Horticulture, College of Agriculture, Atatürk Univ., 25240 Erzurum – Turkey

Key words: organic biostimulants, seaweed extract, grape, quality

Abstract: Effects of two different organic biostimulants containing mineral and organic compounds (based on seaweed) on the quality of table grape cv. Karaerik in field conditions were investigated in 1995 and 1996. Both of the biostimulants decreased the content of soluble solids(TSS) and reducing sugars of the must and total dry matter of the berry in 1995 and 1996. However, the effects of 300 and 900 mg kg^{-1} Proton and 1250 mg kg^{-1} Maxicrop on TSS of must and dry matter of berry were not statistically significant in 1996. When the effects of biostimulator applications were compared to that of the control, all treatments decreased the ratio of spilled berry and all applications had similar effect in 1995. In 1996, there was no effect of Proton application on the ratio of spilled berry whereas Maxicrop treatments excepting 1250 mg kg^{-1} decreased the ratio of spilled berry.

1. Introduction

Owing to the fact that the world's population increases, demand for agricultural products especially food is great. New insights and techniques are required in order to achieve sufficient and sustainable yields to meet global food demand and prevent world hunger.

One approach to increasing crop productivity is the development of non–polluting organic biostimulants that increase plant growth, vigor, crop yield and quality through increased efficiency of nutrient and water uptake.

Definitions for biostimulants vary greatly. However, the biostimulants are loosely defined as non-fertilizer products which have beneficial effects on plant growth. Many of these materials are natural products that contain no added chemicals or synthetic plant growth regulators (Russo and Berlyn, 1990).

The application of seaweed extract which contains minerals, several organic compounds, enzymes, vitamins and plant growth regulators (Verklij, 1992), as an organic biostimulant is fast becoming an accepted practice in horticulture because it has some beneficial effects.

A wide range of beneficial effects including increased crop yield, uptake of inorganic constituents from the soil, resistance of plant to frost and stress conditions, reduced incidence of fungal and insect attack, lesser storage losses of fruit, improved seed germination and increased crop quality have been reported from the use of seaweed extracts (Berlyn and Russo, 1990). On the other hand, seaweed extract being organic and biodegradable is important in sustainable agriculture (Cassan et al., 1992) and recently use of seaweed products in organic farming widely increased (Rader et al., 1985).

In this study, the potential role of two different organic biostimulants (based on seaweed) on quality of table grape cv. Karaerik was investigated.

2. Materials and Methods

This study was carried out in Üzümlü district of Erzincan during 1995 and 1996. Karaerik vines were trained with Baran system which is a prostrated system. Two different biostimulants (based on seaweed) was sprayed on foliage at concentrations of Proton 300, 600, 900 mg kg^{-1} and Maxicrop 1250, 2500 and 3750 mg kg^{-1} (+ 2.5 ml 10 l^{-1} Agral as a surfactant). Control received 2.5 ml 10 l^{-1} Agral (Villiers et al., 1983). The trial was laid out according to a nested design with two replications and three vines per treatment (Düzgüneş et al., 1987).

Biostimulants were sprayed three times in a year on foliage: at first bloom, fruit set and unripe grape stage. In order to determine the effects of these products on quality of Karaerik vine, total soluble solids (TSS), reducing sugars and tartaric acid contents of must, total dry matter of berry, number of seeds per berry, berry weight, bunch weight, number of berries per bunch and the ratio of spilled berry (Villiers et al., 1983; Delas, 1994) were determined.

3. Results and Discussion

The effects of biostimulant treatments on all parameters are presented in Table 1. Maxicrop and Proton applications decreased content of soluble solid (TSS) as compared with control in 1995. Treatments except for 300 mg kg^{-1}, 900 mg kg^{-1} Proton and 1250 mg kg^{-1} Maxicrop applications decreased TSS in 1996. 600 mg kg^{-1} Proton in 1995 and 3750 mg kg^{-1} Maxicrop in 1996 caused the greatest losses in TSS. Total dry matter of berry was decreased by all biostimulant applications in 1995. While 600 mg kg^{-1} Proton, 2500 mg kg^{-1} and 3750 mg kg^{-1} Maxicrop decreased total dry matter of berry, other treatments had effects similar to control in 1996. In both years, the least total dry matter content was obtained with 3750 mg kg^{-1} Maxicrop.

Biostimulant applications except for 900 mg kg^{-1} Proton, 1250 mg kg^{-1} Maxicrop in 1995 and 1250 mg kg^{-1} Maxicrop in 1996 decreased reducing sugar content of must. Application of 3750 mg kg^{-1} Maxicrop caused the greatest losses of reducing sugar content in both years. This treatment decreased reducing sugar content of the must by 29.9 % and 27.7 % in 1995 and 1996, respectively. While the biostimulant treatments except for 1250 mg kg^{-1} Maxicrop increased the acid content of the must in 1995, only 2500 mg kg^{-1} and 3750 mg kg^{-1} Maxicrop increased the acid content and the other applications had no significant effect on this parameter in 1996.

Proton applications did not have any effects and Maxicrop treatments decreased berry weight when the effects of seaweed extract applications on berry weight were compared to that of the control in 1995. Both Maxicrop and Proton applications had no significant effect on this parameter in 1996.

The effects of Proton treatments on the number of berries per bunch were not significant in 1995, but Maxicrop applications increased the number of berries in the same year. In 1996 only effects of 600 mg kg^{-1} Proton and 3750 mg kg^{-1} Maxicrop applications on the number of berries per bunch were significant. The highest number of berries per bunch was obtained from 3750 mg kg^{-1} Maxicrop application in both years. This treatment increased the number of berries per bunch by 26.5 % and 30.6 % in 1995 and 1996, respectively. The biostimulant treatments increased the number of seeds per berry except for 1250 mg kg^{-1} Maxicrop in 1995, 1250 mg kg^{-1} and 2500 mg kg^{-1} Maxicrop in 1996.Application of 3750mg kg^{-1} Maxicrop produced the highest number of seeds per berry in both years.

The effects of Maxicrop applications on bunch weight were not significant in both years. While all Proton treatments increased

Table 1. Effects of biostimulants on total soluble solid (TSS), reducing sugars (RS) and acid content (as tartaric acid) of must (AC), total dry matter content of berry (TDM), berry (BW) and bunch weight (Bun. W), number of berries per bunch (NB), number of seeds per berry (NS) and the ratio of spilled berries (RSB).

Parameters	Years	Control	300 mg kg⁻¹	Proton 600 mg kg⁻¹	900 mg kg⁻¹	1250 mg kg⁻¹	Maxicrop 2500 mg kg⁻¹	3750 mg kg⁻¹	LSD 0.05
TSS (%)	1995	18.41 a	17.40 b	16.60 c	17.46 b	17.32 b	17.02 bc	16.85 bc	0.616
	1996	17.60 a	16.50 abc	16.10 bc	17.25 ab	16.75 ab	16.60 bc	15.40 c	1.239
TDM (%)	1995	20.45 a	19.18 b	18.19 d	19.30 b	19.03 bc	18.69 c	17.44 e	0.344
	1996	19.93 a	18.96 abcd	18.31 bcd	19.77 ab	19.15 abc	18.07 cd	17.39 d	1.496
RS (g/100 ml)	1995	14.36 a	11.45 bc	10.83 bc	12.49 ab	12.61 ab	11.82 bc	10.06 c	1.966
	1996	14.24 a	11.41 cd	10.74 de	12.42 bc	13.38 ab	11.43 cd	10.30 e	0.993
AC (g/100 ml)	1995	1.041 c	1.158 ab	1.198 a	1.152 ab	1.084 bc	1.192 a	1.220 a	0.099
	1996	1.150 c	1.250 bc	1.280 abc	1.175 bc	1.230 bc	1.330 ab	1.430 a	0.156
BW (g)	1995	6.35 a	6.66 a	6.85 a	6.77 a	5.59 b	5.48 b	5.31 b	0.469
	1996	6.08 ab	6.58 a	6.74 a	6.66 a	5.92 b	5.63 b	5.54 b	0.629
Bun. W (g)	1995	532.0 c	608.2 b	660.2 a	632.7 ab	538.0 c	543.1 c	554.6 c	41.86
	1996	507.0 b	616.3 ab	720.8 a	659.1 ab	526.7 b	543.9 b	608.8 ab	147.7
NB	1995	83.7 c	91.2 bc	96.1 abc	93.3 abc	93.4 abc	101.2 ab	105.9 a	12.61
	1996	83.3 b	93.8 ab	107.3 a	98.8 ab	82.6 b	96.3 ab	108.8 a	22.04
NS	1995	1.31 c	1.59 b	1.81 a	1.69 b	1.28 c	1.68 b	1.87 a	0.107
	1996	1.30 d	1.95 abc	2.30 ab	2.08 ab	1.50 cd	1.75 bcd	2.35 a	0.519
RSB (%)	1995	38.14 a (38.12)	33.42 b (35.30)	32.00 b (34.45)	32.84 b (34.94)	31.40 b (34.08)	31.37 b (34.08)	30.52 b (33.52)	2.564
	1996	43.33 a (41.15)	37.54 abc (37.76)	34.58 abc (36.03)	39.18 ab (38.76)	29.19 abc (32.71)	24.63 bc (29.73)	21.65 c (27.76)	9.791

Data in parenthesis were transformed

bunch weight in 1995, only 600 mg kg⁻¹ Proton application increased bunch weight in 1996. The highest bunch weight was obtained from 600 mg kg⁻¹ Proton in both years.

While all seaweed extract applications decreased the ratio of spilled berry in 1995, only 2500 mg kg⁻¹ and 3750 mg kg⁻¹ Maxicrop treatments had significant effects on the ratio of spilled berry in 1996. The least ratio of spilled berry was obtained from 3750 mg kg⁻¹ Maxicrop application in both years. This treatment decreased the ratio of spilled berry by 19.9 % and 50.0 % in 1995 and 1996, respectively.

The reasons for the effects of seaweed extracts are not fully understood because the compositions of these products are complex . But, losses of the TSS, reducing sugar content of must and total dry matter content of berry may be attributed to shade effect because of the vegetative growth of the plants treated with seaweed extracts. The vines treated with seaweed extracts had a greater leaf area and more lateral shoots indicating that seaweed extracts may include cytokinins (Featonby-Smith and Staden, 1984; Avijit and Debashish, 1994).

Our results on berry weight were similar to results reported by Delas (1994). Delas determined that effects of seaweed extracts were different because of the fact that their contents were different and seaweed extracts had different effects from year to year. Furthermore, seaweed is also a plant and its composition should also vary for different times of the year (Villiers et al., 1983). On the other hand, the reason why Maxicrop applications decreased berry weight in 1995 may be that these treatments increased the number of berries per bunch in 1995 (Childers et al., 1995).

Biostimulant applications increased the number of berries per bunch and the number of seeds per berry. The results may be attributed to an increase in enzyme activity of pollen grain and/or in style during pollination because of the fact that seaweed extracts include enzyme and boron (Filiti et al., 1986; Verklej, 1992; Crouch and Staden, 1994).

Decreasing ratio of spilled berries may be attributed to increasing Ca content of berry and decreasing sugar content of must. Some researchers reported that seaweed extract applications increased Ca uptake (Villiers et al., 1983; Köse, 1997). However, many authors noted that ratio of spilled berries with a high Ca content was much lesser than in berries with a low Ca content (Dickinson et al., 1964). Considine and Kriedeman (1972) reported such negative relationship obtained between spilling of berry and sugar content of must. Furthermore, our results on the ratio of spilled berry were similar to Austin (1977) who determined that seaweed extract treatments decreased fruit cracking on apples, pears and plums.

More recent researches showed that seaweed extracts contain certain plant hormones, minerals, enzymes, vitamins and numerous unknown organic compounds. So, how effects of seaweed extracts do occur is not fully understood. Research should be planed because benefits can reasonably be expected from seaweed extracts.

4. References

Austin M (1977) Foliar feeding gave results in 1976. Hort. Abst., 47(9):Nr. 8108.

Avijit S and Debashish S (1994) Effect of Biozyme-a seaweed extract and N P K on the expression of agromorphological traits in wheat. Orissa Jour. Agric. Res.,7(1):46-48.

Berlyn GP and Russo RO (1990) The use of organic biostimulants in nitrogen fixing trees. Nitr. Fixing Trees Res. Reports 811- 2

Cassan L Jeannin I Lamaze T and Gaudry J F (1992) The effect of the A. nodosum extract Goemar GA14 on the growth of spinach. Bot. Marina, 35: 437-439.

Childers NF Morris JR and Sibbet GS (1995) Modern Fruit Science: Orchard and small fruit culture. Horticultural publications, Florida, pp. 632

Considine JA and Kriedemann PE (1972) Fruit splitting in grapes: Determination of critical turgor pressure. Aust. J. Agric. Res. 23: 17-24.

Crouh IJ and Staden JV (1994) Commercial seaweed products as biostimulants in horticulture. J. Home and Cons. Hor., 1 (1): 19-76.

Delas J (1994) Effets des biostimulants sur la physiologie de la vigne. Prog. Agric. et Viticole, 111 (18): 407-410.

Dickinson DB and McCollum JP (1964) The effect of Calcium on cracking in tomato fruits. Amer. Soc. Hort. Sci. 84: 485-490.

Düzgüneş O Kesici T Kavuncu O and Gürbüz F (1987) Araştırma ve Deneme Metotları (İstatistik metotları II) A.Ü. Ziraat Fak. Yay No: 1021, Ankara. pp. 381

Featonby-Smith BC and Staden JV (1984) Identification and seasonal variation of endogenous cytokinins in Ecklonia maxima. Botanica Marina, 27: 527- 531.

Filiti N Cristoferi G and Maini P (1986) Effects of biostimulants on fruit trees. Acta Horticulturae 179: 277-278.

Köse C (1997) Proton biostimulatörünün karaerik üzüm çeşidinin vejetatif ve generatif gelişimi üzerindeki etkileri (M.Sc.)

Rader JS Walser RH, Williams CF and Davis T (1985) Organic and conventional production and economic. Biological Agriculture and Horticulture 2:215-222.

Russo RO and Berlyn GP (1990) The use of organic biostimulants to low input sustainable agriculture. J. of Sust. Agriculture 1 (2): 9-42

Verkleij FN (1992) Seaweed extracts in agriculture and horticulture: a review. Biological Agriculture and Horticulture 8:309-324.

Villiers JD Kotze WAG and Joubert M (1983) Effect of seaweed foliar sprays on fruit quality and mineral nutrition. The Deciduous Fruit Grower, March 1983 97-101.

Chapter 50

Effects of compost material on yield and quality of glasshouse tomatoes grown in different textured soils

B.Okur[1], Y.Tüzel[2], S.Toksöz[1], D.Anaç[1]

[1]Ege University. Fac. of Agric. Soil Dept. 35100 Bornova-Izmir/Turkey

[2]Ege University. Fac. of Agric. Horticulture Dept. 35100 Bornova-Izmir/Turkey

Key words: organic farming, compost, tomato

Abstract: This research was established to study the effects of compost material, a supplement of organic matter, on tomato yield and fruit quality. Tomato plants were grown up in plastic bags (25 kg) containing different textured soils (sandy and clay) and different amounts of compost material (15-30 and 60 ton ha[-1]). The highest yield was determined in the clay soil+I dose compost parcel as 3.91 kg m[-2]. From fruit characteristics, total soluble solids, dry matter content and titratable acidity increased by compost applications, but pH and Vitamin C values were not affected by the treatments.

1. Introduction

Reutilization of plant residues in agricultural soils will be an important organic matter gain for soils in future. Since manure is very expensive due to its scarcity, farmers often search for other alternative organic sources. In the long run, plant production and soil fertility for sure will benefit from the use of organic soil additives (Brito and Santos, 1995).

Although the natural sources of organic matter in soils are indigenous plant and animal debris, cropping wastes and animal manures have been traditionally used as the only means of maintaining and increasing soil fertility and crop production. Mature compost is the end product of the stabilization stage and can be considered an organic fertilizer for general purposes, being fully suitable for application on the soils even in the presence of standing crops (Senesi, 1989).

This investigation was conducted to determine the effect of compost material on yield and fruit quality of glasshouse tomatoes grown in two different textured soils.

2. Materials and Methods

This research was carried in the glasshouses of Horticulture and Soil Science Department, Ege University.

Soil samples of the research were sandy and clay in texture. The compost material was obtained by composting the plant residues from agricultural lands. Some physical and chemical properties of the experimental soils and the compost material are given in Table 1.

Compost material was applied to a sandy and a clay soil in different doses (15-30 and 60 ton ha[-1]) and the mixtures were placed in plastic bags of 25 kg, seperately.

The experiment was conducted in 172 plastic bags totally; 8 applications, 4 replications, 6 plants making up each replication formed the layout of the randomized parcel design. Treatments were as follows:

Sand I   and Clay I   1.25 kg compost/bag
Sand II and Clay II   2.50 kg compost/bag
Sand III and Clay III   5.00 kg compost/bag
Sand IV and Clay IV (control) 25 kg soil/bag

Table 1. Physical and chemical properties of the experimental soils and the compost material.

	Menemen Soil	Bornova Soil	Compost
Texture	Sandy	Clayloam	-
pH	8.25	8.12	7.60
Total soluble salt %	0.045	0.090	1.15
$CaCO_3$ %	4.12	5.32	14.11
Organic matter %	1.00	2.30	68.7
CEC me 100g^{-1}	7.30	22.00	65.85
Total-N %	0.029	0.097	1.430
Available P mg kg^{-1}	0.6	5.5	92.07
Available K mg kg^{-1}	66	594	12400
Available Ca mg kg^{-1}	2097	3589	6050
Available Mg mg kg^{-1}	80	200	780

Tomato (cv.Elif) seeds were sown on the 16th of September, 1996 and were transplanted to the glasshouse on the 25th of October, 1996. Some physical and chemical properties of the soils were determined according to conventional methods. Also, the fruit quality parameters (pH, Vitamin C, titratable acidity, total soluble solids-TSS) were measured following Pearson (1970) and Hortwirth (1960).

Tomato harvests were made every 3 weeks and the total yield was determined in cumulative (kg m^{-2}).

3. Results and Discussion

Tomato harvest began on the 3rd of May 1997 and the total yield was evaluated by successive additions. In the first 3 weeks, the highest yield was obtained in the sandy control as 1.46 kg m^{-2} and the lowest yield in the clay II dose compost parcel as 0.649 kg m^{-2}. In this period the effect of compost doses on yield was significant at a level of 5 % (Table 2).

Table 2. The effect of compost material on early tomato yield.

Compost Dose	Sandy Soil	Clay Soil	Average
I	0.992 ns	1.052 ns	1.022 b
II	0.895 ns	0.649 ns	0.772 b
III	0.721 ns	0.825 ns	0.773 b
Control	1.466 ns	1.037 ns	1.252 a
Average	1.019 ns	0.891 ns	

ns: not significant

At the end of the 2nd, 3 weeks (6 weeks later) the highest yield (0.849 kg m^{-2}) was in the clay II dose compost parcel and the lowest yield (0.317 kg m^{-2}) in the sandy control. The effects of soil textures and compost doses on tomato yield were significant at a level of 5 % in this period too.

When average yield was considered, maximum amount was in the II dose compost application and the minimum in the control parcels (Table 3).

Table 3. The effect of compost material on tomato yield between 3 to 6 weeks.

Compost Dose	Sandy Soil	Clay Soil	Average
I	0.619 ns	0.722 ns	0.671 ab
II	0.579 ns	0.849 ns	0.714 a
III	0.499 ns	0.644 ns	0.571 abc
Control	0.317 ns	0.551 ns	0.434 c
Average	0.503 b	0.692 a	

ns: not significant

The effect of soil texture and compost levels on the yield regarding the 6 to 9 week period was statistically significant. If the average yield in relation to compost applications are considered, maximum yield was obtained in the II dose as 1.155 kg m^{-2} and the minimum yield in control parcels.

Tomato yield obtained between 6 to 9 weeks increased 24 % in the clay soil and 128 % in the sandy soil by the compost applications compared to the control parcels. This result markedly showed that compost applications to sandy soils increased the yield (Table 4).

Table 4. The effect of compost material on tomato yield between 6 to 9 weeks.

Compost Dose	Sandy Soil	Clay Soil	Average
I	0.982 ns	1.260 ns	1.121 a
II	1.140 ns	1.170 ns	1.155 a
III	1.037 ns	1.262 ns	1.150 a
Control	0.500 ns	1.019 ns	0.759 b
Average	0.915 b	1.178 a	

ns: not significant

A 52 % increase in average total yield was obtained by the compost applications compared to control. In recent years, composts of different organic wastes are used as the primary sources to prepare potting substrates (Verdonck, 1988). Average tomato yield obtained between 9 to 12 weeks was the highest in the clay II dose compost application (0.84 kg m⁻²) and minimum in the sandy III dose (0.406 kg m⁻²). Taking into consideration the average yields, minimum yield was determined in the control parcel (Table 5).

Table 5. The effect of compost material on tomato yield between 9 to 12 weeks.

Compost Dose	Sandy Soil	Clay Soil	Average
I	0.440 ns	0.870 ns	0.650 ns
II	0.620 ns	0.740 ns	0.680 ns
III	0.400 ns	0.840 ns	0.620 ns
Control	0.430 ns	0.492 ns	0.460 ns
Average	0.470 b	0.730 a	

ns: not significant

When the total yield of all the considered weeks was evaluated, the highest yield was determined in clay soil I dose compost parcel as 3.9 kg m⁻² and the lowest yield in sandy III dose compost parcel as 2.6 kg m⁻². II dose compost applied to the sandy soil lead to a maximum yield of 3.24 kg m⁻². When the average of two textures were evaluated, the increase due to the compost applications compared to control was 20 %. Abdel Sabour et al., (1996) showed that a single application of good quality organic compost to a sandy soil is beneficial with no potential risk of heavy metal contamination.

In this study, the quality parameters of fruits were also measured. The effect of compost doses on pH of tomato fruits was significant at the level of 5 % (Table 6).

Table 6. The effect of compost material on pH of tomato fruit of the 3ʳᵈ period.

Compost Dose	Sandy Soil	Clay Soil	Average
I	4.24 ns	4.22 ns	4.23 ns
II	4.26 ns	4.23 ns	4.24 ns
III	4.27 ns	4.27 ns	4.27 ns
Control	4.24 ns	4.22 ns	4.23 ns
Average	4.25 ns	4.23 ns	

ns: not significant

It was reported that the pH of tomato fruits is frequently around 4.30 (Gould, 1983) and should be maintained between 4.20 to 4.50 (Matthews and Denby, 1966).

Total soluble solids of the fruits were analysed in every harvest period. After the first harvest, this value in III dose was 10.62 % for the sandy soil and 10.79 % for the clay soil. After the second harvest, this parameter was the highest in the II and III dose compost applications. When the data of the 3ʳᵈ week was studied, maximum TSS was determined in the control parcel. Hanna (1961) reported that in 5 studied tomato varietes the highest measurement was 6.54 %. Lower and Thomson (1966) stated that a 2 % increase in TSS was of economic importance. According to the experiment results, TSS of tomato fruits was fairly high (Figure 1).

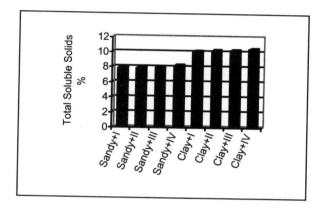

Figure 1. The effect of compost material on average TSS of tomato fruits obtained at the end of the 3ʳᵈ periods.

Titratable acidity of the fruits in relation to the compost applications was found highest in the II dose compost parcel as 23.32 %. Compost material applied to the clay soil increased the titratable acidity 10-40 % compared to the sandy soils. The lowest values were determined in the sandy control parcels (Figure 2). Vitamin C content of tomato fruits was generally high in the control parcels with the exception of the first harvest. While Vitamin C content of the treated fruits ranged from 12.5 to 22.67 mg 100g⁻¹, these values reached 18.19 to 26.35 mg 100g⁻¹ in the control parcels. Tolkybaev (1975) claimed that Vitamin C content of fresh tomatoes is generally 26.05 mg 100g⁻¹. The effects of compost applications on Vitamin C content of fruits were not significant. Warman and Havard (1996) report an experiment of 3 years in a sandy loam soil with potatoes and

222

sweetcorn grown under organic and conventional farm management. They state that the yield and Vitamin C content of the potatoes were not affected by the treatments.

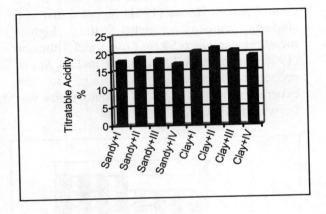

Figure 2. The effects of compost material on average titratable acidity of tomato fruits obtained at the end of the 3rd periods.

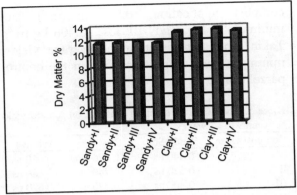

Figure 3 The effect of compost material on average dry matter content obtained at the end of the 3rd periods.

The effect of soil texture on dry matter content of tomatoes was significant at the level of 5 %. Especially, dry matter content of tomatoes grown in the clay soils was 7.4 % higher than those grown in the sandy soils. The highest measurement was obtained in the clay II dose compost application as 12.96 %. According to the results of final week, dry matter content of tomatoes grown in the clay soil was 42 % higher than those grown in the sandy soils. Maximum dry matter content of tomatoes was measured in the clay III dose compost application as 15.10 % (Figure 3).

It is concluded that, compost materials applied to different textured soils in different doses increased the tomato yield. In addition, the fruit characteristics, TSS (%), dry matter content and titratable acidity increased by compost applications however pH and Vitamin C were not affected.

4. References

Abdel Sabour, M. F., A.A.G. Abdalla, M.F.Z. Emara, 1996. Residual Effects of Single Organic Compost Application to Sandy Soil on Trace Elements and Heavy Metals Content of Common Bean Seeds. Bulletin of Faculty of Agric. Univ. of Cairo. 49:1, 153-168.

Brito, L.M.C.M., J.Q.D. Santos, 1995. Cellulose Sludges and Urban Wastes as Organic Soil Additivies: Quantitative Studies on Their Effects on Lettuce (*Lactuca sativa L.*) and Cabbage (*Brassica oleracea var. capitata*). Revista de Ciencias Agrarias (18) 3:65-78 Lisboa, Portugal.

Hortwirth, W., 1960. Official Methods of Analysis A.O.A.C. Chapter 29 Sugar and Sugar Products. A.O.A.C. Benjamin Fraklin Station. Washington.

Pearson, D., 1970. The Chemical Analysis of Foods. Auxill, London.

Senesi, N., 1989. Composted Materials as Organic Fertilizers. The Science of the Total Environment 81/82 (521-542), The Netherlands.

Tolkybaev, Z.H., 1975. Tomato Fruit Quality in Relation to Rates of Nitrogen Fertilizers in Karakalpakia. *Hort. Sci* 45(6).

Verdonck, O., 1988. Compost From Organic Waste Materials as Substitutes for the Usual Horticultural Substrates. Biological Wastes 26, 325-330. Great Britain.

Warman, P.R., K.A. Havard 1996. Yield, Vitamin and Mineral Contents of Organically and Conventionally Grown Potatoes and Sweetcorn. Agric., Ecosystems and Environment. 68:207-216. Canada.

Chapter 51

Soil pollution by nitrates using sewage sludge and mineral fertilizers

J.M.C.Brito, D.Ferreira, C.A.C.Guerrero, A.V.Machado, and J. Beltrão
Unidade de Ciências e Tecnologias Agrárias, Universidade do Algarve - Campus de Gambelas
Faro – Portugal

Key words: nitrate-leaching, nitrogen use efficiency, sewage sludge, fertigation, and mineral fertilizers

Abstract: With the objective of evaluating soil pollution by nitrates, due to nitrogen leaching, several treatments using urban sludge and two mineral nitrogen fertilizers – ammonium sulfate applied as basic fertilization and potassium nitrate, applied by fertigation – were applied to fertilize cucumber plants (*Cucumis sativus*, L.). An alluvial soil was used; pH was 7.9 and its organic matter content was a middle value. Using fertigation (potassium nitrate), the best yield and the highest nitrogen use efficiency were obtained and only an insignificant amount of nitrogen was leached. Amounts of ammonium sulfate of 3.5 ton.ha^{-1} were toxic to the plants, and their nitrogen use efficiency was about 10%. On the other hand, the urban sludge treatment showed a nitrogen efficiency of 45% and there was practically not any nitrogen loss by leaching.

1. Introduction

To maximize profits and increase yields farmers use more and more fertilizers, namely those containing nitrogen since this element can produce spectacular effects on plant growth. For this reason there is an increase of the use of those fertilizers, which can cause soil and water pollution.

In areas where animal production is absent, farmers generally do not apply manure, and if they want it they must buy it far away from their sites, which leads to increasing costs. Simultaneously, human activities produce high amounts of organic residues, which can be used, after the necessary treatment, in the agriculture activity (Carrasco de Brito, 1986). Examples are, for instance, the urban solid residues and urban sludges.

An irrational application of fertilizers to crops is one of the main causes of soil and water pollution. An example of this is the nitrate contamination of soil and groundwater. This trend is associated with a growing population and with more intensive agriculture. High levels of nitrates in groundwater are known to result mainly from fertilizers applied to crops, manure generated by concentrated animal production, septic systems and sewage treatment plants. Additionally there are some other contamination factors as follows (Franco *et al.*, 1993): a) soil characteristics; b) irrigation practices that lead to large volumes of deep percolation increasing the leaching of nitrates; c) crops needing heavy nitrogen fertilization and frequent irrigation; d) concentrated heavy rains and mild temperature lead to more leaching of nitrates; e) small distance from the root zone to groundwater means a more immediate problem; f) population density; g) and availability of an alternate water supply.

The last two factors appear to be the most important problems since nitrate concentration is critical if groundwater is used for human and animal drinking supplies. One well-known potential threat is the relationship between high nitrate levels in drinking water and an infant disease called methemoglobinemia – blue baby syndrome. Also gastric cancer in human and birth defects in human and animals have been subject of concern in relation to high levels of nitrates in drinking water. However, there is yet no firm link established (Bruning-Fann & Kaneene, 1993).

High levels of nitrates in water are highly important in areas where no public water supply exists. This situation is typical on rural areas, in which the main activity is agriculture, and there are no adequate septic systems or sewage treatment plants. An example of this is the rural area called "Campina de Faro", surrounding Faro, in the south of Portugal – Algarve – where a study carried out in 1994-1995 showed that 94% of the monitored wells had nitrate levels over the Portuguese legal limit of 50 mgl^{-1} NO_3^- (Guerrero, 1996).

Since agricultural practices, namely fertilization and irrigation management, appear to be one of the problems of nitrate pollution, this study might give a small contribution for the mentioned subject. The mean goal was to study the effects of the use of sewage sludges and two inorganic fertilizers (ammonium sulfate and potassium nitrate) on the production of cucumber plants in pots. Parameters as crop yield, nitrogen use efficiency, and nitrate leaching are studied here.

2. Materials and Methods

Soil The experimental work was realized in randomized pots. These were filled with an Eutric Fluvisol (FLe) according to the FAO-UNESCO Soil Classification, for soils composed of sand without calcareous material.

The soil was divided into 30 sub-samples, each one of 7.0 kg for growth tests. One soil sample of 0.5 kg was collected for some physical and chemical analyses (Table 1).

Urban sludge and inorganic fertilizers The urban sludge was collected at the water treatment of plant Vale do Lobo, Loulé, dried at an average temperature of 25°C for 1 week and ground in a laboratory grinder to an average size of 10 mm. One sample of 0.5 kg was collected for physical and chemical analyses (Table 2).

The inorganic fertilizers used were ammonium sulfate, with 20.5% of N, as basic

fertilization, and potassium nitrate, with 13% of N, applied using fertigation.

Table 1. Soil properties

Parameters	Method	Values
Texture	Sodium hexam.	sandy-loam
pH	Potentiometry	7.9
Elec. Con. (dS.m^{-1})	Potentiometry	1.1
Organic Matter (%)	Walkley-Black	2.0
N (%)	Kjeldhal	0.1
P_2O_5 (mg.kg^{-1} soil)	Olsen	81.8
K_2O (mg.kg^{-1} soil)	Ammon. Acetate	215.8

Table 2. Sewage sludge properties

Parameters	Method	Values
pH	Potentiometry	6.7
Elect. Cond. (dS.m^{-1})	Potentiometry	5.2
Organic Matter (%)	Calcination	73.1
C/N		9.0
N (%)	Kjeldahl	4.0
P_2O_5 (%)	Photometry	1.7
K_2O (%)	Flame photometry	0.4

Urban sludge and inorganic fertilizer applications Urban sludge was separately added to 9 sub-samples of the soil. To another 9 sub-samples were added the ammonium sulfate as basic fertilization; 9 more sub-samples of the soil were prepared for fertigation during vegetative cycle of the cucumber culture. The urban sludge and inorganic fertilizers were added to those samples according to an increasing scale of nitrogen (Table 3). The last three sub-samples were used as control. Each experimental amount of urban sludge and inorganic fertilizers was assayed in triplicate. After this, all soil samples were placed in Mitsherlich pots, containing 10 cm of limestone to allow water drainage, if necessary.

Table 3. Amounts of urban sludges and inorganic fertilizers applied by treatment (g per pot)

Treat.	N	Ur. Slud (US)	$(NH_4)_2SO_4$ (AS)	KNO_3. (PN)
Control	0		0	
I	1.2	30	5.8	9.2
II	2.4	60	11.7	18.4
III	3.6	90	17.5	27.6

Planting conditions After plantation (February 1997), all pots were irrigated at field capacity, and maintained in this condition during the experiment. If some water was lost through the bottom of the pot, it was immediately replaced on the surface of the soil to avoid water and nutrient losses. The pots were placed in a greenhouse at the Experimental Field of the Algarve University, and kept at an average temperature of 25°C.

Fresh weight of fruit and leaves, nitrogen use efficiency and nitrate losses During April and May 1997 all cucumbers fruit and leaves were harvested and weighed (fresh weight). Afterwards, the samples were dried separately at 105 ±2°C to constant weight (dry weight). Those vegetable samples were ground in laboratory for chemical analyses of nitrogen (the Kjeldahl method was used). Total dried weight of leaves and fruits multiplied by the respectively nitrogen contents gives the approximate amount of nitrogen absorbed and the use efficiency of this element from both plant components. At the end of the harvest period all pots were irrigated with 1L of water. This amount produced a determined volume of drainage water that was collected. In these the nitrate content was determined by HPLC. This value and the volume of leaching give the losses of nitrate. Fresh weight results, N use efficiency and losses of nitrates were submitted to variance analysis (ANOVA), with a significant level of 95%, and Duncan test to define the significant differences.

3. Results and Discussions

From Duncan test results it can be concluded that the average crop yield obtained with the highest level of potassium nitrate was different from all other treatments. Also the average crop yields obtained in all treatments were different from the control, with the exception of the highest level of ammonium sulfate treatment (ASIII) (Figure 1). With the exception of PNIII and ASIII treatments, and control, the increased application of fertilizers did not cause significant differences in the cucumber crop yield. The increasing urban sludge and fertigation applications generally caused significant differences between average productions of cucumber leaves (Figure 1). Thus the highest production of leaves was obtained in the PN III treatment (as it occurred for crop yield) showing that cucumber plant should absorb preferentially the nitrate-nitrogen form. In contrast the highest level of ammonium sulfate may have caused some toxic effects on cucumber plants.

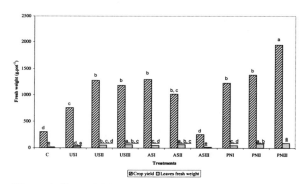

Figure 1. Crop yield and cucumber fresh weight of leaves

The amounts of N absorbed in fruits increased in the urban sludge and fertigation treatments. Although there were not significant differences on the urban sludge treatment results. Also USI, ASIII and NOI treatments did not differ from the control (Figure 2). Analyzing the ammonium sulfate treatments, the nitrogen absorbed decreased with the highest application of this fertilizer. The toxic effects produced by the ammonium-nitrogen form resulted in a decrease of fruit production and obviously in a decrease of the amounts of nitrogen absorbed. The average nitrogen absorption in leaves was significantly different from the control only in PNII and PNIII treatments (Figure 2).

226

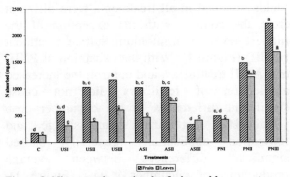

Figure 2. Nitrogen absorption by fruits and leaves

Figure 3 shows the effects of the increasing application of fertilizers on the nitrogen use efficiency and losses of the same element in the nitrate form. The average nitrogen absorbed in fruits and leaves of the control was deducted to the average nitrogen absorbed in the other treatments. According to the amounts of nitrogen applied by the urban sludge, ammonium sulfate and potassium nitrate treatments, the percentage losses of nitrogen differed from 0.33% to 56.2% (Figure 3).

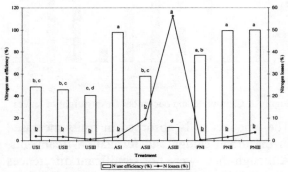

Figure 3. Nitrogen use efficiency and percentage of nitrogen losses

With the exception of the highest level of ammonium sulfate (3.2 g/pot N), which lead to the highest loss of nitrogen (56.2%), no significant differences occurred in the average nitrogen loss results. In the ASIII treatment the ammonium-nitrogen form, which have caused toxic effects, was not absorbed by the plants. If this phenomenon occurred the ammonium-nitrogen form might have been oxidized to nitrate, which is soluble and could have been leached through the soil.

Analyzing the average nitrogen use efficiency we can conclude that: the best values were obtained with the inorganic fertilizer application with exception of the ASIII treatment (toxic effects). Fertigation treatments obtained the highest average nitrogen use efficiency (75% - 97%.). The urban sludge treatment showed an average nitrogen use of 40% to 50%, approximately.

4. Discussion

The applications of nitrate-nitrogen in fertigation promoted a good production of leaves and the best crop yield. The nitrogen use efficiency was from 75% to 97% with losses of nitrate at these controlled plant conditions. Ammonium sulfate at 3.6 g.pot^{-1} N (an equivalent of 3.5 ton.ha^{-1}) might be toxic to cucumber plants; this nitrogen form remaining in the soil must be converted to nitrate which can be submitted to leaching process. In fact these results show that fractionated fertilization and controlled water supply are two important procedures to reduce the risks of nutrient losses and the probability by groundwater contamination of nitrates and other soluble contaminants. The nitrogen use efficiency from urban sludge was near 45%. The organic mineralization of the sludge slowly releases the nitrogen to the culture, avoiding nutrient losses through the soil profile.

As concluding remarks it was shown that nutrient and water management are two important conditions to reduce groundwater pollution. Nevertheless, according to the requirements of the plants, there are a few other requisites that must not be forgotten in agriculture, namely soil characteristics, climate, according to the above descriptions.

Acknowledgements

The authors are grateful to Dr. Graça Miguel and Dr. Denise Martins from the Algarve University for the dispensed time in HPLC nitrate determination.

This work were partially supported by the Portuguese Project PRAXIS XXI/3/3.2/ HORT/2146/95, entitled "Novos" Fertilizantes Orgânicos em Horticultura Intensiva.

5. References

Brunning-Fann, CS & Kaneene, JB (1993). The effects of Nitrate, Nitrite and N-Nitroso compounds on human health: A review. Veterinary and Human Toxicology, 35 (5): 521-538.

Carrasco de Brito, JM (1986). As lamas pretas como fertilizante: contribuição para o seu estudo. PhD thesis, Instituto Superior de Agronomia, Lisboa, Portugal.

Franco, J, Schad, S & Cady, CW (1993). California's experience with a voluntary approach to reducing nitrate contamination of groundwater: The Fertilizer Research and Education Program (FREP). Journal of Soil and Water Conservation: 76-81.

Guerrero, CAC (1996). Estudo sobre o comportamento do azoto na cultura do pimenteiro (*Capsicum annum*, L.) – Aspectos relacionados com a poluição das águas subterrâneas. Master thesis, Instituto Superior de Agronomia, Lisboa, Portugal.

Chapter 52

Macro and micro nutrient contents of tomato (*Lycopersicon esculentum*) and eggplant (*Solanum melongena* var. *Esculentum*) seedlings and their effects on seedling growth in relation to humic acid application

A. Dursun[1], İ. Güvenç[1], M. Turan[2]

[1]*University of Atatürk, Faculty of Agriculture, Deparment of Horticulture, 25240-Erzurum/Turkey*
[2]*University of Atatürk, Faculty of Agriculture, Deparment of Soil Science, 25240-Erzurum/Turkey*

Key words: macro and micro elements, tomato, eggplant, seedling growth, humic acid

Abstract: Plant performance mainly depends on characteristics of photosynthetically active organs. Many of these characteristics such as growth performance and quality of seedlings are modified by ecological conditions and growing techniques. Production of vegetable crops from seedlings is commonly practiced by vegetable growers. This study was conducted to determine macro and micro nutrient contents of tomato (*Lycopersicon esculentum*) and eggplant (*Solanum melongena* var. *esculentum*) and their effects on seedlings in relation to humic acid applications. Different levels of humic acid (50, 100, 150, 200 ml l^{-1}) were applied to growth media (peat) after transplanting of seedlings of the species every ten days by the time of planting. Macro and micro plant nutrient contents of both species increased with humic acid applications compared to the control.

1. Introduction

Tremblay and Senecal (1998), Melton and Dufault (1991) and Nicola and Bassoccu (1994) reported on vegetable seedling growth and its effect on yield in response to different fertilization levels in nursery. Effects of fertilizers on yield, quality and nutrient content of some vegetables were also studied by Csizinszky (1986) on tomato, Csizinszky et al. (1990) on pepper and tomato, Alan and Güvenç (1994) on bean, Padem et al. (1995) on lettuce, Eris et al. (1995) on pepper, Güvenç and Padem (1995) on tomato and Böhme and Thi Lua, (1997) on tomato.

Böhme and Thi Lua (1997) reported that humic acids had beneficial effects on the nutrient uptake by plants, and they were particularly important to the transport and availability of micro elements in the plant.

Humic substances (humic and fulvic acids) comprise 65-70 % of the organic matter in soils. These compounds are products of the decomposition of plant tissues, and they are predominantly derived from lignified cell walls. The major functional groups of humic acids include carboxyles, phenolic hydroxyles, alcoholic hydroxyles, ketones, and quinones (Russo and Berlyn, 1990). The mechanism of humic acid action in promoting plant growth is not completely known, but several explanations have been given by some authors (Vaugham, 1974, Cacco and Dell Agnolla, 1984 and Russo and Berlyn, 1990) such as: increasing of cell membrane permeability, oxygen uptake, respiration and photosynthesis, phosphorus uptake, root and cell elongation, ion transport and acting as cytokinin like substances.

An experiment was carried out with the aim

of finding the effects of different levels of humic acid on macro and micro nutrient contents of tomato and eggplant seedlings

2. Materials and Methods

This study was carried out to determine the effects of different levels of humic acid on macro and micro nutrient contents of tomato (cv. Sandoz-182) and eggplant (cv. Pala Yalova-49) seedlings under greenhouse conditions at the Department of Horticulture of the College of Agriculture, Atatürk University, Erzurum, in 1997.

Different levels of humic acid (50, 100, 150 and 200 ml l^{-1}) were applied to the growth media. This humic acid (HA) contained 5 % N, P_2O_5 and K_2O; 1 % Ca, Fe, Zn, and Mg; 200 mg kg^{-1} Mn, 90 mg kg^{-1} B, 14 mg kg^{-1} Cu, 5 mg kg^{-1} Co, and 8 mg kg^{-1} Mo.

Peat in 1 liter pots was used as growth medium. Experimental layout was a completely randomized design with three replicates and each replication had 30 plants. Data were subjected to analysis of variance and the means were compared according to Duncan Test (Düzgüneş al., 1987) except for application of 200 ml l^{-1} HA in tomato seedling because of its inhibition effect (Table 1).

Seeds of both species were sown in the first week of March, and seedlings were transplanted in the first week of April in 1997. Humic Acid was applied four times to the peat after transplanting of seedlings with ten day intervals, starting from transplanting.

Macro and micro (N, P, K, Ca, Mg, Na, Cu, Fe, Mn, Zn, and Cu) plant nutrient contents of seedlings were determined in both species (Kacar, 1972).

3. Result and Discussion

Nutrient contents

N content: Treatments and cultivars significantly affected N content in tomato and eggplant leaf tissues (Tables 1 and 2).The highest N contents (5.34, 7.64 and 8.10%) were obtained from 100 ml l^{-1} HA application in tomato and 150 and 200 ml l^{-1} HA applications in eggplant, respectively. All humic acid treatments in both cultivars significantly increased N contents of leaves in comparison to the controls.

P content: Phosphorus contents of tomato (0.83%) and eggplant (0.37%) were higher in the plants treated by the 100ml l^{-1} and 200 ml l^{-1} HA applications than the other applications and controls, respectively (Tables 1 and 2). There were statistically significant differences among them.

K content: The highest K content in both species occurred in the plants treated with 100 and 150 ml l^{-1} HA application in tomato (3.52%) and 200 ml l^{-1} HA application in eggplant (2.99%) as compared to the controls, and there were significant differences among the other applications and controls (Tables 1 and 2).

Na content: All HA applications in tomato and 200 ml l^{-1} HA application in eggplant significantly increased the content of Na according to the control in tomato and the other applications and control in eggplant (Tables 1 and 2).

Mg content: Mg content of tomato and eggplant significantly increased in the plants treated by HA and the highest Mg contents of tomato and eggplant (0.59% and 1.15%) were obtained from 100 ml l^{-1} HA and 200 ml l^{-1} HA applications, respectively, as compared to the controls, (Tables 1 and 2).

Ca content: The Ca contents of tomato and eggplant were significantly influenced by HA applications and the highest Ca contents were determined in 150 ml l^{-1} HA application in tomato (3.52%) and 200 ml l^{-1} HA in eggplant (6.68%) when they were compared to the controls (Tables 1 and 2).

Fe content: HA applications significantly affected the contents of Fe in tomato and eggplant when compared to the controls. The highest values were obtained from 150 ml l^{-1} HA application in tomato (305.7 mg kg^{-1}) and

Table 1. The effects of humic acid on macro and micro elements in seedlings of tomato.

Treatments	N (%)	P (%)	K (%)	Na (%)	Mg (%)	Ca (%)	Fe (mg kg⁻¹)	Mn (mg kg⁻¹)	Zn (mg kg⁻¹)	Cu (mg kg⁻¹)
50 ml l⁻¹	3.52b	0.60b	2.52b	0.94a	0.57ab	3.31a	222.2b	96.15ab	207.1c	16.67^{NS}
100 ml l⁻¹	5.34a	0.83a	3.52a	1.08a	0.59a	3.51a	287.6b	106.24a	222.5bc	18.54
150 ml l⁻¹	4.08b	0.60b	3.45a	0.94a	0.51c	3.52a	305.7ab	120.47a	277.4a	18.55
Control	1.14c	0.43b	2.31b	0.36b	0.56b	2.63b	132.7c	76.45b	189.8c	14.81

NS: Not Significant (LSD p=0.05 level).

Table 2. The effects of humic acid on macro and micro elements in seedlings of eggplant.

Treatments	N (%)	P (%)	K (%)	Na (%)	Mg (%)	Ca (%)	Fe (mg kg⁻¹)	Mn (mg kg-1)	Zn (mg kg⁻¹)	Cu (mg kg⁻¹)
50 ml l⁻¹	5.40c	0.29c	2.27d	0.56d	0.98b	3.65c	153.5b	25.23bc	165.1bc	14.25ab
100 ml l⁻¹	6.89b	0.31bc	2.50bc	0.77bc	1.01b	4.81b	146.1b	34.53abc	203.3bc	15.21a
150 ml l⁻¹	7.64a	0.32b	2.58b	0.96b	1.02b	5.57b	195.0a	38.56ab	221.1ab	15.19a
200 ml l⁻¹	8.10a	0.37a	2.99a	1.49a	1.15a	6.68a	227.5a	50.81a	246.1ab	14.08ab
Control	2.49d	0.27d	2.08d	0.38d	0.85c	2.85c	135.9b	16.38c	117.96c	10.05ab

200 ml l⁻¹ HA application in eggplant (227.5mg kg⁻¹) (Tables 1 and 2).

Mn content: Mn content of tomato and eggplant significantly increased in the plants treated by HA applications. The highest Mn contents of tomato and eggplant (120.47mg kg⁻¹ and 50.81 mg kg⁻¹) were obtained from 150ml l⁻¹ HA and 200ml l⁻¹ HA applications as compared to the controls, respectively (Tables 1 and 2).

Zn content: The highest Zn contents of tomato (277.4mg kg⁻¹) and eggplant (246.1mg kg⁻¹) were determined in the plants treated by 150 ml l⁻¹ and 200 ml l⁻¹ HA applications, respectively (Tables 1 and 2) and the applications significantly affected the contents of Zn in both cultivars when compared to the controls.

Cu content: The highest Cu content in tomato (18.55mg kg⁻¹) was determined from 150 ml l⁻¹ HA application. However, no significant effect of HA applications on the Cu content was obtained in tomato. HA applications significantly affected the contents of Cu in eggplant when compared to the control. The highest value was obtained from 100 ml l⁻¹ HA application in eggplant (15.21mg kg⁻¹) (Tables 1 and 2).

Seedling growth in relation to nutrient content
Depending on the levels of HA applications, N, K, Na, Fe, Mn, Zn, Ca and Cu contents of

plants significantly increased. Increasing N contents of the plants affected P, Na, Ca and Fe contents of plants and also significantly affected K, Mn, Cu contents and the number of leaves of plants. Na and Ca contents of the plants affected P contents of the plants and there were relations between K, Fe and the number of leaves of the plants. Based on Mg contents of the plants, the number of leaves, leaf width, stem diameter, stem fresh and dry weight, and root fresh and dry weight of plants significantly increased. Zn contents of plants negatively affected stem diameter, stem and root lengths and root fresh and dry weights of the plants.

Humic acid applications affected plant nutrient elements N, P, K, Na, Mg, Ca, Fe, Mn, Zn and Cu of tomato and eggplant seedlings. There were significant effects of humic acid applications on plant growth and nutrient contents as compared to the controls. Similar results were reported by some authors such as Csizinszky (1990) on bell pepper, Csizinszky et al. (1990) on pepper and tomato, Verkleij (1992) on horticultural crops, Sanchez-Andreu et al. (1994) and Böhme and Thi a Lua (1997) on tomato.

In conclusion, the results of this study indicated that humic acid applications to peat increased plant growth and nutrient contents in both species. Therefore, it was recommended to the vegetable growers.

232

4. References

Alan R and Güvenç • (1994) Influence of foliar sprays and application time of day on pod yield and yield components of snap beans (*Phaseolus vulgaris* L.) cv. K•z•lhaç and Yalova-17. Tr. J. of Agriculture and Forestry, 18:27-32.

Böhme M and Thi Lua H (1997) Influence of mineral and organic treatments in the rhizosphere on the growth of tomato plants. Acta Horticulturae, 450:161-168.

Cacco G and Dell Agnolla G (1984) Plant growth regulator activity of soluble humic complexes. Canadian J. of Soil Science, 64:225-228.

Csizinszky AA (1986) Response of tomatoes to foliar biostimulant sprays. Proc. Fla. State Hort. Soc., 99:353-358.

Csizinszky A A (1990) Response of two bell pepper (Capsicum annum L.) cultivars to foliar and soil-applied biostimulants. Soil and Crop Sci. Soc. Fla. Proc. 49:199-203.

Csizinszky AA Stanley CD and Clark GA (1990) Foliar and soil-applied biostimulant studies with micro irrigated pepper and tomato. Proc. Fla. State Hort. Soc., 103:113-117.

Düzgüne• O Kesici T Kavuncu O and Gürbüz F (1987) Deneme Metotlar• Ankara Üniv. Ziraat Fakültesi Yay., No:1021, Ankara.

Eris A Sivritepe HÖ and Sivritepe N (1995) The effects of seaweed (*Ascophyllum nodosum*) extract on yield and quality criteria in peppers. Acta Horticulturae, 412:184-192.

Güvenç • and Padem H (1995) The effect of foliar application of nitrogen sources on yield and yield components of tomatoes. Soil Fertility and Fertilizer Management 9th International Symposium of CIEC, 22-30 September, Ku•adas•, p 243-246.

Kacar B (1972) Bitki ve Topra••n Kimyasal Analizleri. Ankara Univ. Bas•mevi, Ankara.

Melton RR and Dufault R.J (1991) Nitrogen, phosphorus and potassium fertility regimes affect tomato transplant growth. HortScience(2):141-142.

Nicola S and Basoccu L (1994) Nitrogen and N, P, K relation affect tomato seedling growth, yield and earliness. Acta Horticulturae, 357:95-102.

Padem H and Alan R (1994) The effect of some substrate and foliar fertilizers on growth and chemical composition of peppers under greenhouse. Acta Horticulture, 366:453-460.

Padem H Alan R and. Güvenç • (1995) The effect of foliar fertilizers on yield, chlorophyll and nutrient contents of lettuce (*Lactuca sativa* L.). Soil Fertility and Fertilizer Management 9th International Symposium of CIEC, 22-30 September, Ku•adas•, p 175-180.

Russo R O and Berlyn GP (1990) The use of organic biostimulants to help low input sustainable agriculture. J. of Sust. Agric., 1(2):19-42.

Sanchez-Andre, J Jorda J and Juarez M (1994) Humic substances incidence on crop fertility. Acta Horticulture, 357:303-316.

Tremblay N and Senecal M (1988) Nitrogen and potassium in nutrient solution influence seedling growth of four vegetable species. HortScience 23(6):1018-1020.

Vaugham D (1974) Possible mechanism for humic acid action on cell elongation in root segments of *Pisum sativum* under aseptic conditions. Soil Biology and Biochemistry, 6(4): 241-247.

Verkleij F N (1992) Seaweed extracts in agriculture and horticulture: a review Biological Agriculture and Horticulture, 8:309-324.

Chapter 53

Calcium cyanamide – a unique source of nitrogen promoting healthy growth and improving crop quality of vegetables

H.-J. Klasse

Skw Trostberg Ag, P.O. Box 1262, D-83303 Trostberg, Germany

Key words: calcium cyanamide, nitrification, nitrogen fertilization, nitrate content, *Phytophthora capsici*, *Plasmodiophora brassicae*, *Sclerotinia sclerotiorum*, tipburn of lettuce

Abstract: Recent research has shown that the quality of vegetable crops can be improved by applying the nitrogen fertilization one or two weeks before sowing or planting in form of calcium cyanamide. Due to its retardent nitrification calcium cyanamide can reduce the nitrate content of the vegetables as well as tipburn on lettuce crops. In addition this fertilizer application has proved to be able to reduce occurrence of several soil borne diseases such as sclerotinia rot of lettuce, root collar rot on red pepper and clubroot on brassicas.

1. Introduction

Calcium cyanamide has been produced since 1905 and was the first industrial nitrogen fertilizer. Nowadays it is sold as a granuled fertilizer under the brandname PERLKA® containing 19.8 % N total nitrogen and having a liming value equivalent to 50 % CaO. Calcium cyanamide differs from all other mineral fertilizers as it first has to undergo a metabolism in the soil to become a source of plant available nitrogen (Figure 1).

The first intermediate formed one or two days after the fertilizer has been in contact with the soil's humidity is hydrogen cyanamide. This is phytotoxic and to some extent has fungicidal, herbicidal and molluscicidal properties. Therefore a waiting period has to be observed between fertilizer application and sowing or planting. Consequently this cyanamide phase can be used to diminish germinating weeds and to decrease the viability of some important soil borne pathogens. Within one to two weeks the cyanamide is converted totally into well

tolerated nitrogen forms such as urea and ammonium.

Another peculiarity of calcium cyanamide is its retarded and even conversion into nitrate. A minor pathway of hydrogen cyanamide in the soil leads to the formation of dicyandiamide which is known to act as an inhibitor of those soil bacteria responsible for nitrification. By this way the nitrogen in calcium cyanamide remains longer in the ammonium form than that of common fertilizers. Ammonium is adsorbed by the soil and is not leached out from the root zone but can be taken up by the crop gradually over a longer period.

In recent years calcium cyanamide has gained increased interest particularly in the production of vegetables. Its slow conversion into nitrate fits perfectly to the need of most vegetable crops. The even and well balanced nitrogen nutrition promotes healthy growth and results in vigorous crops. In addition the suppressing effects of calcium cyanamide on soil borne diseases can help to minimize the use of fungicides and can increase the marketable yield as well as the quality of vegetable crops. Thus fertilization with calcium cyanamide is regarded as an important cultural practice in the system of Integrated Pest Management (IPM).

2. Results

Lettuce

Scientists in Denmark have investigated the influence of different fertilizer applications on yield and quality of butterhead lettuce (Willumsen and McCall, 1997). 80 kg of nitrogen per hectare were applied either as calcium cyanamide 10 days before planting or with other fertilizers in split applications.

The results in table 1 indicate that there were no significant differences in the yield (fresh weight and dry matter) but significant differences in the quality depending on the fertilizer application. Compared to all other treatments calcium cyanamide significantly reduced the undesireable nitrate content in the lettuce heads. Similar results had been achieved earlier in studies on lettuce and kohlrabi at the Technical University of Munich-Weihenstephan (Venter, 1978; Venter and Fritz, 1979).

In addition the Danish trial revealed that the preplant application of calcium cyanamide significantly reduced the occurrence of tipburn on lettuce compared to split applications of calcium nitrate. Tipburn is a major quality problem in lettuce production and reduces the marketability of the crop. Tipburn is not caused by a disease but has to be regarded as a nutrient disorder resulting from a temporary calcium deficiency. Several factors are known to promote the occurrence of tipburn: Uneven supply of water, excessive or uneven nitrogen supply, insufficient root growth and delayed harvest. The reduction of tipburn through calcium cyanamide can be explained by the sustaining and even nitrogen supply to the plant and the stimulation of root growth through the ammonium based nitrogen nutrition in the first weeks after planting. In addition reasonable quantities of plant available calcium are

released into the soil water during the transformation of calcium cyanamide.

The soil borne fungus *Sclerotinia sclerotiorum* can cause severe crop losses (bottom rot, white mould) and is hard to control with fungicides. The resting bodies (sclerotia) of this pathogen remain viable in the soil for several years. From the literature it is known that calcium cyanamide can inhibit the germination of these sclerotia (Huang and Sun 1991; Mattusch, 1984).

Several studies were performed on lettuce to evaluate the influence of calcium cyanamide on the occurrence of bottom end rot. Applications of 60 and 100 kg.ha^{-1} of nitrogen in the form of calcium cyanamide two weeks before planting reduced the number of infected lettuce plants from 13.3 % (control, standard nitrogen fertilizer) to 2.2 and 0.8 % respectively. At the same time the percentage of marketable lettuce heads increased from 59.7 % (control) to 74.2 % for the lower and 79.7 % for the higher application rate of calcium cyanamide (Ilovai and Ceglarska-Hódi, 1992). These results were confirmed by other studies: In Western Australia calcium cyanamide reduced Sclerotinia infection on iceberg lettuce from 4.8 % on the standard fertilizer treatment to 1.6 % (Klasse, 1998). In Scotland applications of calcium cyanamide were also associated with substantial reductions in the levels of Sclerotinia disease in field trials conducted with iceberg lettuce, further experimentation is in progress (McQuillken, 1998).

Red peppers

In a two year study on red peppers in the Kahramanmaras region of Turkey it has been found that applications of 100 kg ha^{-1} nitrogen in form of calcium cyanamide two weeks before sowing can reduce root collar rot caused by *Phytophthora capsici* even more effectively than repeated fungicide applications and resulted in significant yield increase. However, excessive application rates or inadequate waiting periods have resulted in yield reductions due to the phytotoxicity of the cyanamide (Koç and Eskalen, 1996).

Brassica crops

The suppressing effect of calcium cyanamide on the fungus *Plasmodiophora brassicae* which causes clubroot on all types of brassica crops like cabbage, cauliflower and broccoli has often been described in literature (Humpherson-Jones et al. 1992, Klasse 1996). Recent research has been carried out in Australia by the Horticultural Research and Development Corporation HRDC. Calcium cyanamide has been included in a field trial programme testing the influence of different fertilizers, soil activants and fungicides on the occurrence of

clubroot (Porter, 1997). Preplant applications of calcium cyanamide not only gave significant reductions in disease severity but in additon resulted in a highly profitable yield increase (Figure 2).

The post-harvest shelf-life of vegetables is strongly dependent on a sufficient calcium content in the plant tissue. As calcium cyanamide reduces the root galling caused by the clubroot disease it improves the crop's ability to take up sufficient calcium from the soil. Reduction of root galling therefore is associated with an improved storeability of the crop.

Table 1: Influence of the nitrogen fertilization on yield and quality of lettuce (according to willumsen and mccall, 1997)

Nitrogen fertilisation	Fresh weight g/plant	Dry weight g/plant	Tipburn % of plants	Nitrate content ppm
Calcium nitrate, split applications	240	14.9	31	2 680
Ammonium sulphate, split applications	256	15.7	0	2 350
Ammonium sulphate plus DIDIN, split applications	256	16.1	13	1 629
Calcium Cyanamide, 10 days preplanting	243	15.5	6	1 670
LSD 0.95	15	1.0	16	210

3. References

Huang HC and Sun SK (1991) Effects of S-H Mixture or Perlka on carpogenic germination and survival of sclerotia of Sclerotinia sclerotiorum. Soil Biol. Biochem. Vol. 23, No. 9: 809 – 813

Humpherson-Jones FM, Dixon GR, Craig MA and Ann DM (1992) Control of Clubroot using calcium cyanamide – a review. Brighton Crop Protection Conference – Pests and Diseases – 9B-4: 1147-1154

Ilovai Z and Ceglarska-Hódi E (1992): Report on fungicide trial. Csongrád County Plant Health and Soil Conservation Station

Klasse HJ (1996): Calcium Cyanamide – an effective tool to control clubroot. Proc. Int. Sym. on brassicas, Acta Hort. 407: 403 – 409

Klasse HJ (1998): Neue Versuchsergeb-nisse mit Kalkstickstoff bei Salat. Gemüse 3/1998: 171 – 172

Koç K and Eskalen A (1996) Untersuchungen über die Wirkung von Perlka (Kalkstickstoff) auf den Krankheitsbefall von Phytophtora capsici beim Paprika. Trial report, not published yet.

Mattusch P (1984) Elimination of the apothecia of Sclerotinia sclerotiorum under field and glasshouse conditions. Acta Hort. 152: 49 – 55

McQuillken M (1998) SAC, Scottish Agricultural College, personal communication

Porter I (1997) Integrated strategies control and prevent the spread of clubroot in brassicas. HRDC Research Report 1996-97: 27

Venter F (1978) Einflüsse auf den Nitratgehalt von Kopfsalat (Lactuca sativa L. var. capitata L.). Landw. Forsch., Sonderheft 35: 616 – 622

Venter F and Fritz PD (1979) Nitrate contents of kohlrabi (Brassica oleracea L. var. Gongylodes Lam.) as influenced by fertilization. Qualitas Plantarum – Plant Foods for Human Nutrition, vol. 29, no. 1-2: 179 – 186

Willumsen AJ and McCall D (1997) Kalkkvaelstof til salat. Gartner tidende, arg 113 (6): 6 – 7

Figure 1: Metabolism of calcium cyanamide in the soil

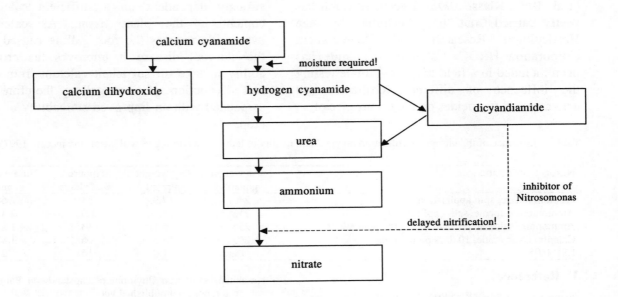

Figure 2: Effect of treatment on marketable yield of broccoli, Werribee (Victoria) 1996 (according to Porter, 1997) Soil type: Medium red brown clay loam; Shirlan was applied as a plant drench in all cases

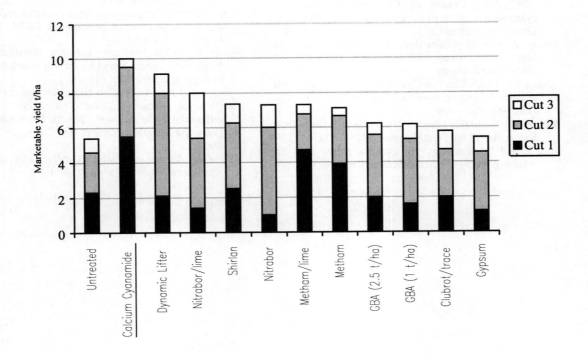

Chapter 54

Effect of manure doses and growth media on the productivity and mineral composition of rocket leaves (*eruca sativa*)

D.Eşiyok[1], B.Okur[2], S.Delibacak[2], İ.Duman[1]
[1]Ege University Faculty of Agriculture, Department of Horticulture 35100 Bornova, İzmir, Turkey
[2]Ege University Faculty of Agriculture, Department of Soil Science 35100 Bornova, İzmir, Turkey

Key words: rocket, growth media, pumice, perlite, manure

Abstract: Rocket (*Eruca sativa*) is a member of the *Brassicaceae* family. It is found endemic in most Mediterranean countries and in Northern and Eastern Europe. According to 1996 records, the annual production of rocket in Turkey was nearly 1100 t green fresh leaves which are mostly used as appetisers with traditional Turkish foods like "pide" (Turkish pizza) or "Kısır" (Turkish wheat meal) or to prepare fresh salads. Rocket was also believed to have aphrodisiac properties. It is used in traditional pharmacopoeia for many different purposes.
In this study, three different levels of farmyard manure (4 t da^{-1}; 8 t da^{-1}; 12 t da^{-1}) were applied to experimental soil; experimental soil mixed with pumice and perlite. Five kg da^{-1} N; 2 kg da^{-1} P_2O_5; 1.5 kg da^{-1} K_2O were applied to growth media as basic fertilisation.
The highest yield of rocket (2717 g m^{-2}) was found at the first cutting in soil + 20 % pumice + 4 t da^{-1} manure treatment. It was followed by 2483 g m^{-2} yield in soil + 20 % perlite + 4 t da^{-1} manure and soil + 4 t da^{-1} manure with 2420 g m^{-2} yield. The increasing levels of manure gave the lowest rocket yield.
Addition of different levels of manure to different growth media was responsible for statistically significant variations in yield and mineral composition of rocket. It can be said that for rocket production the optimum level of manuring is 4 t da^{-1}.

1. Introduction

Rocket (*Eruca sativa*) is a member of the Brassicaceae family. It is found endemic in most Mediterranean countries and in Northern and Eastern Europe. It thrives in many kinds of soils (preferably calcareous) in fields. Green fresh leaves are mostly used as appetisers with traditional Turkish foods like "pide" (Turkish pizza) or "kısır" (Turkish wheat meal), or to prepare fresh salads (Eşiyok, 1996).

Rocket was believed to have aphrodisiac properties (Fernald, 1993). It is used in traditional pharmacopoeia for many different purposes: antiphlogistic, astringent, depurative, diuretic, digestive, tonic, laxative, stomachic, anti-inflammatory for colitis, antiscorbutic and rubefacient (Bianco, 1994). An infusion at 4-8 % is used against itching, chilblains, scalds and

urticaria. Among other applications is the preparation of a lotion to enhance hair regrowth and to fight against greasy scalps (Ellison et al. 1980), as a tonic for the face (Anonymous, 1988) and to cure catarrh and hoarseness (Mascagno, 1987). Oil and leaf extracts showed also a good potential as insect repellents (Sharma and Watta, 1982).

According to 1996 records, the production of rocket is nearly 1100 t per annum in Turkey. It is not possible to say exact status of rocket production, because a great amount of rocket is produced on small parcels and home gardens, which are generally not included in official records (Eşiyok, 1996).

In this research, effect of different levels of manure and growth media on the productivity and mineral composition of rocket leaves were taken into account.

237

238

2. Materials and Methods

The experiment was conducted in the experimental fields, which are characterised by sandy clay loam texture soils (Entisol-Typic Xerofluvent) with slight alkaline reaction. On volume basis, 20 % pumice, 20% perlite was mixed with soil separately to 10 cm depth.

Three different levels of manure (4 t da^{-1}; 8 t da^{-1}; 12 t da^{-1}) were applied to experimental soil, experimental soil mixed with pumice and experimental soil mixed with perlite. Five kg da^{-1} N; 2 kg da^{-1} P$_2$O$_5$; 1,5 kg da^{-1} K$_2$O were applied to growth media as basic fertilisation at the sowing time. The experiment was designed according to randomised blocks with three replications. In this study, thirteen different growth media were used. These media were symbolised with numbers: 1- Treated soil+20 % pumice; 2- Treated soil + 20 % pumice + 4 t da^{-1} manure; 3- Treated soil + 20 % pumice + 8 t da^{-1} manure; 4- Treated soil + 20 % pumice + 12 t da^{-1} manure; 5- Treated soil + 20 % perlite; 6- Treated soil + 20 % perlite 4 t da^{-1} manure; 7- Treated soil + 20 % perlite + 8 t da^{-1} manure; 8- Treated soil + 20 % perlite + 12 t da^{-1} manure; 9- Treated soil; 10- Treated soil + 4 t da^{-1} manure; 11- Treated soil + 8 t da^{-1} manure; 12- Treated soil + 12 t da^{-1} manure ; 13- Untreated soil (control).

Rocket seeds (Eruca sativa) of 2 g m^{-2} were sown to the experimental parcels of 1 m^2 size. Seeds were sown on September 23rd, 1997 to 1cm of soil depth, in rows 20 cm apart, without tilling operations. Irrigation was provided when essential. First (November 5th, 1997) and second harvests (December 10th, 1997) were made manually with knives. Fresh yield of rockets leaves were weighed, and dried in a forced air oven at 70 °C for 48 hours. Total N, P, K, Na, Ca (Kacar, 1972), Mg, Fe, Cu, Zn and Mn (Munoz, 1968) analysis were done in dried leaves. Data were analysed according to Tarist statistical program (Açıkgöz et al. 1994).

3. Result and Discussion

Some physical and chemical properties and mineral composition of the experimental soil are shown in Table 1.

The marketable yield and mineral composition of rocket from 13 different growth media are given in Table 2.

As shown in Table 2, the highest yield (2717 g m^{-2}) of rocket was found at the first cutting with treated soil + 20 % pumice + 4 t da^{-1} manure. It was followed by treated soil + 20 % perlite + 4 t da^{-1} manure (2483 g m^{-2}) and treated soil + 4 t da^{-1} manure (2420 g m^{-2}). The lowest yields of rocket were determined from mixtures of 12 t da^{-1} manure. The increasing levels of manure decreased rocket yields. There was no statistical relationship between the second cutting yields of rocket of different growth media. Yield of rocket decreased in second harvest. Results obtained are corroborated with Bianco and Boari (1996), Pimpini and Enzo (1996). Total yield of rocket

Table 1. Some physical and chemical properties of the experimental soil

Sand %	Silt %	Clay %	Texture	pH	CaCO$_3$ %	Soluble salt %	Org.mat. %	Total N %
52.56	24.00	23.44	SCL	7.42	2.53	0.075	1.93	0.143
P mg kg^{-1}	K mg kg^{-1}	Ca mg kg^{-1}	Mg mgkg^{-1}	Na mg kg^{-1}	Fe mg kg^{-1}	Cu mg kg^{-1}	Mn mg kg^{-1}	Zn mg kg^{-1}
5.44	250	4100	258	40	2.80	5.64	19.98	5.74

in the treated soil + 20 % pumice + 4 t da⁻¹ manure, like the results of first cutting. Other growth media mixed with 4 t da⁻¹ manure

mg kg⁻¹). Increasing levels of manure caused to accumulate N, P, Na, Mg, Mn and Zn in rocket leaves. The decrease in the rocket yield might

Table 2. The marketable yield and mineral composition of rocket from different growth media

Treatments no	Marketable rocket (g m⁻²)			N %	P mg kg⁻¹	K mg kg⁻¹
	1st cut.	2nd cut.	Total yield			
2	2717	1716	4433	4.444	5097	56202
6	2483	1625	4108	4.464	5120	53047
10	2420	1730	4150	4.430	5139	53638
1	2243	1630	3873	4.153	4825	53836
5	2117	1435	3551	4.389	4540	52456
9	1967	1710	3676	4.404	4916	52061
3	1913	1440	3520	4.487	5255	52850
7	1820	1360	2847	4.583	5239	53639
11	1617	1785	3402	4.445	5310	54230
13	1567	1736	3303	4.201	4916	52258
4	1550	1370	2920	4.586	5550	56005
8	1517	1550	3067	4.629	5628	54033
12	1417	1480	2897	4.620	5563	56991
LSD						
(0.05):*	-	ns	-	-	-	ns
(0.01)**	671.3	ns	1367.5	0.33	508.9	ns

Table 2. Continued

Na mg kg⁻¹	Ca mg kg⁻¹	Mg mg kg⁻¹	Fe mg kg⁻¹	Cu mg kg⁻¹	Mn mg kg⁻¹	Zn mg kg⁻¹
1472	25854	3081	617	16.6	89.0	47.3
1374	24658	3081	593	15.3	92.3	55.0
1407	24958	3157	543	17.6	94.3	59.0
949	25555	3419	637	13.3	86.6	47.6
982	23463	2630	583	13.6	72.6	41.6
982	24509	3081	579	16.6	88.3	45.3
1538	23613	3119	503	15.0	88.3	49.3
1996	23613	3457	572	15.0	104.0	51.0
2061	25406	3420	523	19.0	106.3	59.0
982	25703	3459	624	18.6	84.3	60.0
2159	24509	3946	517	17.0	101.3	55.0
2159	24808	3833	573	20.3	112.3	50.6
2258	26004	3607	440	18.3	105.0	64.3
-	1521.7	-	ns	-	-	-
407.6	-	685.84	ns	3.44	15.63	14.15

followed it. Increasing levels of manure decreased total yield of rocket. Comparison of first and second and total yields of rocket in different growth media is shown at Figure 1.

It was determined that the increasing levels of manure (Table 2) increased mineral compositions of rocket leaves (N %, P mg kg⁻¹, Na mg kg⁻¹, Mg mg kg⁻¹, Mn mg kg⁻¹ and Zn

be due to accumulated minerals in rocket leaves. It can be concluded that different growth media can be used to produce rocket. It was found that addition of different levels of manure to growth media may cause statistically significant variations in respect to yield and mineral composition of rocket. Especially,

240

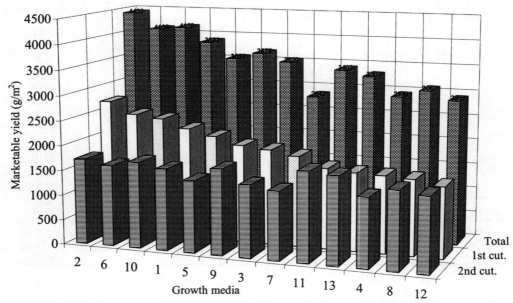

Figure 1. Comparison of first and second cutting and total yields of rocket in different growth media.

rocket yield was negatively effected with the highest levels of manure. Application of 4 t da^{-1} manure can be an optimum level for rocket production.

4. References

Açıkgöz, N. et al., 1994. TARİST: Tarımsal Araştırmalar İçin Bir İstatistik Paket Program (A Statistical Packet Program For Agricultural Research) E.Ü.Z.F. Tarla Bitkileri Bölümü, Bornova, İzmir.

Anonymous, 1988. Il grande Libro delle erbe, 1 st. End. , Peruzzo Editore, Italy.

Bianco, V.V., 1994 Rocket, an Ancient Underutilized Vegetable Crop and It's Potential. Rocket Genetic Resources Network, Report of the First Meeting, 13-15 November 1994, Lisbon, Portugal.

Bianco, V.V. and F.Boart, 1996. Up To Date Developments on Wield Rocket Cultivation, Rocket: A Mediterranean Crop for The World. Report of a workshop 13-14 December 1996, Legnaro (Padova), Italy.

Ellison, J.A., P.Hylands, A.Peterson, C.Pick, K.Sanecki and M.Stuart, 1980,Enciclopedia delle erbe. 1 st. edn. A.Mondadori, Verona

Esiyok, D., 1996. Marketing and Utilisation of Rocket in Turkey. Rocket A Mediterranean Crop

For The World. Report of a Workshop 13-14 December 1996, Legnaro (Padova), Italy.

Fernald, M.L., 1993. Gray's Manual of Botany. 1 st. edn., 2.Dioscorides Press Portland, Oregon.

Kacar, B., 1972. Bitki ve Toprağın Kimyasal Analizleri. 2. Bitki Analizleri, Ankara Üniv. Ziraat Fak. Yayın no: 453. Ankara.

Mascagno, V., 1987. Coltiva o selvatica la rucola e ottima in insalata. Vita in campagna 5(12): 42-43.

Munoz, J., 1968. Atomic Absorption Spectroscopy and analysis by Atomic Absorption Flame Photometry. Elsevier Publishing Company, Amsterdam, London, New-York

Pimpini, F. And M. Enzo, 1996. Present Status and Prospects for Rocket Cultivation in the Veneto Region. Rocket: A Mediterranean Crop For The World. Report of a Workshop 13-14 December 1996, Legnaro (Padova), Italy.

Sharma, S.K. and B.L.Wattal, 1982. Further studies on Mosquito Larvicidal potential of mucilaginous seeds. J. Entomol. Res. 6: 159-165.

Chapter 55

Sugar beet and durum wheat quality characteristics as affected by composted urban waste

G. Convertini, D. De Giorgio, D. Ferri, P. La Cava, L. Giglio

Istituto Sperimentale Agronomico (Ministry of Agriculture) Via C. Ulpiani 5 - 70125 Bari, Italy

Key words: municipal solid waste, sugar beet, durum wheat, sucrose, α-amino nitrogen, protein

Abstract: The research, started 1993, was carried out at Foggia (Southern Italy) on a silty-clay soil with "accentuated thermo-Mediterranean" climate. The field experiment was laid out in randomized blocks with three replications, comparing three types of fertilizer (A1: mineral fertilizer; A2: organic + mineral fertilizer; A3: municipal solid waste compost) and three N doses (40, 80, 120 kg N ha^{-1}) on Autumn-sown sugar beet, and durum wheat, both cropped each year on adjacent plots. Product quality was evaluated determining polarization angle and α-amino N, Na$^+$ and K$^+$ concentrations of sugar beet pulp and the protein content of durum wheat grain. The best yield responses and quality characteristics of sugar beet and durum wheat in MSW-amended soil depend on differing previous crops and agronomical inputs used over many years in the experimental field. Agronomical benefits of MSW-compost applied to the vertisol type soil used were observed also in relation to mineralization of organic N contained in MSW-compost that gradually released N during the cropping season. When MWS-compost was applied for 4 years the α-amino N and Na$^+$ contents of sugar beet pulp decreased but the supply of mineral N to durum wheat plants provided adequate requirements for grain d.m. accumulation.

1. Introduction

The recycling of municipal solid wastes (MSW) serves to recover plant nutrients and organic matter that might otherwise have a dangerous environmental impact in the absence of adequate controls on chemical characteristics of the MSW compost and on variations in soil properties. Some authors (Martel, 1988) suggested a need to evaluate nutrients, heavy metals, organic matter, water content and degree of maturity of MSW compost. This should be coupled to soil evaluation from heavy metals, total organic carbon (TOC), total extracted carbon (TEC) and humified carbon(C_{HA} + C_{FA}) (Sequi et al., 1986; Ministry of Agricultural Resources, 1992).

The encouraging results obtained in previous studies carried out on MSW compost (De Bertoldi et al., 1987; Businelli et al., 1988;

Senesi et al., 1989; Goldberg Federico L. et al., 1989; Sequi et al., 1991; Giusquiani et al., 1992), did not focus on qualitative characteristics of agricultural products as affected by organic matter and/or the N supplied by the compost in Southern Italy. In fact, some qualitative characteristics of agricultural products (e.g. α-amin nitrogen, in the case of sugar beet) depend on the type of fertiliser applied (Venturi, 1982). To determine yield and qualitative characteristics of sugar beet and durum wheat as affected by MSW compost , an investigation started in 1993.

2. Materials and Methods

The research was carried in 1993 (Southern Italy) on a silty-clay soil (Typic Chromoxererts, Soil Taxonomy classification; OM: 2.07%; clay: 31.1%; pH: 8.3; C/N: 10; CaCO3: 7.3).

The climate is "accentuated thermo-mediterranean", with rain mainly concentrated in the winter months and summer temperatures which can rise above 40°C.

Autumn sown sugar beet (SB cv. Suprema) and wheat (DW -cv. Simeto), cropped in a 2-year rotation were treated with three types of fertilizer (A1: mineral fertilizer; A2: organic + mineral fertilizer; A3: municipal solid waste compost) and three N application rates (B1=40 kg N ha^{-1}; B2=80 kg N ha^{-1}; B3=120 kg N ha^{-1}) arranged in a completely randomized block design with 6 replications.

Yields and qualitative characteristics of sugar beet roots and durum wheat grain were examined. On the sugar beet roots, polarisation degree (sucrose %), α-amino N concentration, potassium and sodium content in the pulp, alkalinity coefficient (AC), molasses sugar (MS), juice purity (JP), extracted sugar, corrected polarization, all according to determination of Biancardi et al. (1994) and other authors (Mantovani, 1977; Barbanti et al., 1994) were all determined. Protein content and mineral composition of durum wheat grain were also quantified.

The compost was obtained by the SLIA Plant (Brindisi, Castel di Sangro - Italy) through aerobic transformation of municipal solid waste.

Soil samples were collected before and after of both cropping cycles (0-40 cm depth) of all the 54 elementary plots. Heavy metal contents, total organic carbon (TOC), total extracted carbon (TEC) and humified carbon(C_{HA} + C_{FA}) were determined (Sequi et al., 1986; Ministry of Agricultural Resources, 1992). Humification parameters were also assessed as well as soil N-NO$_3$, exchangeable N-NH$_4$, NaHCO$_3$-extracted P, NH$_4$Ac extracted-K and soil aggregate sizes.

3. Results and Discussion

The application of MSW-compost during three years seemed to increase SB yield, when compared with mineral fertilizer (mean of three N doses 40, 80,120 kg N ha-1). It is probable that application of MSW-compost supplied organic N for gradual mineralisation more synchronized to plant N requirements. The observed differences between the two treatments seem poor, but enough to show an interesting effect of recycling biomass in "thermomediterranean" conditions of Southern Italy.

Observed variations in sucrose percentages of pulp (Figure 1) were not significant, but trended to decrease, especially in the last year.

Among the quality characteristics of sugar beet, an important role is played by the α-amino-N content (Figure 2). It may lead to deterioration in juice purity and other technological parameters when it exceeds 7 meq.100 g^{-1} of sugar beet pulp. The recorded decrease in α-amino N content of sugar beet cropped in MSW treated plots, point to its potential usefulness in improving juice quality. Slow mineralization of MSW-compost, organic N reduce the mineral N substrate for translocation and enzymatic transformation to accumulate aminoacids.

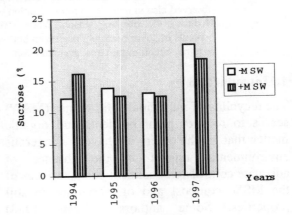

Figure1. Sucrose content (%) of sugarbeet pulph as affected by MSW applications during the years.

Figure 3 illustrates variations in SB pulp Na$^+$ content grown on MSW-compost. Values were low in comparison to maximum values indicated by some authors of up to 3 meq100g^{-1} of pulp and they decreased significantly during the trial period. This finding shows that conventional N fertilizer supply seemingly releases from the soil higher quantities of Na⁺ , in contrast to the effects recorded here with MSW-compost application.

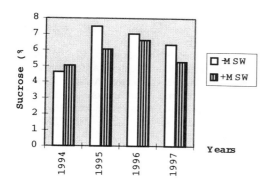

Figure2. Alfa amino N content (meq $100g^{-1}$) of sugarbeet pulp as affected by MSW applications during the years.

In Table 1 the main effects observed on durum wheat in response to MSW application are reported. It improved grain yield but protein content remained comparitively higher with mineral N fertilizer (-MSW), as means of four years. Fertilizer mineral N may have exceeded plant requirements such as to accumulate as crude protein N in grain. On the contrary, mineral N supplied by MSW-compost was probably lower, but adequate to improve grain yield. In this case, the reduced mineral N did not facilitate an increase of grain protein content.

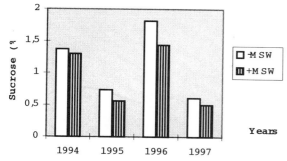

Figure3. Na content of sugarbeet pulp (meq 100g-1)as affected by MSW applications during four years.

Durum wheat cropped with MSW compost, seemed to maintain yield performances after 4 years, but showed a slight decrease in grain

protein content. This result should be evaluated comparing the costs of mineral N fertilizers with MSW-compost, taking in to consideration any variations in soil properties the two different agronomical inputs may have induced, e.g. mineral N fertilizer increases soil $N-NO_3$ that can be leached, while MSW-compost improves also soil structure, in addition to providing N.

Table 1: Main yield and quality characteristics of durum wheat during the trial period.

Years	+MSW		-MSW	
	Grain yield t.ha⁻¹	Protein %	Grain yield t.ha⁻¹	Protein %
1994	6.47	14.5	5.92	17.5
1995	4.20	13.5	4.50	12.8
1996	4.70	12.9	4.60	12.0
1997	3.22	12.4	3.33	13.6
Mean	**4.65**	**13.3**	**4.59**	**14.0**
C.V. (%)	**47.4**	**11.0**	**39.5**	**17.7**

4. Conclusions

Agronomic benefits of MSW-compost applied to a vertisol type soil were observed mainly in relation to the following aspects: (i) α-amino N of sugar beet decreased probably because the reduced rate of mineralization of MSW-compost organic N facilitated gradual transformation by enzymes without the accumulation of α-aminoacids; (ii) lower quantities of Na⁺ were supplied or released from the soil, contrary to the effects recorded with N fertilizer; (iii) the N supplied to durum wheat by MSW-compost sufficed for the accumulation of grain dry matter, but the same quantities supplied by fertilizer N seemed excessive.

5. References

Barbanti L., Bimbatti M., Peruch U., Poggiolini S., Rosso F. (1994) Factors affecting sugar beet quality in the Po river valley. Part 2: Nitrogen fertilization. Proc. 2nd ESA Congress, Abano-Padova, 574-575.

Biancardi E., Colombo M., Graf A., Negrini G. (1994) La barbabietola da zucchero nell'Italia centro-settentrionale. L'Industria Saccarifera Italiana, 1, 87: 7-20.

Businelli M., GiusquianiP.L., Gigliotti G.(1988)Evoluzione dei residui solidi urbani durante il compostaggio e loro influenza sulla fertilità del suolo.In E. Santucci (ed.) Proc. Int.Congress Energy

244

and Materials Recovery from Wastes. Perugia, Italy. 6-9 June. Litograf, Perugia, Italy,pp.233-242.

De Bertoldi M., Ferranti M.P.,L'Hermite P.,Zucconi,F.(eds.) (1987) Compost: production, quality and use. Elsevier Applied Science Publishers, London, New York.

Giusquiani P.L., Gigliotti G., Businelli D.(1992) Mobility of heavy metals in urban waste-amended soils. J. Environ. Qual. 21: 330-335.

Goldberg Federico L., Rossi N., Spallacci P. (1989) Agricoltural use of organic wastes (livestock slurries, sewage sludges, composts): the situation in Italy. Chimica Oggi. 7: 29-32.

M.A.F. (1992) Approvazione dei "Metodi ufficiali di analisi chimica del suolo". D.M. n. 79 del 11/5/92 S.o. G.U. n. 121 del 25/5/92-S.g.

Mantovani G. (1977) Sul valore tecnologico della barbabietola da zucchero. L'Industria Saccarifera Italiana, 3: 51-64.

Martel L. (1988) Quality and marketing of urban sludge composts in the EEC. Commission of the EEC, Dir. Genn. XI and XII, Brussels-Agro-Development S.A., St. Quentin Yvelines (F).

Senesi N., Sposito G., Holtzclaw K.M., Bradford G.R. (1989) Chemical properties of metal-humic fractions of a sewage sludge-amended aridisol. J. Environ. Qual., 18: 186-194.

Sequi P., De Nobili M., Leita L., Cercignani G. (1986) A new index of humification. Agrochimica, 30: 175-179.

Sequi P.,Ciavatta C., Vittori Antisari L.(1991) Organic fertilizers and humification in soil. In Organic Substances and sediments in water. Vol. 1. Humic and soil (Baker R.E.D.), Lewis Pub. Chelsea, MI., 351-367.

Venturi G. (1982) Barbabietola da zucchero (*Beta vulgaris* L.). Da "Coltivazioni erbacee", Patron Editore, pp. 467-514.

Chapter 56

Leachability of phosphorus and mineral nitrogen of the soils amended with solid phase from pig slurry

E. Vasconcelos, F. Cabral and C.M.d.S. Cordovil

Department of Agriculture and Environmental Chemistry, Instituto Superior de Agronomia, Tapada da Ajuda, 1399 Lisboa Codex, Portugal.

Key words: soil solid phase; pig slurry; wheat; phosphorus; mineral nitrogen; nutrient leaching

Abstract: An outdoor pot experiment was conducted to evaluate the risk of contamination of drainage water by leaching of phosphorus and mineral nitrogen as a result of increasing applications (5, 15 and 25 t ha^{-1}) of the solid phase from pig slurry (SP) to a Cambic arenosol (soil A) and a Dystric Cambisol (soil B). A control treatment consisted of a basic dressing of NPK fertiliser. Results showed that increasing additions of SP enhanced the available P content on both soils although this increase was much less in soil A due to it is lower P fixing capacity. Leaching losses of P significantly increased with the level of SP applied and were about ten times greater on soil A than on soil B. In contrast, water soluble phosphorus (P_{ws}) supplied as fertiliser (superphosphate) was much more resistant to leaching than P_{ws} supplied as SP. Leaching loss of mineral nitrogen increased with the level of SP applied during the first stages of crop development. Soil properties were also important in this process since the amount of mineral nitrogen leached was significantly greater from soil A than soil B.

1. Introduction

Application of organic wastes, can recycle nutrients back onto agricultural land and thus reduce the need for mineral fertilisers. Other potential benefits from animal waste application arise from improvement to soil structure, infiltration rate and water holding capacity due to the addition of organic matter. This ensures a better soil environment for root development and nutrient uptake (Mbagwu et al., 1994). Pig slurries in particular have been shown to enhance the nutrient status and water stability of soil aggregates and to improve overall productivity of soils. Pig slurries are usually evaluated in terms of their ability to supply N to the crops as their N content is rich in readily available forms (Smith et al.,1994). In particular the solid phase from pig slurry (SP), is rich in organic matter, organic N and P

(Smith and Chambers, 1995). In the long term, P contained in SP can be regarded as equivalent to P supplied by mineral fertilisers and high application rates can lead to enrichment of P in the soils (Smith and van Djick, 1987). Leaching of nutrients, such as N, P and K to surface and ground waters can pose animal and human health problems, varying for example from ingestion of excess nitrate or lead to eutrophication (Siegenthaler and Stauffer, 1991). Such losses not only represent a danger to the environment but also an agronomic loss. Losses of nutrients following waste application to soils depend mostly on the rate of application. Nevertheless, at rates of typical farm practice, significant losses of P and K from the rooting zone are unlikely to occur except on soils with a low fixing capacity (Christie, 1987).

According to Vasconcelos et al. (1997), increasing application rates of SP led to a significant accumulation of P in the surface layer of both soils tested (Cambic arenosol and a Dystric cambisol) after the second year of a wheat experiment conducted in lysimeters. This suggested a possible contamination from P of surface water after successive applications of SP to these soils. In addition, nitrate leaching significantly increased with the level of SP applied. Thus, a key objective of this research was to evaluate the risk of drainage water contamination by inorganic phosphorus (P) and mineral nitrogen (N) compounds associated with increases in the application rate of SP on both soils tested.

2. Materials and Methods

Winter wheat (*Triticum aestivum* L. cv.Lotti) was grown in an outdoor pot experiment (9/11/95 - 31/05/96) with a completely randomised design. The PVC pots (0.038 m^2 in surface area and 0.25 m in depth) were filled with 12.0 kg of the upper layer (0-25 cm) of a Cambic arenosol (soil A) or 11.0 kg of the upper layer (0-25cm) of a Dystric cambisol (soil B) (Table 1). These soils were taken from the lysimeters where a two-year wheat experiment (1993 to 1995) had been conducted using the same basic experimental design as in the present research (Vasconcelos et al., 1997). In the present experiment, four treatments were replicated four times and consisted of: control (NPK) and 5, 15, and 25 t ha^{-1} of SP on a fresh weight basis (Table 2). The basic dressing of a NPK fertiliser consisted of: 0.5g N pot^{-1} (ammonium sulphate, 20.5% N), 0.83g K pot^{-1} (potassium chloride, 49.8% K)and 0.44g P pot^{-1} (superphosphate, 7.8% P). Two N top-dressings (13/12/95 and 29/01/96) were applied in every treatment and consisted of 2 x 0.25g N pot^{-1} (ammonium nitrate, 26% N) were applied. At the beginning of the experiment, pots were watered with demineralised water sufficient to allow seed germination. After emergence a selection was made in order to leave the twenty most uniform plants in the pots (20/11/95). During the experiment, four collection of leachates were taken from each pot, measured and analysed (6/12/95; 11/12/95; 23/01/96 and 15/05/96).

Table 1. Some characteristics of experimental soils at the beginning of the experiment

Determ.	pH (H$_2$O)	O.M. (%)	N$_{min}$ mg kg^{-1}	P$_{avail}$ mg kg^{-1}	P$_{ws}$ mg kg^{-1}
Soil A					
NPK	6.30	0.83	7.06	25.28	12.11
5 t SP	6.90	0.85	7.30	41.00	14.74
15 t SP	6.80	0.98	7.79	60.57	18.11
25 t SP	6.80	1.20	10.40	140.82	31.76
Soil B					
NPK	6.20	1.02	8.86	22.18	2.72
5 t SP	6.60	1.06	9.01	48.47	6.31
15 t SP	6.65	1.29	10.81	115.70	15.51
25 t SP	6.55	1.34	10.90	174.60	23.59

Table 2. Chemical composition of solid phase from pig slurry (fresh weight basis).

Determ.		Determ.	
Moisture (%)	82.70	K total (%)	0.16
O. Matter (%)	7.76	K soluble (%)	0.06
N$_{Kj.}$ (%)	0.58	Ca (%)	0.95
NH$_4$-N (%)	0.02	Mg (%)	0.66
NO$_3$-N(mg kg^{-1})	0.44	Na (%)	0.03
P$_{total}$ (%)	0.83	pH	7.92
P$_{ws}$ (%)	0.33	C/N ratio	7.76

Organic matter in the SP was determined by loss-on-ignition at 350-400°C. Available P in the soil was determined by Egner-Riehm extraction. N in SP was determined by Kjeldahl digestion and N$_{min}$ in leachates was determined using a continuous flow segmented auto-analyser. P in leachutes and P$_{ws}$ in the soils was determined according to Rod and Payot (1983). N$_{min}$ in the soils was also determined by continuous flow segmented analysis, after soils were leaching with 0.01 mol l^{-1} CaCl$_2$ (Houba et al., 1990). NO$_3$-N and NH$_4$-N in SP were determined after extraction with water or 2 mol l^{-1} KCl respectively (Rod and Payot, 1983). Results from the study were subjected to a one-way ANOVA, followed by a Scheffe F-test at p< 0.05.

3. Results and Discussion

In each soil there was an increase in available P over control dependant on the SP treatment (Table 3). However, comparing the amounts of available P at the end of the experiment with those present in the soils at the beginning, a different pattern occurred on each of the two soils.

Table 3. Some characteristics of both soils at the end of the experiment

Determ.	pH (H$_2$O)	O.M (%)	N$_{min.}$ mg kg^{-1}	P$_{avail.}$ mg kg^{-1}	P$_{ws}$ mg kg^{-1}
Soil A					
NPK	6.00 b	0.81 a	5.91 a	34.23 c	13.10 c
5 t SP	6.54 a	0.82 a	6.33 a	29.40 c	16.30 c
15 t SP	6.59 a	0.83 a	6.91 a	85.77 b	22.40 b
25 t SP	6.59 a	0.93 a	8.86 a	140.13 a	31.50 a
Soil B					
NPK	5.92 b	1.17 a	7.27 a	41.97 d	3.60 c
5 t SP	6.53 a	1.15 a	6.35 a	74.30 c	11.20 b
15 t SP	6.66 a	1.32 a	6.67 a	149.13 b	21.60 a
25 t SP	6.62 a	1.29 a	9.93 a	221.50 a	27.60 a

Different letters in the same column and for each soil means that treatments differ significantly (p<0.05)

Thus, for soil B the available P was significantly higher at the end of the experiment whereas for soil A it was much closer to the initial values because of the high P$_{ws}$ leaching (Table 4). For both soils tested pH values were always higher for SP treatments than for control. As shown in Table 4 losses of P by leaching significantly increased with increasing rates of SP applied on both soils, whose responses however differed considerably. The total amount of P leached from soil A was about 10 times greater than that obtained from soil B. Results given on Table 5 confirm the different behaviour of each soil, in soil A 45-50% of P$_{ws}$ was leached while in soil B it was significantly lower (3-4%). Losses of P by leaching were significantly less when P$_{ws}$ was supplied as superphosphate than in SP.

Table 4. Phosphorus (P) leaching losses from both soils (mg pot^{-1})

Soil A	1st	2nd	3rd	4th	Total
NPK	118.80c	13.30c	13.90d	6.50c	152.50c
5 t SP	96.90d	7.10d	6.00c	4.30c	114.30d
15 t SP	134.30b	20.60b	37.70d	12.70b	205.30b
25 t SP	184.30a	42.10a	63.50a	22.60a	312.50a
Soil B					
NPK	8.31b	0.84c	2.10c	0.04b	11.29c
5 t SP	2.80c	0.18d	1.40c	0.05b	4.43d
15 t SP	8.33b	1.60b	4.60b	0.09b	14.62b
25 t SP	13.90a	2.90a	7.80a	0.24a	24.84a

Different letters in the same column and for each soil means that treatments differ significantly (p<0.05)

Table 5. Relation between P$_{ws}$ applied and P leached on both soils.

Treat	Soil A P$_{ws}$ (mg pot^{-1})	P$_{leached}$ (mg pot^{-1})	(P$_{leach.}$/P$_{ws}$) x 100 (%)
NPK	582.00 *	152.60	26.20
5 t SP	240.60 **	114.30	47.50
15 t SP	406.60 **	205.20	50.50
25 t SP	696.50 **	312.10	44.80
	Soil B		
NPK	466.60 *	11.30	2.40
5 t SP	132.50 **	4.40	3.30
15 t SP	359.80 **	14.60	4.10
25 t SP	574.40 **	24.80	4.30

*(Pws in the soilx soil weight) + (P from superphosphate)
**(Pws in the soilx soil weight)+ (Pws from SPxSP weight)

For both soils the N$_{min}$ lost by leaching was much greater in every treatment where SP was applied than where mineral fertiliser alone was applied (Table 6). This was likely due to high rainfall in the initial phases of the crop growth period that led to the runoff of most of the N$_{min}$ applied as basic dressing fertilisation. As observed for P loss by leaching, the N$_{min}$ leached were much greater on soil A due to its low fixing capacity. Most of the N$_{min}$ was leached during the first three weeks of the experiment.

Table 6. Mineral nitrogen leaching losses from two experimental soils (mg pot^{-1})

Soil A	1st	2nd	3rd	4th	Total
NPK	233.5a	5.21a	0.55a	0.15a	239.41a
5 t SP	23.80d	1.31b	0.19b	2.03ab	27.33d
15 t SP	38.40c	1.30b	0.17b	2.34a	42.21c
25 t SP	87.50b	1.60b	0.17b	1.42ab	90.96b
Soil B					
NPK	94.46a	25.10a	0.75a	1.57a	121.88a
5 t SP	7.12d	1.09b	0.10b	1.65a	9.96d
15 t SP	41.44c	1.54b	0.11b	1.30a	44.39c
25 t SP	75.76b	1.96b	0.11b	1.76a	79.59b

Comparing control (NPK) treatment between soils showed retarded leaching losses of N$_{min}$ on soil B. Hence leaching losses of N$_{min}$ on the 2nd collection (11/12/95) represented 23 % of the total leached, but on soil A only 3 %. Leaching of N$_{min}$ increased with increasing application rates of SP on both soils. Considering that the N$_{min}$ content in SP very low, leaching observed on SP treatments for both soils must have proceeded mainly from the amounts of N$_{min}$

initially existent in that soil as well as from the mineralisation of organic N in the SP. On both soils N top-dressings were quite efficient since no significant N leaching occurred after they were applied (13/12/95 and 29/01/96).

4. Conclusions

Increasing application rates of SP enhanced the content of available P on both soils although that increase was much more substantial on soil B due to the smaller amounts of P leached. In fact, the leaching of P significantly increased with increasing doses of SP applied but was about ten times greater on soil A due to its low P fixing capacity. P_{ws} supplied as superphosphate was much more resistant to leaching than P_{ws} supplied as SP. Mineral nitrogen leaching losses increased with the level of SP applied especially during the first stages of crop development. In this case soil characteristics played a decisive role since N_{min} was significantly greater in soil A. N top-dressing was found to be highly conservative of nitrogen since small amounts of N_{min} were leached after its application.

The data indicate that successive applications of high rates of SP could cause contamination of surface waters with P, particularly in soils of a low P- fixing capacity under climatic features favourable to leaching.

5. References

Christie, P (1987) Some long term effects of slurry on grassland. J. Agric. Sci. Camb., 108, 529-541.

Houba VJG, Novozamsky, I, Lexmond, TM and van der Lee, JJ (1990) Applicability of 0.01 M CaCl$_2$ as a single extractant for the assessment of the nutrient status of soils and other diagnostic purposes. Commun. Soil Sci. Plant Anal., 21, 2281- 2290.

Mbagwu, JSC, Unamba-Oparah, I and Nevoh, GO (1994) Physico chemical properties and productivity of two tropical soils amended with dehydrated swine waste. Biores. Technol., 49, 163-171.

Rod, PH and Payot JM (1983) Methodes d' analyses de terre, substrats et végétaux. Document interne RAC Changins/1260 Nyon.

Siegenthaler, AF and Stauffer, W (1991) Environmental effects of long-term slurry and sewage sludge application: reasons and measures. In: Treatment and use of sewage sludge and liquid agricultural wastes. P. L'Hermité (ed.). Elsevier Applied Science Publishers, London, England. pp. 82-89.

Smith, KA. and van Djick, TA (1987) Utilisation of phosphorus and potassium from animal manure on grassland and fodder crops. In: Animal Manure on Grassland and Fodder Crops. van de Meer, H. G. et al.(eds.) Martinus Nijhoff Publ., Dordrecht, Boston and Lancaster, 87-101.

Smith, KA, Chambers, BJ and Jackson, D R (1994) Solid manure and animal waste slurries as a source of nitrogen in arable crops rotations. In: Solid and liquid wastes: their best destination (II) Proc. III Congr. Int. Quimica de la Anque, Tenerife, 285-294.

Smith, KA and Chambers, BJ (1995) Muck: from waste to resource utilisation: the impacts and implications. Inst. Agric. Eng., 33-38.

Vasconcelos E, Cabral, F and Cordovil, CM.d.S (1997) Effects of solid phase from pig slurry on soil chemical characteristics, nitrate leaching, composition and yield of wheat. J. Plant Nutr. 20 (7), 939-952.

Chapter 57

Effect of fulvic+humic acid application on yield and nutrient uptake in sunflower (*Heliantus annuus*) and corn (*Zea mays*)

A.Aydın , M. Turan, Y. Sezen

University of Atatürk, Faculty of Agriculture, Department of Soil Science, 25240 Erzurum/Turkey

Key words: macro and micro nutrients, sunflower, corn, humic acid, fulvic acid

Abstract: This study was undertaken to determine the effects of K-Humate application on dry matter content, elemental composition and uptake of plant nutrients in corn and sunflower. K-Humate was applied to soil or leaves at 150, 300 and 450 g da^{-1} doses. Results indicated that the effects of K-Humate application on dry-matter content and mineral composition of both plants were significant.

1. Introduction

Improving soil conditions and establishing the equilibrium among plant nutrients are important for soil productivity and plant production. Soil organic matter increases agricultural production by improving soil physical, chemical and biological properties. Organic matter content of soils in all over Turkey is generally low, because of lack of crop rotation, insufficient rainfall and limited irrigation systems.

In order to solve the problems caused by low organic matter some natural and sufficiency organic amendments have been started applying to soil, especially in the western-countries, recently.

Among these amendments many researches reported that humic acid increased availability of plant nutrients in soils and crop production by improving soil physical, chemical and microbiological properties (Jalali and Takkar, 1979; Foortun and Lopez-Fondo, 1982; Bowen et.al, 1985; Vaughan and Malcolm, 1985; Sözüdoğru and et. al., 1996).

The objective of this study was to determine the effects of K-Humate application on dry matter content, elemental composition and nutrient uptake in corn and sunflower.

2. Materials and Methods

Soil samples were taken at 0-20 cm depth in the agricultural experimental station of Atatürk University, Erzurum. The experimental design was a factorial design with 4 K-Humate doses (0,150,300 and 450 g da^{-1}), 2 application methods (to soil, foliage), 2 plant species and 3 replications (4x2x2x3=48 pots).

Before sowing, 10 kg N da^{-1}, 10 kg P$_2$O$_5$ da^{-1} and 10 kg K$_2$O da^{-1} were applied to soil as basic fertilizer. Four seeds were planted in each pot. After germination, only two plantlets were left in the pots. K-Humate was applied to the soil at once, but to the foliage at 15 days intervals during growing season (4 times). Soil moisture was kept between field capacity and wilting point during the experiment.

About 10 weeks later, plants were harvested and dried at 65°C for 72 hours. Dried plant samples were analysed and the results were statistically evaluated .

Soil analyses: Soil texture was determined using Bouyoucos hydrometer method (Demiralay, 1993), pH using glass pH meter on 1:2.5 soil-water mixture, organic matter by Smith-Weldon method, exchangeable cations and cation exchange capacity and plant available P by sodium bicarbonate blue-colour method (Sağlam, 1994), plant available Zn, Fe, Mn and Cu by DTPA method (Lindsay and Norwell, 1969).

Plant analyses: Total nitrogen was determined using micro – Kjeldahl method, P by spectrophotometry of the P-Vanadomolibdoyellow - colour method K, Ca, Mg, Fe, Mn, Zn ve Cu by atomic adsorption spectrophotometry (Bayraklı, 1987).

3. Results and Discussion

The soil used in this study was a clay-loam (38.4 % clay, 39.9 % silt and 21.7 % sand), slightly alkaline (pH 7.8), moderate in organic matter (2.90 %), and rich in $CaCO_3$ content (11.5 %), its cation exchangeable capacity was 41.4 cmol kg^{-1}, exchangeable Ca+Mg, K, Na were 38.3, 2.2, and 0.5 cmol kg^{-1}, plant available P, Fe, Mn, Zn and Cu were 9.4, 3.5, 3.5, 1.1 and 1.8 mg kg^{-1}, respectively.

Effects of K-Humate application on dry matter contents of corn and sunflower:

Dry matter contents of corn and sunflower after K-Humate application are shown in Table 1. Dry Matter contents of sunflower and corn increased by increasing rates of K-Humate application. With soil application, the maximum increasing rate of dry-matter content was obtained at the highest dose (KH_3= 450 kg da^{-1}) with rates of 24.17 % and 28.32 % in sunflower and corn, respectively. Average increasing rates were 11.66 % for sunflower and 13.97% for corn as compared to the control. However, with foliar application , the maximum increasing rate of dry-matter content was obtained at the highest dose application (KH_3= 450 kg da^1) with rates of 2.58 % and 3.03 % in sunflower and corn, respectively.

Average increasing rates were 1.06 % for sunflower and 1.63 %, for corn as compared to the control. Effects of K-Humate application methods and doses on dry matter content of sunflower and corn were statistically ($p < 0.01$) significant (Table2).Increase in dry-matter content was more clear with soil application than foliar application. This may be due to the positive effect of K-Humate on soil physical , chemical and biological properties. Similar result was reported by Tok et al., (1988).

Effects of K-Humate application on mineral composition of sunflower and corn:

Mineral composition of sunflower and corn is given in Table 3. There were slight differences in mineral composition of sunflower and corn with respect to K-Humate application methods, doses and plant species. N, P, K, Ca, Mg, Fe, Mn, Zn and Cu contents of both plants generally increased by increasing rates of K-Humate applications. Mineral contents of plants were generally higher with application to soil than to leaf. Similar results were reported by Jalali and Takkar (1979), Sözüdoğru et.al., (1996).

Effects of K-Humate applications on nutrient uptake:

Nutrient uptakes from 100 g soil by sunflower and corn are given in Table 4. In general, nutrient uptake by both plant increased by increasing rates of K-Humate application. It was more clear with soil application than with foliar application. This may be due to the effect of K-Humate application on the availability of plant nutrients in soil.

4. Conclusion

Dry Matter contents of sunflower and corn increased by increasing rates of K-Humate application to soil and foliage. The soil application was more effective on increasing dry matter content. Similarly, mineral contents of plants and nutrient uptake increased by increasing rates of K-Humate application.

Table 1. Effects of K-Humate application on dry matter contents of corn and sunflower

Application Method	K-Humate Doses, g da^{-1}	Sun-flower Dry Matter g pot^{-1}	Dry Matter Incre. Rate,%	Corn Dry Matter g pot^{-1}	Dry Matter Incre.Rate, %	Doses Average	Application Average
Soil	0	17.50	-	9.18	-	13.34 c	
	150	17.98	2.74	9.83	7.08	13.47 c	
	300	20.95	19.71	11.06	20.48	14.73 b	15.00 a
	450	21.73	24.17	11.78	28.32	15.23 a	
	Average	19.54	11.66	10.46	13.97		
Foliage	0	17.45	-	9.23	-		
	150	17.68	1.32	8.40	1.84		
	300	17.51	0.03	9.38	1.63		13.51 b
	450	17.90	2.58	9.51	3.03		
	Average	17.64	1.06	9.38	1.63		
	Plant Aver.	18.59 a		9.32 b			

Table 2. ANOVA table for dry-matter contents of sunflower and corn

V.S	Df	MS	F
Plant species(P)	1	901.25	1140.8**
Application Methods (A)	1	26.78	33.9**
Doses (D)	3	9.17	11.6**
PxA	1	2.03	2.6
PxD	3	0.73	0.9
AxD	3	7.06	8.9**
PxAxD	3	0.71	0.9
Error	32	0.79	

**: $p < 0.01$

Table 3. Effects of K-Humate application on mineral composition of sunflower and corn.

Plant	Applic. Meth.	Doses, g da^{-1}	N (%)	P	K	Ca	Mg	Fe (mg kg^{-1})	Mn	Zn	Cu
S U N F L O W E R	Soil	0	0.87	0.233	2.57	0.57	0.14	171.9	15.7	58.4	15.3
		150	0.96	0.255	3.01	0.74	0.17	193.5	21.8	77.8	17.9
		300	1.03	0.312	2.72	0.65	0.15	188.1	19.9	77.8	18.3
		450	1.09	0.323	3.10	0.81	0.16	198.5	24.0	93.4	15.3
		Average	0.99	0.281	2.85	0.69	0.16	188.0	20.6	76.9	16.7
	Foliage	0	0.88	0.237	2.52	0.61	0.15	178.1	16.1	61.2	15.3
		150	1.02	0.250	2.82	0.62	0.14	185.4	19.8	74.7	17.3
		300	1.02	0.248	2.74	0.75	0.17	183.3	22.3	81.1	18.5
		450	0.99	0.253	2.83	0.71	0.18	201.5	23.8	97.3	13.4
		Average	0.98	0.247	2.73	0.67	0.16	187.1	20.5	78.6	16.1
C O R N	Soils	0	1.57	0.235	3.01	0.43	0.16	228.4	65.1	75.9	9.9
		150	1.60	0.255	3.07	0.39	0.14	238.4	60.5	76.8	8.9
		300	1.66	0.273	3.29	0.50	0.17	236.5	67.6	87.8	10.1
		450	1.75	0.265	3.09	0.61	0.18	240.2	72.1	102.2	12.8
		Average	1.65	0.257	3.12	0.48	0.16	235.9	66.3	85.7	10.4
	Foliage	0	1.55	0.236	3.05	0.42	0.17	236.5	63.6	74.7	9.9
		150	1.62	0.247	2.99	0.51	0.17	231.8	63.0	82.2	10.2
		300	1.60	0.238	3.17	0.50	0.19	235.3	67.6	87.2	10.4
		450	1.72	0.248	3.21	0.57	0.18	240.2	71.3	89.7	11.8
		Average	1.62	0.242	3.11	0.50	0.18	236.0	66.4	83.5	10.6

Table 4. Effects of K-Humate application on the mineral composition of sunflower and corn (mg $100g^{-1}$).

Plant species	Applic. Meth.	Doses, g da^{-1}	N	P	K	Ca	Mg	Fe	Mn	Zn	Cu
S U N F L O W E R	Soil	0	7.61	2.40	22.49	4.99	1.23	0.150	0.014	0.051	0.013
		150	8.63	2.29	27.06	6.65	1.53	0.174	0.020	0.070	0.016
		300	10.79	3.27	28.49	6.81	1.57	0.197	0.021	0.081	0.019
		450	11.84	3.51	33.68	8.80	1.74	0.216	0.026	0.0101	0.017
		Average	9.67	2.75	27.84	6.74	1.51	0.184	0.020	0.075	0.016
	Foliage	0	7.68	2.07	21.99	5.32	1.31	0.155	0.014	0.053	0.013
		150	9.02	2.21	24.93	5.48	1.24	0.164	0.018	0.066	0.015
		300	8.93	2.17	23.99	6.57	1.49	0.160	0.020	0.071	0.016
		450	8.86	2.26	25.33	6.35	1.61	0.180	0.021	0.087	0.012
		Average	8.64	2.18	24.07	5.91	1.41	0.165	0.018	0.069	0.014
C O R N	Soils	0	7.21	1.08	13.82	1.97	0.73	0.105	0.030	0.035	0.005
		150	7.86	1.25	15.09	1.92	0.69	0.117	0.030	0.038	0.004
		300	9.18	1.51	18.19	2.77	0.94	0.131	0.037	0.049	0.006
		450	10.31	1.56	18.20	3.59	1.06	0.141	0.042	0.060	0.008
		Average	8.48	1.34	16.32	2.51	0.84	0.123	0.035	0.045	0.005
	Foliage	0	7.15	1.09	14.08	1.94	0.78	0.109	0.029	0.034	0.005
		150	7.61	1.16	14.05	2.40	0.80	0.109	0.030	0.039	0.005
		300	7.50	1.12	14.87	2.35	0.89	0.110	0.032	0.041	0.005
		450	8.18	1.18	15.26	2.71	0.86	0.114	0.034	0.043	0.006
		Average	7.60	1.13	14.59	2.35	0.84	0.111	0.031	0.039	0.005

5. References

Bayraklı F (1987) Toprak ve Bitki Analizleri. 19 Mayıs Üniv. ayın No: 17. Samsun

Bowen JE Krotjy BA and Perreira P (1985) Adapting crops to troubled soils is key to solving words food problems. Plant Breeding, 3-10.

Demiralay İ (1993) Toprak Fiziksel Analizleri. Atatürk Üni. Yayınları No:143. Erzurum.

Foortun C and Lopez-Fondo C (1982) Influence of humic acid on the mineral nutrition and development of maize roots cultivated in normal nutrient solutions and lacking Fe and Mn. Anales de Edafolog Agrobilogia XLI :335-349.

Jalali VK and Takkar PN (1979) Evaluation of parameters for simultaneous determination of micro nutrient cations available to plants from soils. Indian J. Agric Sci. 49:622-626.

Lindsay WL and Norwell WA (1969) Development of DTPA Soil Test for Zinc, Iron, Manganese and Copper. Soil Sci. Soc. Amer. Proc. Vol:33, p:49-54.

Sağlam T (1994) Toprak ve Suyun Kimyasal Analiz Yöntemleri. Trakya Üniv. Tekirdağ Ziraat Fak. Yayınları No: 189.

Tok HH Sağlam MT and Ekinci H (1988) Toprağa Konsantre Hümik Asit Bileşikleri İlave Etmenin Ayçiçeği ve Soya Fasulyesinin Bazı Verim Özellikleri Üzerine Etkileri. T. G. S. A.Ş. Dergisi Sayı 6.s:13-16

Vaughan YD and Malcolm RE (1985) Influence of humic substances on growth and physical processes. P:37-75. In: D. Vaughan and R.eE. Malcolm (ed.) Soil organic matter and biological activity. Martinus Nijhoff/Dr W.Junk Publ. Dordrec.

Sözüdoğru S Kütük C Yalçın R and Usta S (1996) Hümik asitin fasülye bitkisinin gelişimi ve besin maddeleri alımı üzerine etkisi. Anakara Üniv. Ziraat Fak. Yayınları No: 1452, Bilimsel Araş. Ve inc. 800 Ankara.

Chapter 58

Response of corn to bio-and organic fertilizers in a newly reclaimed sandy soil

S.M.A. Radwan and M. Saber
Agricultural Microbiology Dept. National Research Centre, Cairo, Egypt

Key words: corn, biofertilizers, organic fertilizers

Abstract: Two field experiments were carried out in a newly reclaimed sandy soil during 1997 and 1998 seasons to evaluate the effect of two rates of composted sawdust (20 or 40 m^3/feddan) in the presence and absence elemental sulphur, the foliar fertilizer (Nofatrein) ,the seed dresser (Coatingen) and/or the multi-strain biofertilizer (Microbein) on chemical composition and yield of corn crop.

The weight of 100 grain increased from 16.8 g under the usual rate of chemical fertilizer (NPK) to 18.9 and 20.4 g under either 20 or 40 m^3/ feddan composted sawdust, respectively. Associating the biofertilizer with either organic or chemical fertilizers led to a marked increase in the grains and straw yields compared to their sole effect under different treatments. The greatest N, P, Zn, Fe and Cu -contents of corn grains were achieved when Microbein was associated with elemental sulphur or Nofatrein in comparison with Coatingen. Generally, combinations of composted sawdust (40 m^3/feddan) and Microbein with elemental sulphur or Nofatrein, led to highly significant differences over the usual rate of chemical NPK fertilization in corn yield and its chemical composition.

1. Introduction

Misuse of agrochemicals in agricultural practices caused significant adverse environmental impacts. During the last two decades, man tried going back to nature through organic farming. However, the recent advances in biotechnology enabled the innovation of a newly farming system known as ecological farming or bio-organic farming. Such system is composed of three major components ; organic manuring, biofertilization and biological control of agricultural pests (Saber 1998)

The over all goal of the present study is to evaluate the application of different rates of composted sawdust (organic manuring) in the presence and absence elemental sulphur, micronutrients, and/ or the multi- strain biofertilizer Microbien on chemical composition and yield of corn grains.

2. Materials and Methods

Two field experiments were carried out at South Tahreer province, Bahira Governorate, Egypt, during 1997 & 1998 seasons. Mechanical and chemical analyses of the experimental soil were carried out according to Jackson (1971), and are presented in Table (1). Split - plot design with four replicates was used. The main plots were organic manure, while the sub- plots were biofertilizer with and without different additives. Sub-plot area was 10.5 m^2 (1/400 feddan).

The treatments were as follows :

I- Main plots
1-Chemical fertilizers (NPK) as recommended rate
2-Composted sawdust at the rate of 20 m^3/ feddan

3-Composted sawdust at the rate of 40 m³/ feddan

II - Sub- main plots

1- Control (without Microbien)

2- Microbien (multi-strain biofertlizer)

3- Microbien + elemental sulphur

4- Microbien + Coatingen (seed dresser)

5- Microbien + Nofatrien (foliar fertilizer)

Composted sawdust as organic manure has been added during soil preparation at two rates (20 and 40 m³/ feddan). The chemical composition of compost used is shown in Table(2).

Microbien biofertilizer contains highly efficient strains of phosphate dissolving bacteria (PDB), *Azotobacter spp.*, *Azospirilum spp.* and *Pseudomonas spp.* These strains were separately grown in specific nutrient broth for 48 hours at 30 C° in a rotary shaking incubator. Equal amounts of each strain broth were collectively mixed for preparation of inocula using peat moss as a carrier. Liquid broth cultures initially containing 1×10^8 ; 5×10^7 ; 3×10^7 and 2×10^7 viable cell/ml of PDB, *Azotobacter spp.*, *Azospirilum spp.* and *Pseudomonas spp* respectively . Microbien was used as a seed dresser just prior to sowing.

Grains of corn (*Zea mays* L.) c.v. single cross 10 were sown on mid June in both seasons . In biofertilized treatments, the grains were mixed with the Microbien using 40% arabic gum as sticker. Coatingen powder was used as a seed dresser during sowing (13 gm/1kg seeds). It is a micronutrient preparation with three chelate elements (Zn, Mn and Fe) and elemental sulphur. Nofatrien was used as a foliar fertilizer (1L/300L water) and contains macro and micronutrients. It was sprayed twice, at appearance of the fifth leaf and two weeks after the first one. Both Coatingen and Nofatrien are commercially produced by Ministry of Agriculture in Egypt.

Elemental sulphur was added during soil preparation at the rate of 300 kg/feddan and mixed with the surface soil. Ammonium sulphate (20.5% N) at the rate of 120 kg N/feddan in four equal doses before the first, second, third and fourth irrigation. Calcium superphosphate (16% P_2O_5) and potassium sulphate (48% K_2O) were applied at 200 and 100 kg/feddan, respectively.Other agricultural practices were carried out as recommended.

At maturity, 120 days of sowing, the plants were harvested. Grain, straw and biological yield per feddan, as well as seed index (100 grain weight per grams) were determined. N, P, Mn, Zn, Fe and Cu- contents of corn grain were estimated according to Jackson (1971).

The obtained results were statistically analysed using LSD at 5% level of significance according to Snedecor and Cochran (1987).

3. Results and Discussion

(A) Macro and micronutrient contents of corn grains

1- N and P contents of corn grains

Results presented in Table (3) revealed that treatments with Microbien in the presence and absence of elemental sulphur, Coatingen or Nofatrien gave always higher values of N and P contents as compared to control (without Micobien) under organic or chemical applications.

The most pronounced increase in N and P contents of corn grains were recorded as a result of the combined action of elemental sulphur and Microbien under either chemical or organic fertilizer applications. In this respect, Kabesh et al.(1988) and Saraf et al.(1997) reported that biofertilizers and elemental sulphur application help in releasing, celating and increasing the availability of plant nutrients in soil.

Nevertheless, significant differences were found for N and P contents of corn grains between composted sawdust treatments at the rates of either 20- 40 m³ / feddan and chemical fertilization treatments. This means that composted sawdust application improved chemical characteristics resulting in the release of more available nutrient elements to be absorbed by plant roots. Confirming trends were obtained by Abd El-Moez (1996) .

2- Micronutrient contents of corn grains

Sole Microbien application showed less increases in Zn, Mn, Fe and Cu contents of cograins, from control, than those recorded under the combined effect of Microbien with elemental sulphur, Coatingen or Nofatrien under organic or chemical fertilizers application. The greatest micronutrients contents were found in grains harvested from plots receiving Microbien associated with Nofatrien under organic manure application. This effect of Nofatrien could be attributed to suppling plant with its requirements of micronutrients, in balanced ratios and effective chemical form, that are easily absorbed by the plant leaves.

It is evident from the results that the bio-organic farming exerted a more pronouncly positive effect on Zn and Fe contents in corn grains compared to Mand Cu-con. Worthme, that the composted sawdust gave higher values for the studied micronutrients content in corn grains at 40 m³ /feddan than at 20 m³ /feddan followed by chemical fertilizers. These results are in accordance with that reported by Amer et al. (1997).

(B) Yield of corn

1- Weight of 100 grain

Significant increases were calculated in the weight of 100 grain as a results of the application of Microbien in the presence and absence the different additives (elemental sulphur, Coatingen or Nofatrien) as compared to control under organic or chemical applications (Table,4). These effects of Microbien (multi strain biofertilizer) could be attributed to four main functions; asymbiotic nitrogen fixation, mineralization of certain macro-and micro-nutrients, secreting a set of growth promoting principles and bio-controlling certain soil born diseases. Similar conclusions have previously obtained by Reiad et al.(1987); El-Shanshoury (1995) and Saber (1998).

Increasing the rate of composted sawdust application from 20 to 40 m³/feddan exhibited distinguished increases in weight of 100 grain. These increases ,however, did not exhibit any significant differences. The present data confirm the finding of Ahmed et al.(1997) and Amer et al. (1997).

2- Grain yield

Results presented in Table(4) revealed that the associative action of Microbien and elemental sulphur, Coatingen or Nofatrien led to marked effects on grain yield of corn as compared the sole Microbien under organic or chemical fertilizers applications. The largest grain yield per feddan was obtained by Microbien+Nofatrien being 16.98 ardab/feddan(Ardab=140 kg) descendingly followed by Microbien + elemental sulphur giving 16.11 ardab/feddan and by Microbien + Coatingen giving 15.73 ardab/feddan under composted sawdust at the rate of 40 m³ / feddan application. Worthmentioning, that significant differences were calculated in grain yield as a results of composted sawdust application at the rates of either 20 or 40 m³/feddan compared to chemical fertilizers application under various treatments. This means that organic manure application enhanced the biological cycles within the farming system; maintenance and increase the long-term sustainable fertility of soils as stated by Abd El-Moez (1996) and Amer et al.(1997).

3- Straw yield

No significant differences were observed in the straw yield of corn between chemical fertilizers and composted sawdust at the rate of 20 m³ / feddan application under Microbien with different additives applications treatments (Table, 4). On the other hand, these differences were significant under the rate of 40 m³/feddan application. Similar results were reported by Abd El- Gawad et al. (1995).It is worthy to state that, the combined effect of Microbien with elemental sulphur on straw yield was superior to its effect with either Coatingen or Nofatrien. These results were in agreement with those obtained by Saraf et al.(1997).

Table 1. Mechanical and chemical analyses of the experimental soil

Soil characters	Values
Mechanical analyses	
Sand	93.4%
Silt	5.9%
Clay	0.7%
Texture	**Sandy**
Chemical analyses	
pH (1-5)	8.2
EC(1-5)	0.42 m mhos cm^{-2}
CaCO$_3$	2.73%
Organic matter	0.29%
Total N	0.04%
Available P	4.5 mg kg^{-1}
Available Zn	0.62 mg kg^{-1}
Available Mn	0.40 mg kg^{-1}
Available Fe	3.70 mg kg^{-1}
Available Cu	0.10 mg kg^{-1}

Table 2 . Chemical characteristics of composted sawdust

Character	pH	EC (mmhos/cm^2)	Organic Carbon %	Total N %	Available P (mg kg^{-1})	C/N ratio
Values	7.68	2.53	29.4	1.18	79.8	24.92

Table 3 . Effect of bio-organic farming with different additives and their interaction on N and P -contents of corn grain (Combined analysis of two seasons)

Treatment	N-content (mg/plant)			P-content (mg/plant)		
	Chemical Fert.	Organic manure 20 m^3/fed.	40 m^3/fed.	Chemical Fert.	Organic manure 20 m^3/fed.	40 m^3/fed.
Control	1516	1615	1718	409	426	452
Microbien(M)	1632	1719	1769	436	512	549
M+Sulphur	1770	1914	2110	572	619	638
M+Coatingen	1681	1851	1872	491	583	592
M+Nofatrien	1709	1892	1921	513	595	621
LSD $_{0.05}$ for : - Organic manures	92			15		
- Bio with additives	117			21		
- Interaction	335			62		

Table 4. Effect of bio-organic farming with different additives and their interaction on corn yield at harvest (Combined analysis of two seasons)

Treatments	Weight of 100 grain (g)			Grain yield (ardab/fed.)			Straw yield (tons/fed.)		
	Chemical Fert.	Organic manure 20m^3 / fed.	40m^3 / fed.	Chemical Fert.	Organic manure 20m^3 / fed.	40m^3 / fed.	Chemical Fert.	Organic manure 20m^3 / fed.	40m^3 / fed.
Control	16.8	18.9	20.4	11.13	13.28	14.65	6.06	6.56	6.76
Microbien(M)	18.1	21.5	22.9	11.67	13.72	15.54	6.26	6.72	6.83
	23.1	25.7	27.2	11.94	14.17	16.11	6.83	7.12	7.31
M+Sulphur									
	21.9	23.8	25.7	11.78	13.87	15.73	5.89	6.59	6.79
M+Coatingen									
M+Nofatrien	24.8	25.4	28.2	12.57	14.79	16.98	6.41	6.85	6.90
LSD $_{0.05}$ for :									
- Organic manures		1.18		0.53			0.59		
- Bio with additives		1.51		0.61			0.37		
- Interaction		2.62		1.19			1.32		

4. References

Abd El-Gawad, A.A.; Saad,A.O.M.; Edris,A.S.A. and El kholy,M.E.(1995). Effect of the stimulative dose of some forms of nitrogen fertilizers and bacterium inoculation on yield and its components and grain quality of maize plants.Bull. NRC,Egypt.20, No.4,493-506.

Abd El-Moez,M.R.(1996). Dry matter yield and nutrient uptake of corn as affected by some organic wastes applied to a sandy soil. Annals of Agric. Sc. Moshtohor, 34(3):1319-1330.

Ahmed, M.K.A.; Saad,A.O.M. ; Alice T.Thalooth and Kabesh, M.O. (1997). Utilization of biofertilizers in field crops production.10-Yield response of groundnut to inorganic, organic and biofertilizers. Annals Agric.Sci, Ain Shams Univ.,Cairo,42(2),365-375.

Amer,A.A.; Badawi,M.A. and El-Banna,A.A.(1997). Effect of organic manuring on wheat plants grown in sandy soil.Annals Agric.Sci. Ain Shams Univ.,Cairo,42 (1),1o7-116.

El-Shanshoury A.R.(1995). Interactions of Azotobacter chroococcum,Azospirillum brasilense and Streptomyces mutabilis, in relation to their effect on wheat development. J. Agron. & Crop Sci. 175:119-127.

Jackson,M.L.(1971)."Soil Chemical Analysis" Printice Hall of India Ltd., New Delhi.

Kabesh,M.O.; El-Baz F.K. and saber M.S.M.(1988). Utilization of biofertilizers in field crops production. 8- Effect of elemental sulphur application in the presence and absence of biofertilization on chemical composition of maize plants. Annals Agric.Sci, Ain Shams Univ.,Cairo,33(2),1057-1066.

Reiad,M.Sh.; Abdrabou R.Th. and Hamada M.A. (1987). Response of maize plant to inoculation with Azotobacter and Azospiriluum, nitrogen and organic fertilization rates. Annals of Agric. Sci. Moshtohor, 25 (1):1-13.

Saber, M.S.M. (1998). Prospective of sustinable farming through microbial biofortification of plant rhizosphere. 8th International Symposium on Microbial Ecolog, Halif, Can.P,288.

Saraf, C.S.; Shivakum, B.G. and Patil R.R. (1997). Effect of phosphorus, sulphur and seed inoculation on performance of chickpea (Cicer arictinum).Indian J. Agron.42(2): 323-328.

Snedecor and Cochran (1987).Statistical Methods. Lowa State Univ. Press, Ames, Lowa, USA.

Chapter 59

The effects of GA_3 and additional KNO_3 fertilisation on flowering and quality characteristics of *gladiolus grandiflorus* 'eurovision'

O. Karagüzel[1] S. Altan[2] İ. Doran[3] Z. Söğüt[2]

1. University of Akdeniz, Faculty of Agriculture, Antalya-Turkey

2. University of Cukurova, Faculty of Agriculture, Adana-Turkey

3. Ministry of Agriculture, General Directorate of Agricultural Researches, Ankara-Turkey

Key words: gladiolus, gibberellic acid, potassium nitrate

Abstract: This study was carried out to determine the combined effects of GA_3 with additional KNO_3 fertilisation on flowering and cut flower quality of *Gladiolus grandiflorus* 'Eurovision' in late autumn planting.The corms (size 10/12) were soaked in solutions of GA_3 at 0 (control), 50 and 100 mgkg^{-1} for one hour 5 days before planting and were planted at 49 corms.m^{-2} density on November 24. Before planting, the basal dose of 30 g.m^{-2} ammonium sulphate (21 %) and 45 g.m^{-2} triple super phosphate (42 %) were mixed with the soil. At the three-four leaf stage, the dose of KNO_3 (43 %, 13 %) at 25 g.m^{-2} was applied in all the experimental plots. Applications of 25 g.m^{-2} KNO_3 as an additional fertiliser in half of the experimental plots at weekly intervals were continued until 2 weeks before the corms were dug. Both the treatment of GA_3 at 100 ppm and additional KNO_3 fertilisation shortened the time from planting to harvest, and increased flowering percentage, the length of flower stem and spike, the number of flowers per spike and the diameter of flower stem in late autumn planting.

1. Introduction

Gladiolus is the third of the top 10 cut flower crops in Turkey with 26 million stems total production and 1 850 000 US$ wholesale value (Ertan et al., 1996). In response to the highest prices in markets in the winter and early spring seasons, majority of the growers in the Mediterranean coastal region have planted corms in greenhouses in the autumn.

As known, the large part of commercial gladiolus cultivars as 'Eurovision' was bred for summer cropping by spring planting. Growing these cultivars in subtropical and Mediterranean climates during the winter months has been frequently resulted in lack of flowering and low-quality flowers (Cohen and Barzilay, 1991). Although breeders have focused their studies on new cultivars which can flower at higher percentage and produce high-quality

flowers under winter conditions, it has been still needed to know growing techniques including the use of growth regulators and fertiliser to obtain desirable flowering percentage and flower quality in the autumn plantings.

Previous studies showed that GA_3 significantly affected flowering percentage, number of flowers or flower stem length in several bulbous species as *Begonia x tuberhybrida* (Tonecki, 1986), *Freesia* (Talia, 1983), *Liatris* (Mor and Berland, 1986) and *Zantedeschia* (Corr and Widmer, 1987). It was also mentioned that GA_3 treatments had a positive effect on earliness and flower quality in some gladiolus cultivars (Roychowdhury, 1989).

Potassium has been considered the base nutrient in gladiolus fertilisation. Comparing with nitrogen and phosphorus the higher rates of potassium have been suggested by several authors (Bushman, .. ; Wilfret, 1980; Altan and Altan, 1984; Moltay et al, 1989). Aouichaoui and Tissaoui (1989) obtained significant increase in flowering and good stem elongation in response to KNO_3 application in gladiolus cultivars planted in November.

This experiment was carried out to determine the combined effect of GA_3 with additional KNO_3 fertilisation on flowering and some quality characteristics of *Gladiolus grandiflorus* 'Eurovision' under plastic greenhouse conditions in late autumn planting.

2. Materials and Methods

Experiment was conducted in a plastic greenhouse at the Horticultural Research Institute (Erdemli) located on the coastal line of east Mediterranean Region of Turkey. The corms of *Gladiolus grandiflorus* 'Eurovision' (size 10/12) were soaked in solutions of GA_3 at 0 (control), 50 and 100 $mgkg^{-1}$ for one hour, and were dried in shade for 5 days before planting. Then the corms were planted at 7 cm depth with a distance of 15x15 cm^2 (49 corms.m^{-2}) on November 24.

The soil of experimental plots was sandy-loam in texture having pH of 8.0. One month before planting 198 g mineral sulphur (80 WP) per square meter was applied to decrease the soil pH around 7.0. As a basal dressing 30 gm^{-2} ammonium sulphate (21%) and 45 g m^{-2} triple superphosphate (42%) were applied before planting.

At the three-four leaf stage, 25 g m^{-2} potassium nitrate (45%, 13%) was applied in all the experimental plots and this dose was assigned as the first level of KNO_3 fertilisation (K1). Application of 25 g m^{-2} KNO_3 as an additional fertiliser (K2) in half of plots at a weekly interval was continued until 2 weeks before the corms were dug.

The greenhouse was not heated, but a heating system was automatically run when greenhouse temperature went down below 4°C to keep minimum greenhouse temperature at 4±2°C. Minimum monthly temperatures measured in the greenhouse were 10.7, 5.8, 6.7, 6.4, 9.6 and 18.2°C in November, December, January, February, March and April, respectively.

During the experiment, the time from planting to harvest, flowering percentage, flower stem and spike lengths, the number of flowers, stem diameter and corm weight were recorded. Also some physical and chemical properties of the greenhouse soil were analysed and determined at the beginning and at the end of the experiment.

3. Results

The results showed that both GA_3 and additional KNO_3 fertilisation had a significant shortening effect on the time from planting to harvest compared to controls (Table 1). The time from planting to harvest was the shortest in plants arisen from corms treated with 100 $mgkg^{-1}$ GA_3 and fertilised additionally with KNO_3. As shown in Table 1, it was determined that the combined effect of 100 $mgkg^{-1}$ GA_3 with additional KNO_3 fertilisation shortened the time from planting to harvest about 10 days under described growing conditions.

Flowering percentage was separately affected by GA_3 applications or additional KNO_3 fertilisation. By increasing the dose of GA_3, flowering percentage significantly increased. Also, flowering percentage of plants which were additionally fertilised with KNO_3 was higher than that of the plants fertilised only once with KNO_3 at tree-four leaf stage. As a result, the flowering percentages in cv Eurovision were enhanced about 9.5 % by GA_3 treatment and additional KNO_3 fertilisation (Table 1).

Only additional KNO_3 fertilisation had significant effect on the flower stem length. With GA_3 treatment a slight elongation in flower stem length was recorded, but it was not found statistically notable (Table 1).

Spike length was affected by both GA_3 and additional KNO_3 fertilisation at 1 % and 5 % levels, respectively. The results revealed that increase in the dose of GA_3 resulted in increase

Table 1: The effects of GA$_3$ and additional KNO$_3$ fertilisation on flowering and quality characteristics of *Gladiolus grandiflorus* 'Eurovision'.

Treatments		Time to harvest (day)	Flowering percentage (%)	Flower stem length (cm)	Spike length (cm)	Number of flowers (flowers.spike^{-1})	Stem diameter (mm)	Corm weight (g.corm^{-1})
KNO$_3$	GA$_3$							
K1	Control	145.3 a[z]	82.3 d	129.3 b	65.2 c	17.0 b	11.69 b	55.3 b
	50 ppm	143.0 ab	87.1 bc	129.3 b	68.5 b	17.7 ab	11.99 b	58.9 b
	100 ppm	141.7 bc	89.8 ab	131.4 b	69.6 b	18.6 a	11.86 b	57.6 b
K2	Control	143.3 ab	85.0 c	135.2 a	71.4 b	17.9 ab	12.68 a	68.3 a
	50 ppm	140.3 c	88.4 b	138.0 a	75.4 a	18.7 a	12.68 a	72.3 a
	100 ppm	135.3 d	91.8 a	138.5 a	76.3 a	18.8 a	12.90 a	71.1 a
Significance								
KNO$_3$		*[y]	*	*	*	*	*	*
GA$_3$		**	**	N.S.	**	*	N.S.	N.S.
GA$_3$ x KNO$_3$		*	N.S.	N.S.	N.S.	N.S.	N.S.	N.S.

(z) : Mean separation within columns by SNK test, % 5 level.
(y): Nonsignificant (NS), or significant at 5 % (*) or 1 % (**) levels.

Table 2: Some physical and chemical properties of the greenhouse soil at the beginning and end of the experiment.

	Texture	pH	CaCO$_3$ (%)	Organic matter (%)	Available K (mg kg^{-1})	P (mg kg^{-1})
At the beginning of the experiment	Sandy-loam	8.0*	23.1 a[z]	0.9 c	81.1 c	63.3 b
At the end of the experiment						
K1	Sandy-loam	7.1	12.1 b	2.6 b	130.8 b	79.9 a
K2	Sandy-loam	7.3	10.6 b	3.8 a	421.6 a	82.0 a

*: Before planting, 198 g.m^{-2} mineral sulphur (80 WP) was applied and mixed in 25 cm depth to reduce the soil pH about 7.0
(z) : Mean separation within columns by SNK test, % 5 level

in spike lengths. Also, additional KNO$_3$ fertilisation provided a significant elongation in spike lengths (Table 1). GA$_3$ applications and additional KNO$_3$ fertilisation significantly increased the number of flowers per spike. As seen in Table 1, increase in the dose of GA$_3$ increased the flower number per spike and additional KNO$_3$ fertilisation resulted in higher number of flowers per spike. However, increases in flower number were found practically slight.

The results indicated that only additional KNO$_3$ fertilisation had a notable increasing effect on the final weight of the corms. There was no significant variation in the final weights between GA$_3$ treated or non treated corms in cv Eurovision (Table 1). Some physical and chemical properties of the greenhouse soil are given in Table 2. Application of 198 g.m^{-2} mineral sulphur in the greenhouse soil reduced the soil pH about 7.0 (Table 2).CaCO$_3$ contents of the soil decreased from 23.1 % to 12.1 or 10.6 % at end of the experiment. According to the base

fertilisation and KNO$_3$ fertilisations, the organic matter content of the soil significantly increased. It was found that KNO$_3$, particularly additional KNO$_3$ fertilisation resulted in considerably increases in the available K content of the greenhouse soil (Table 2). Also, the base fertilisation with triple superphosphate at the beginning of the experiment increased the available P content of the soil, but the notable variations did not recorded in the available P content with respect to different KNO$_3$ fertilisation levels.

4. Discussion and Conclusion

In the previous studies, it was found that GA$_3$ increased flowering percentage, flower number or flower stem length in several bulbous flowering plants (Talia, 1983; Mor and Berland, 1986; Tonecki, 1986 and Corr and Widmer, 1987). In gladiolus, GA$_3$ and KNO$_3$ fertilisation were found effective to improve flowering, flower quality and corm yield in separate studies (Aouichaoui and Tissaoui, 1989; Roychowdhury (1989).

In the light of previous results, to determine the combined effect of GA$_3$ with additional KNO$_3$ fertilisation on flowering and cut flower quality of *Gladiolus grandfilorus* 'Eurovision' was aimed in this study. Similar results were obtained with slight differences. However, additional KNO$_3$ fertilisation was found to be more effective than GA$_3$ on the flower stem length and corm weight in cv Eurovision.

In this study, the reason of continuing additional KNO$_3$ fertilisation until 2 weeks before placing the corms was to determine the effect of KNO$_3$ on the final corm weight. It can be concluded that soaking the corms in 100 mgkg^{-1} GA$_3$ solution for one hour before planting and fertilising the plants with 25 g.m^{-2} KNO$_3$ 5 or 6 times in a weekly interval after three-four leaf stage can be accepted as a suitable growing method for cv Eurovision in late autumn plantings.

5. References

Altan T and Altan S (1984) Glayol ve Gerbera Yetiştiriciliği. TAV Yayınları No.6, Yalova, pp. 54.

Aouichaoui S and Tissaoui T (1989) Mineral nutrition effect on the flowering of hybrid gladioli cv.s under plastic greenhouse. Acta Hortic. 246:213-218.

Bushman JCM (..) Gladiolus as cutflower in subtropical and tropical regions. International Flowerbulb Center, Hillegom, Holland, 31 pp.

Cohen A and Barzilay A (1991) Miniature gladiolus cultivars bred for winter flowering. HortScience 26(2):216-218.

Corr BE and Widmer RE (1987) Gibberellic acid increases flower number in *Zantedeschia elliottiana* and *Z. rahmanii*. HortScience 22(4):695-697.

Ertan N; Karagüzel O; Kostak S; Gürsan K and Özçelik A (1996) Kesme Çiçek Raporu (Süs Bitkileri). DPT,Yay No.DPT 2464-ÖİK:515, pp. 63-88.

Moltay İ; Genç Ç and Gürsan K (1989) Yalova Bölgesinde Yetiştirilen Bazı Kesme Çiçeklerin Ticari Gübre İstekleri II. Gladiol (Sonuç Raporu). ABKME, Yalova, pp. 20.

Mor R and Berland M (1986) Effect of various corm treatments on flowering of *Liatris spicata* WILLD. Acta Hortic. 177:197-201.

Roychowdhury N (1989) Effect of plant spacing and growth regulators on growth and flower yield of gladiolus under polyethylene tunnel. Acta Hortic. 246:259-263.

Talia MC (1983) Effects of gibberellin upon Freesia flowering. Acta Hortic. 137:225-228.

Tonecki J (1986) Effect of short photoperiod and growth regulators on growth and flowering and tuberization of *Begonia x tuberhybrida*. Acta Hortic. 177:147-156.

Wilfret GJ (1980) Gladiolus. In:R. A. Larson (Editor), Introduction to Floriculture. Academic Press Inc., New York, USA, pp.166-181.

Chapter 60

Grass response to municipal wastewater reuse as compared to nitrogen and water application

J. Beltrão, P. Gamito, C. Guerrero, A. Arsénio and J.C. Brito

Unidade de Ciências e Tecnologias Agrárias, Universidade do Algarve,Campus de gambelas, 8000 Faro, Portugal

Key words: wastewater irrigation, nitrogen fertilization, grass appearance, grass yield

Abstract: Bermuda grass (*Cynodon dactylon*, L. Pers) is used very often in the fairways of golf courses, due to its tolerance to drought, high temperatures and treading damage. The objective of this work was to study the response of this grown cultivar with Dallis grass (*Paspalum dilatatum*, Poiret), to several levels of municipal wastewater. An experimental setup known as sprinkle point source was used to simulate the various levels of water application, expressed by the crop coefficient kc and the crop evapotranspiration rate. Results were compared with those obtained for sprinkle irrigated plots under nitrogen fertilization. At lower water application rates (kc < 0.5), grass yield response to wastewater application was comparable to potable water application without added nitrogen. At higher application rates (kc > 1), grass yield response to wastewater was comparable to potable application combined with nitrogen application between 0 and 25 kg ha^{-1}.month^{-1}. Visual appearance of the lawngrass was good for kc values > 1.2, when wastewater was applied.

1. Introduction

The reuse of treated wastewater is considered as an alternative to the use of potable water in the Mediterranean agriculture and landscape (Oron *et al.*, 1992), especially in golf courses (Arsénio *et al.*, 1998). An important growth and crop quality factor is nutrient content which may be influenced by this practice (Marecos do Monte, 1996). The intense use of effluent for sprinkle irrigation has attracted public awareness in respect of environmental pollution and water quality impact (Asano and Mills, 1990). Secondary and tertiary effluent reuse is subject to various environmental and health criteria currently being defined for each region (Angelakis and Tchobanoglous, 1997). Effluent applied under sprinkle irrigation, while exposed to solar radiation, may experience dieaway of the contaminating pathogens (Oron and Beltrão, 1993). The present work compares grass response, at different yields and quality levels, to wastewater application and to the combined effects of nitrogen and potable water application.

2. Material and methods

Irrigation was studied on two mixed cultivar grass swards: Bermuda grass (*Cynodon dactilon*, L. Pers) and Dallis grass (*Paspalum dilatatum*, Poiret). The experiments were conducted in Campus da Penha, University of Algarve, Portugal, during Spring and Summer. The water reclamation plant serves this Campus of the University, including laboratories and a canteen. Raw wastewater flows through a comminutor where the course solids are ground up. Afterwards, it flows to a septic tank to experience a pre-anaerobic treatment to reduce suspended solids. Later it permeates through an anaerobic filter. The final effluent is discharged into a photosynthetic pond system.

Subsequently, the effluent is filtrated through sand filters.

Table 1 gives soil physical parameters and Table 2 shows chemical and microbiological parameters of the applied effluent.

Table 1. Soil physical parameters to a depth of 0.3 m

Texture	Sand (%)	89
	Silt (%)	5
	Clay (%)	6
Classification: Loamy sand (organic soil)		
θ_w (kg kg^{-1}) at ψ_m = - 10 kPa		0.30
θ_w (kg kg^{-1}) at ψ_m = - 32 kPa		0.27
θ_w (kg kg^{-1}) at ψ_m = - 100 kPa		0.25
Hydraulic conductivity Kf (cm h^{-1})		1,200
Bulk density (kg dm^{-3})		0.85

Samples of final effluents were collected twice monthly. The most probable number (MPN) of fecal coloforms and *E. coli* were determined by cultivation on enrichment medium and the selective medium "Billis Green". Enterococci were cultivated in "Rothe medium". The identification of *E.coli* was done by the IMVC test. The detection of *Salmonella* was done in the selective media "Rappapor-Vassiliadis" and "Selenite". The isolation process used the media "Rambach Agar" (Merck), and "Kliger". Identification was done by the use of Api system "Rapid 20E".

Chemical parameters were determined by spectrophotometry (ELE – Paqualab TM Photometer EL430-550). The pH and the electrical conductivity of wastewater were determined by a portable potentiometer and by a conductivity meter, respectively.

An experimental design known as sprinkle point source (Or and Hanks, 1992) was used. It is characterised by the assumption that a point application creates a linear irrigation gradient from source, to produce a gradual change in water application, so a high degree of irrigation uniformity must be obtained in parallel isohyets. The wettest zone occurs near the sprinkler and zones near its output fringes are the driest. The irrigation system was stopped when wind speed was greater than 1 m s^{-1}. It comprised a 323/92 Naan sprinkler with 2.5 x

4.5 mm diameter nozzles, using a 300 kPa sprinkler pressure, with a wetting radius about 10m. The Christiansen Coefficient CUC (1942) was used to determine the uniformity of water distribution at 2; 4; 6; and 8 m from the sprinkle point source. Respective CUC values of 92.5; 92.7; 91.9; and 73.2 % were obtained. The wastewater plot was irrigated once a day. Crop daily evapotranspiration rates for each controlled isohyet were compared to crop coefficient times the daily potential evapotranspiration (Penman method), according to Doorenbos and Kassam (1979).

Table 2. Chemical and microbiological parameters of the applied effluent

Parameter	Symbol	Unit	Value
pH			7.1-7.2
Elect. cond	ECw	dS m^{-1}	1.8-2.5
Nitrate	NO$_3^-$	mgkg^{-1}	0-10
Phosphate	P$_2$O$_5$	mgkg^{-1}	0
Potassium	K$_2$O	mgkg^{-1}	0-165
Ammonium	NH$_4^+$	mgkg^{-1}	0
Bicarbonate	HCO$_3^-$	mgkg^{-1}	340-365
Boron	B	mgkg^{-1}	0
Calcium	Ca$_2^+$	mgkg^{-1}	200-217
Chloride	Cl$^-$	mgkg^{-1}	100
Copper	Cu	mgkg^{-1}	0.00-0.01
Iron	Fe	mgkg^{-1}	0
Magnesium	Mg	mgkg^{-1}	23-33
Manganese	Mn	mgkg^{-1}	0
Fecal colif.	MPN 100 ml^{-1}		0.0-1.5x10^3
E.coli	MPN 100 ml^{-1}		0
Enterococeis	MPN 100 ml^{-1}		0.0-4.5x10^3
Salmonella	MPN 100 ml^{-1}		0

The potable water plot was divided into four subplots for nitrogen treatments, similiar to the methodology of Beltrão and Ben Asher (1993). Nitrogen application levels were: 0, 25, 50g and 100 kgha^{-1}month^{-1} (N$_0$, N$_1$, N$_2$, N$_3$).

3. Results

Figures 1 and 2 show grass responses to wastewater application and to combined potable water and nitrogen, during Spring and Summer. Figure 1 expresses grass response in absolute yield (Y) as a function of water consumption as crop coefficient kc. In figure 2 yield is presented as a function of crop evapotransprations

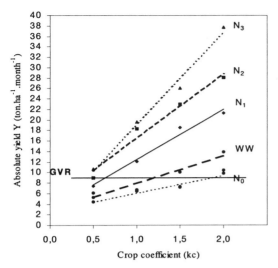

Fig.1. Comparative grass responses to either wastewater or combined potable water and nitrogen fertilizer. Water consumption is expressed by the crop coefficient kc.

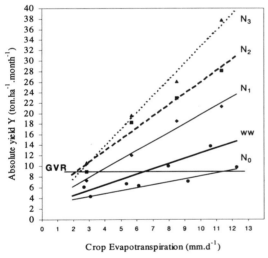

Fig. 2. Comparative grass responses to either wastewater or combined potable water and nitrogen fertilizer. Water consumption is expressed by the daily crop evapotranspiration rate ETc.

The relationship between absolute yield (Y=kg ha^{-1} month^{-1}) and the crop coefficient (kc) of potable water + N (kg ha^{-1} month^{-1}), during Summer, is given by the following equations:

July - N_1 $Y_1 = 9.02$ kc+2.61 $R^2=0.98$ (1)
N_2 $Y_2 = 9.96$ kc+5.29 $R^2=0.93$ (2)
N_3 $Y_3 = 17.87$kc+2.61 $R^2=0.99$ (3)

Aug.-N_1 $Y_1 = 9.26$kc+3.50 $R^2=0.97$ (4)
N_2 $Y_2 = 12.00$kc+ 2.61 $R^2=0.96$ (5)
N_3 $Y_3 = 17.25$kc+ 1.79 $R^2=0.98$ (6)

Sept.-N_1 $Y_1 = 12.00$kc+ 2.29 $R^2=0.97$ (7)
N_2 $Y_2 = 15.13$kc+ 2.33 $R^2=0.98$ (8)
N_3 $Y_3 = 17.68$kc+1.32 $R^2=0.98$ (9)

4. Conclusions

According to Figures 1 and 2, it may be seen that 1; for lower water application rates (kc<0.5), yield response to wastewater is comparable to that of potable water without nitrogen 2) for larger application rates (kc > 1), yield response to wastewater is comparable to that of potable water combined with nitrogen between 0 and 25 kg ha^{-1} month^{-1}. Visual appearance of the lawngrass for kc > 1.2 was good when wastewater was applied. On the other hand, the minimal absolute yield for good usual appearance (9,000 kg ha^{-1} month^{-1}) was obtained for kc = 1.9 using potable water without nitrogen. The corresponding kc values for N_1, N_2 and N_3 were 0.6, for N_1 and 0.5 for both N_2 and N_3

Acknowledgements

This study was partially supported by the CEC, progamme Avicenne: Integrated management of reclaimed wastewater resources in the Mediterranean region, Project RTD, n° 93AVI076.

5. References

Angelakis AN and Tchobanoglous G (1997) Necessity of establishing EU-guidelines for wastewater reclamation and reuse: withemphasis the EU-mediterranean countries. Acta Horticulturae 449(2):673-682.

Arsénio A Gamito P Faleiro ML Brito JC and Beltrão J (1998) Wastewater quality for reuse on irrigation in Algarve (Portugal). 1st Inter-Regional Conference on Environment- -Water: Innovative Issues in Irrigation and Drainage. ICID, Lisbon, Portugal.

Asano T and Mills RA (1990) Planning and analysis for water reuse projects. J. Am. Water Works Association, 38-47.

Beltrão J and Ben Asher J (1993) Simulation of maize yield response to combined effects of nitrogen fertilization versus irrigation and population. MAC Fragoso and ML van eusichem (eds.), Optimization of Plant Nutrition, 153-155. Kluwer Academic Publishers.

Christiansen J (1942) Irrigation by sprinkling Calif. Agric. Esp. Bull. 670.

Doorenbos J and Kassam AH (1979) yield response to water. FAO Irrig. Drain. Pap. 33:1-193. Rome, Italy.

266

Marecos do Monte, MH (1996) Contribution to the use of treated wastewater for irrigation in Portugal. Ph.D. Thesis. Technical University of Lisbon, Portugal

Or D and Hanks RJ (1992) A single point source for the measurements of irrigation production functions. Irrig. Sci. 13:55-64.

Oron G de Malach Y Hoffman Z and Manor Y (1992) Effect of effluent quality and application method on agricultural productivity and environmental control. Water Sci. Technol. 26, 1593-1601.

Oron G and Beltrão J (1993) Complete environmental effluent disposal and reuse by drip irrigation. MAC Fragoso and ML van Beusichem (eds.), Optimization of Plant Nutrition, 589-592. Kluwer Academic Publishers.

Chapter 61

The influence of wastewater treatment on irrigation water quality

P. Gamito, A. Arsénio, M. L. Faleiro, J.C. Brito and J. Beltrão

Unidade de Ciências e Tecnologias Agrárias, Universidade do Algarve, Campus de Gambelas, 8000 Faro, Portugal

Key words: wastewater, irrigation, crop coefficient, soil and plant contamination

Abstract: Reclaimed wastewater has been used extensively as a source of irrigation water for centuries. In addition to provide a low water source cost, other side benefits include increases in crop yields and decreased reliance on chemical fertilizers. One of the main aspects of the wastewater related to its quality is the kind of treatment used. Thus, three wastewater plants were studied. The quality of the final effluents of three different reclamation plants was analysed, in order to apply them on irrigated crops. From these final effluents, it was shown that two of them had good quality to be reused on irrigation, and the characteristics of the one were not good enough to be reused, due to the large concentration of pathogens. Those effluents of better quality were applied to lawngrass, composed by Bermuda grass and Dallis grass, in order to study the response of this lawngrass to the application of this final effluent. It was concluded that for a crop coefficient around 1.2, the lawngrass had a good visual rating, though a lower yield; additionally a very low soil and plant contamination was verified.

1. Introduction

The reuse of municipal wastewater is a practice which maximizes profits and increases yields (Oron and de Malach, 1987). However, the intense use of these effluents can contribute to increase soil, plant and groundwater pollution (Asano and Mills, 1990). A possible remedy to combat this conflict is to use the treated wastewater with minimal pollution risk (Oron and Beltrão, 1993). Secondary and Terciary wastewater disposal and reuse is subject to several environmental and health criteria which are regularly defined for each region (Angelakis and Tchobanoglous, 1997). The advantage of wastewater reuse in agriculture is, beside the irrigation, the soil fertilization (Marecos do Monte, 1996). The objective of this work is 1) to study the bacteriological and chemical characteristics of reclaimed wastewater, from three reclamation plants (A, B and C); and 2) to study the response of soil and crop (lawngrass) to the reuse of wastewater, simultaneously as a yield factor and as a contamination factor.

2. Materials and Methods

The monitored parameters included chemical constituents and bacteria concentrations in the three final effluents and bacteria concentration in the soil and grass leaves of the experimental plots, irrigated with final effluents of the reclamation plant C. Samples of final effluents, crop (lawngrass described below) and soil samples, were collected approximately twice a month, according to Arsénio *et a*l. (1998). The most probable number (MPN) of fecal colforms and *E. coli* were determined by cultivation on enrichment medium and selective medium "Billis Green". Enterococci were cultivated in "Rothe medium". The identification of *E.coli* was done by the IMVC test. The detection of

Salmonella was done in selective media "Rappapor-Vassiliadis" and "Selenite". It was used to the isolation process the media "Rambach Agar" (Merck) , and "Kliger". This identification was done by the use of Api system "Rapid 20E".

Chemical parameters were determined by spectophotometry (kit: ELE-Paqualab TM Photometer EL430-550). The pH and conductivity of wastewater were determined by a portable conductivimeter and a potenciometer, respectively.

Irrigation was studied to two mixed grass cultivars: Bermuda grass (*Cynodon dactilon*, L. Pers) and Dallis grass (*Paspalum dilatatum*, Poiret). The irrigation experiments were conducted in Campus da Penha, University of Algarve, Portugal, during Spring and Summer (Reclamation plant C). This reclamation plant serves this Campus of the University, including laboratories and a cantine.

An experimental design known as sprinkle point source was used (Or and Hanks, 1992). It is characterized by the assumption that a point creates a linear irrigation gradient from the water point source, producing a gradual change in water application, and a high degree of irrigation uniformity must be obtained in parallel isohyets. The wettest zone was near the sprinkler and the treatments near the borders were the driest. The irrigation system was stopped when wind speed was larger that $1 ms^{-1}$. A 323/92 Naan sprinkler, 2.5 x 4.5 mm diameter nozzles, using a 300 kPa sprinkler pressure, with a wetting radius about 10 m was used. Christiansen coefficient CUC (1942) was used to determine the uniformity of water distribution at 2; 4; 6; and 8 m from the sprinkle point source; it was obtained, respectively, CUC's values of 92.5; 92.7; 91.9; and 73.2 %.

3. Results

Table 1 shows the maximum admitted values for irrigation water, according to the Portuguese Law (74/90). Tables 2, 3, and 4 show the results of the effluents quality, respectively obtained to the reclamation plants A, B and C.

Table 1. Maximal admitted values for irrigation water quality, according to the Portuguese Law (74/90)

Parameter	Symbol	Unit	Value
pH			4.5-9.0
Elect. cond	ECw	dS m^{-1}	3
Nitrates	NO$_3$$^-$	mgkg^{-1}	30
Bicarbonate	HCO$_3$$^-$	mgkg^{-1}	520
Boron	B	mgkg^{-1}	0.75
Chloride	Cl$^-$	mgkg^{-1}	150
Copper	Cu	mgkg^{-1}	5
Iron	Fe	mgkg^{-1}	20
Manganese	Mn	mgkg^{-1}	10
Fecal colif.		MPN 100 ml^{-1}	10^3
Salmonella		MPN 100 ml^{-1}	0

Table 2. Chemical and microbiological parameters of the final effluent (Reclamation Plant A)

Parameter	Symbol	Unit	Value
pH			6.7-8.7
Elect. cond	ECw	dS m^{-1}	2.0-2.7
Nitrates	NO$_3$	mgkg^{-1}	0
Phospphate	P$_2$O$_5$	mgkg^{-1}	0.0-2.3
Potassium	K$_2$O	mgkg^{-1}	0-167
Ammonium	NH$_4$$^+$	mgkg^{-1}	0
Bicarbonate	HCO$_3$$^-$	mgkg^{-1}	317-465
Boron	B	mgkg^{-1}	0.3-0.8
Calcium	Ca$_2$$^+$	mgkg^{-1}	100-200
Chloride	Cl$^-$	mgkg^{-1}	150-200
Copper	Cu	mgkg^{-1}	0.00-0.22
Iron	Fe	mgkg^{-1}	0
Magnesium	Mg	mgkg^{-1}	12-26
Manganese	Mn	mgkg^{-1}	0
Fecal colif.		MPN 100 ml^{-1}	0.0-3.5x10^2
E.coli		MPN 100 ml^{-1}	0
Enterococos		MPN 100 ml^{-1}	0
Salmonella		MPN 100 ml^{-1}	0

Tables 5 and 6 show, respectively, the results of the soil and grass leaves, irrigated with the final effluent of plant reclamation C.

Fig. 1 shows the grass response to wastewater application to the crop coefficient kc.

Table 3. Chemical and microbiological parameters of the final effluent (Reclamation Plant B)

Parameter	Symbol	Unit	Value
pH			6.9-8.1
Elect. cond	ECw	$dS\ m^{-1}$	1.7-1.9
Nitrates	NO_3^-	$mgkg^{-1}$	0
Phosphate	P_2O_5	$mgkg^{-1}$	0.0-13.7
Potassium	K_2O	$mgkg^{-1}$	0-223
Ammonium	NH_4^+	$mgkg^{-1}$	0
Bicarbonate	HCO_3^-	$mgkg^{-1}$	285-390
Boron	B	$mgkg^{-1}$	1.05-1.56
Calcium	Ca_2^+	$mgkg^{-1}$	150
Chloride	Cl^-	$mgkg^{-1}$	100
Copper	Cu	$mgkg^{-1}$	0.00-0.03
Iron	Fe	$mgkg^{-1}$	0
Magnesium	Mg	$mgkg^{-1}$	7.5-10.5
Manganese	Mn	$mgkg$	0
Fecal colif.		$MPN\ 100\ ml^{-1}$	$1.3x10^5$-$7.0x10^6$
E.coli		$MPN\ 100\ ml^{-1}$	$1.0x10^3$-$5.1x10^6$
Enterococos		$MPN\ 100\ ml^{-1}$	$1.7x10^5$-$1.7x10^7$
Salmonella		$MPN\ 100\ ml^{-1}$	0

Table 4. Chemical and microbiological parameters of the final effluent (Reclamation Plant C)

Parameter	Symbol	Unit	Value
pH			7.1-7.2
Elect. cond	ECw	$dS\ m^{-1}$	1.8-2.5
Nitrates	NO_3^-	$mgkg^{-1}$	0-10
Phosphate	P_2O_5	$mgkg^{-1}$	0
Potassium	K_2O	$mgkg^{-1}$	0-165
Ammonium	NH_4^+	$mgkg^{-1}$	0
Bicarbonate	HCO_3^-	$mgkg^{-1}$	340-365
Boron	B	$mgkg^{-1}$	0
Calcium	Ca_2^+	$mgkg^{-1}$	200-217
Chloride	Cl^-	$mgkg^{-1}$	100
Copper	Cu	$mgkg^{-1}$	0.00-0.01
Iron	Fe	$mgkg^{-1}$	0
Magnesium	Mg	$mgkg^{-1}$	23-33
Manganese	Mn	$mgkg^{-1}$	0
Fecal colif.		$MPN\ 100\ ml^{-1}$	$0.0-1.5x10^3$
E.coli		$MPN\ 100\ ml^{-1}$	0
Enterococos		$MPN\ 100\ ml^{-1}$	$0.0-4.5x10^3$
Salmonella		$MPN\ 100\ ml^{-1}$	0

Table 5. Bacterial contamination of the grass leaves after the irrigation

Parameter	Symbol	Unit	Value
Fecal colif.		$MPN\ 100\ ml^{-1}$	$0.0-2.3x10^3$
E.coli		$MPN\ 100\ ml^{-1}$	$0.0-2.3x10^3$
Enterococos		$MPN\ 100\ ml^{-1}$	$0.0-9.2x10^3$
Salmonella		$MPN\ 100\ ml^{-1}$	0

Table 6. Bacterial contamination of the soil (0.0 - 0.30 m) after the irrigation

Parameter	Symbol	Unit	Value
Fecal colif.		$MPN\ 100\ ml^{-1}$	0.0-6.0
E.coli		$MPN\ 100\ ml^{-1}$	0.0-6.0
Enterococos		$MPN\ 100\ ml^{-1}$	$0.0-0.6x10^3$
Salmonella		$MPN\ 100\ ml^{-1}$	0

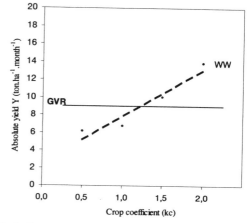

Fig.1. Grass response to wastewater application and to the crop coefficient kc

4. Conclusions

It is possible to consider partially treated final effluents as an alternative disposal to potable water in the golf courses of the Mediterranean region, since we take all cares, according to the various environmental and health criteria, which are regularly defined for each region. The following main aspects are concluded: 1) The bacteriological and chemical analysis demonstrated that the final effluents from reclamation plant A and C can be used on irrigation and the quality of final effluent from reclamation plant B is not good enough for irrigation purposes; 2) Although sampling was conducted after irrigation termination, limited or no bacteria were detected in the soil and grass leaves; 3) For a crop coefficient kc around 1.2, the lawngrass had a good visual rating, though a lower yield obtained; 4) These findings coincide with previous findings, and consequently the wastewater irrigation is a promising technology to satisfy nutrient and water demands and avoid pollution problems.

Acknowledgements

This study was partially supported by CEC programme Avicenne: Integrated management of reclaimed wastewater resources in the Mediterranean region, Project RTD, n° 93AVI076.

5. References

Angelakis, A.N. and Tchobanoglous G. (1997) Necessity of establishing EU-guidelines for wastewater reclamation and reuse: with emphasis the EU-Mediterranean countries. Acta Horticulturae 449(2): 673-682.

Arsénio, A. Gamito, P. Faleiro, M.L Brito, J.C and Beltrão, J. (1998) Wastewater quality for reuse on irrigation in Algarve (Portugal). 1st Inter-Regional Conference on Environment-Water: Innovative Issues in Irrigation and Drainage. ICID, Lisbon, Portugal.

Asano, T. and Mills, R.A (1990) Planning and analysis for water reuse projects. J. Am. Water Works Association, 38-47.

Christiansen, J. (1942) Irrigation by sprinkling. Calif. Agric. Esp. Bull. 670.

Doorenbos, J. and Kassam, A.H (1979) Yield response to water. FAO Irrig. Drain. Pap. 33:1-193. Rome, Italy.

Marecos do Monte, M.H (1996) Contribution to the use of treated wastewater for irrigation in Portugal. Ph.D. Thesis. Technical University of Lisbon, Portugal

Or, D. and Hanks, R.J (1992) A single point source for the measurements of irrigation production functions. Irrig. Sci. 13: 55-64.

Oron, G. and Beltrão , J. (1993). Complete environmental effluent disposal and reuse by drip irrigation. MAC Fragoso and ML van

Beusichem (eds.), Optimization of Plant Nutrition, 589-592. Kluwer Academic Publishers.

Oron, G. and de Malach, Y. (1987) Response of cotton to treated domestic wastewater applied through trickle irrigation. Irrig. Sci.

8: 291-300.

7- Crop quality – nutrient management in general

Chapter 62

S.I.Mul.Fer. - A Multimedia information system for citrus fertilisation

G. Lacertosa[1], G. Basile[1], R. Crudele[2], G. Storelli[2], F. Intrigliolo[3], C. Mennone[4], D. Palazzo[1]

[1]Metapontum Agrobios, SS. 106, Km. 448.2, I-75010 Metaponto (MT), Italy

,[2] Tecnopolis, Valenzano (BA), Italy,

[3] Istituto Sperimentale per l'Agricoltura, Acireale (CT), Italy

[4]A.A.S.D. Pantanello, SS. 106, Km. 448.2, I-75010 Metaponto (MT), Italy

Key words: citrus, fertilisation, multimedia application

Abstract: Multimedia techniques are innovative tools for processing, supplying and diffusing information to be used as a support for land planning and resources management in agriculture. In this framework S.I.MUL.FER has been developed, a prototype of a multimedia information system for Citrus fertilisation. The software has been developed on Personal Computer platform using the Microsoft Fronte Page 98 software. It has been implemented on CD-ROM and is also available on an Internet Web site. It contains an integrated network of knowledge on the different aspects of Citrus nutrition. Therefore it can be satisfactorily used by advisory services for understanding Citrus nutritional problems and, thus, optimising the use of fertiliser inputs. The agronomic knowledge is organised as: macro and micronutrient schedules; information and pictures of nutritional deficiency symptoms on different plant organs; a fertiliser database; algorithms for fertilisation scheduling in the most representative cultural and soil conditions; a guide to the knowledge of Citrus nutrition and cultivation. Thanks to the multimedia technology, an intuitive interaction with the software is possible as well as an easy navigation in the various structures.

1. Introduction

Improper application of fertilisers can cause nutritional imbalances with a consequent depression of crop vegetative and productive responses, and side-effects of environmental pollution. The use of more efficient fertiliser programmes has been halving the cost of fertilisers maintaining production and improving fruit quality in southern Australia (Gallasch, 1992).

Information on soil, climatic and agronomic conditions as well as knowledge of plant nutrition and soil chemistry must be integrated for the formulation of rational fertilisation plans. In order to improve crop yields and quality, a proper utilisation of leaf and soil analysis is necessary, looking at the factor or the factors that restrict profit the most (Embleton, 1996).

Recent technological advances in multimedia applications offer great development opportunities for a more efficient knowledge/information transfer in agriculture (Carrascal et al., 1995). These environments offer many advantages: they let the user "navigate" freely through information, data, images, tables and graphics; they include different levels of explanation according to the user specialisation; they provide natural interfaces, the writer shaping the document to fit the information instead of forcing the information into an arbitrary structure.

274

In this framework S.I.MUL.FER, a prototype of a multimedia information system for Citrus fertilisation, has been developed. It can be satisfactorily used by advisory services for understanding Citrus nutritional problems and, thus, optimising the use of fertiliser input. Moreover the possibility of diffusing S.I.MUL.FER through the Internet allows a fast data update, giving chance to build up an interactive customised service which can reach a higher number of users.

S.I.MUL.FER is the final product of the first phase of a project aimed at developing a multimedia expert system for Citrus fertilisation starting from other works (Palazzo et al. 1993, Resina et al. 1992).

2. Materials and Methods

S.I.MUL.FER. contains an integrated network of knowledge on the different aspects of Citrus nutrition with particular emphasis on regional aspects. The agronomic knowledge and information have been derived from experimental results, survey information, literature data, advisory and expert experience.

The local and regional informations are supported by a monitoring network of reference orchards scattered throughout Metaponto area and a data bank including results from leaf and soil analyses (Palazzo et al. 1997, Lacertosa et al. 1997).

S.I.MUL.FER. has been developed on Personal Computer platform using the Microsoft Front Page 98 and Java Script software. It has been implemented on CD-ROM and is also available on an Internet Web site (www.tno.it/Agrobios).

3. Results and Discussion

At the moment the user can: require information on the functions of plant nutrients; navigate among picture and textual explanations on symptoms of the most important nutritional imbalances; obtain the amounts of each nutrient to be distributed according to the phenological stage, orchard age and/or expected productivity, soil properties and fertilisation method.

The agronomic knowledge is organised as: macro and micronutrient schedules; information and pictures of nutritional deficiencies symptoms on different plant organs; a fertiliser database; a guide to the knowledge of Citrus nutrition and cultivation; algorithms for fertilisation scheduling in the most representative cultural and soil conditions (fig. 1). It is possible to receive information by choosing nutrient, plant organ and nutritional status (fig. 2).

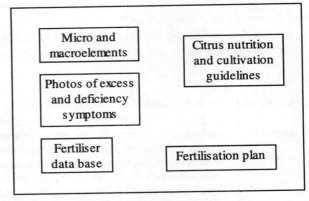

Fig. 1. The agronomic knowledge of the system.

- Nitrogen
- Phosphorus
- Potassium
- Calcium
- Magnesium   Type of plant   Nutritional
- Sulfur   • Old leaf   status
- Iron   • Top leaf   • Excess
- Manganese   • Fruit   • Normal
- Zinc   • Flower   • Deficit
- Boron   • The plant
- Copper   • The branch
- Sodium
- Chlorine

Fig. 2. Information to select

Macro and micronutrient schedules contain different sections: functions and plant requirement of each element, plant and soil content, interaction with other elements.

A proper utilisation of leaf and soil analyses is based on the availability of reference values suitable for a particular agricultural area. S.I.MUL.FER. utilises and spreads local and regional information deriving from a network of reference orchards, where every year leaf, soil and fruit analyses are carried out. For example one of the informations available in S.I.MUL.FER. is the most recent graphical processing of nutritional status of Citrus in Metaponto area (fig. 3).

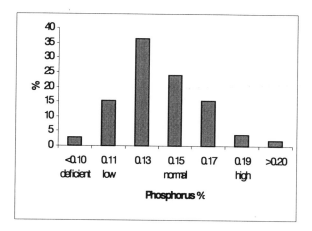

Fig. 3. Bar chart showing citrus leaf P levels measured in 1997 in the reference orchard network of the Metaponto area.

Information on nutritional deficiency and excess symptoms, subdivided for different plant organs, contains: pictures of the symptoms, diffusion of the symptoms and techniques to prevent or to alleviate nutrient imbalances.

The fertiliser database is a catalogue where informations about agricultural fertilisers are represented with a good level of detail: percentage, application amount, producer.

The fertilisation plan provides amounts of each nutrient to be distributed according to the nutritional status and phenological stage. It is possible to receive information about fertilisation scheduling choosing one of the available nutrient elements, the nutritional status, the orchard age, the expected productivity, the type of soil and the

fertilisation method. For each nutrient element 54 fertilisation schedules are available.

A guide to the knowledge of Citrus nutrition and cultivation is always available together with a bibliographical section.

In the second part of the work an expert system for adapting the fertilisation plan to the peculiarities of every citrus orchard will be integrated in the software, exploiting Internet capability to diffuse information.

4. References

Carrascal MJ, Pau L F and Reiner L (1995) Knowledge and information transfer in agriculture using hypermedia: a system review. Comput. Electon. Agric. 12: 83-119.

Embleton TW, Coggins CW Jr. and Witney GW (1996) What is the most profitable use of Citrus leaf analysis? Proc. Int. Soc. Citriculture, Vol. 2: 1261-1264.

Gallasch PT (1992) The use in Southern Australia of Citrus Leaf Analysis to develop more efficient fertiliser programs Proc. Int. Soc. Citriculture, Vol. 2: 560-563.

Lacertosa G, Montemurro N, Martelli S, Santospirito G and Palazzo D (1997) Indagine sulle caratteristiche chimico fisiche dei suoli agrari nella regione Basilicata. L'Informatore Agrario, 26:2-4.

Palazzo D, Lacertosa G, Rondinone T, Mennone C and Montemurro N (1997) Fertilità dei terreni e stato nutrizionale degli agrumi. L'Informatore Agrario, 46: 67-71.

Palazzo D, Basile G, D'Agostino R, Intrigliolo F, Chiriatti K and Resina C (1993) An expert system for diagnosing citrus nutritional status and planning fertilization. In Optimization of Plant Nutrition, Kluwer Academic Publishers, 161-166.

Resina C, Chiriatti K, Palazzo D, Coniglione L and Intrigliolo F (1992) SEFEAG: an expert system for citrus fertilisation. Proc. Int. Soc. Citriculture, 580-583.

Chapter 63

Effects of Cu and Zn added to an agricultural soil in laboratory open system and in Mitcherlich pots

F.A. Biondi[1], C. Di Dio[1], L. Leita[2], S. Socciarelli[1] and A. Figliolia[1]

[1]Istituto Sperimentale per la Nutrizione delle Piante Via della Navicella 2, 00184 Roma, Italy
[2]Istituto Sperimentale per la Nutrizione delle Piante Via Trieste 23, 34170 Gorizia, Italy

Key words: Cu, Zn soil exchangeable cations soil-plant system

Abstract: The purpose of this research was to evaluate the influence of different amounts of cupper and zinc applied to an agricultural soil on the behavior of exchangeable cations (Ca^{++}, Mg^{++}, Na^+ and K^+), in relation to two different open systems: one carried out in a laboratory with use of chromatography columns and one using a pot cultivation of *Hordeum vulgare*. As regards the laboratory experience, the highest degree of cation leaching was found in the treatments receiving only Cu. Oppositely, the lowest degree of leaching was found in the control. On the other hand, the application of both Cu and Zn resulted in an amount of cation leaching lower than the amounts resulting from each of the two heavy metals alone. As regards the Mitcherlich experiment, all the treatments resulted in a decrease of the CEC in comparison to the control, due to the addition of these two metals. Besides, all the treatments with both Cu and Zn showed a decrease in biomass yield relatively to the control, particularly in the treatments in which the metals had been applied at sowing. In both experiment it was found a negative synergism of the two metals added to the systems: as regards the laboratory experience, the treatment receiving Cu + Zn together, showed the lowest concentration of cations leached in the percolates; as regards the Mitcherlich pots experiment, the treatment receiving the two metals together at sowing showed the lowest biomass yield. .

1. Introduction

Heavy metals and other undesirable elements are continuously released into the environment in increasing amounts, due to the common agricultural practice of using manure and biomasses such as sludge, compost and other animal waste, so as to use in the better way the whole amount of human waste (Ottavi et al., 1995).

The literature evidences that Cu, which represents an essential microelement for plant growing, due to its important role in the enzymatic oxidative functional groups, is shown to be particularly present in the root zone. The Cu deficiency threshold in plant has been found to be nearly 4 mg kg^{-1}, while its toxicity threshold has been assessed around 20 mg kg^{-1} (Vismara et al., 1983; Lindsay, 1979; Ross, 1994).

As regards Zn, it is well known that it represents an essential microelement for plant nutrition, particularly for the monocotyledon species which are the most tolerant for its accumulation in their tissues (Robson, 1993). In the literature one can also find that normal amounts of Zn in plants range between 10 and 100 mg kg^{-1}, even though sometimes its phytotoxicity threshold has been found to be about 400 mg kg^{-1}, whereas its deficiency threshold has been considered to be around 20 mg kg^{-1} (Chapman, 1996).

The purpose of this research was to give an updated assessment of the effects of Cu and Zn as applied to an agricultural soil and in particular to evaluate their influence on the yield and uptake by a barley crop, and their effects on the dynamics of Ca, Mg, Na and K present in the soil.

Moreover this research aimed to assess whether these two metals bring about positive or negative effects to the characteristics of the plant-soil system.

2. Materials and Methods

The soil used for both experiments of this study was the surface horizon (Ap) of a soil collected in Latium (Italy) and classified as Mollic Calcic Cambisol (FAO classification, 1971-81).

A laboratory experiment in open system was carried out using 30 cm chromatographic columns (3 cm $\varnothing$), equipped with taps for the methodical collecting of percolates. Each column was filled with 100 g of soil, mixed with a chemically inert material in order to obtain the most suitable drain conditions. Cu and Zn were applied to the columns under four treatments plus a control, laid out as shown. There were three replications. The columns were treated monthly with 100 ml of $H_2O + CO_2$, pH 4.5, to simulate the rainwater acidity conditions. The treatments were: C = no addition(control); C1 = 200 mg kg^{-1} Cu + 300 mg kg^{-1} Zn; C2 = 400 mg kg^{-1} Cu + 600 mg kg^{-1} Zn; C3 = 400 mg kg^{-1} Cu and C4 = 600 mg kg^{-1} Zn

In a second experiment, *Hordeum vulgare* was grown in Mitcherlich pots containing 7 kg of soil, with three replication in random blocks. These pots were kept outdoors, protected from rainfall and maintained to the field capacity. Two series of each treatment plus a control were prepared:

1st series: Cu and Zn, as sulfate, were applied at sowing: S1 = 200mg kg^{-1} Cu+400 mg kg^{-1} Zn; S2 = 600 mg kg^{-1} Zn and S3 = 400 mg kg^{-1} Cu.

2nd series: Cu and Zn, as sulfate, were applied at tillering: V1 = 200mg kg^{-1} Cu+400 mg kg^{-1} Zn; V2 = 600 mg kg^{-1} Zn and V3 = 400 mg kg^{-1} Cu.

At the end of the experience the barley yield was determined and the soil characteristics of each treatment were analyzed: pH (measured in a soil water ratio of 1:2.5), Ca, Mg, Na and K concentrations, CEC values, total and available amounts of Cu and Zn.

Ca, Mg, Na and K concentrations in columns percolates and in the soil of the pots were determined by atomic absorption spectrometer analysis (Mi.R.A.A.F., 1994), while the amount of Cu and Zn were measured by ICP (Martin T., Martin E. and Mckee, 1987).

The obtained results were submitted to Duncan's statistical test.

3. Results and Discussion

As regards the laboratory open system experiment, at the end of the six leachings, the treatments added with Cu and Zn, showed the maximum release of exchangeable cations in the percolates, from 0.1 to 3.3 times that of the control. The highest values of exchangeable cations at the end of the experiment were found in C3, which had received 400 mg kg^{-1} of Cu. No relevant differences of Cu and Zn concentrations in the percolates of all the treatments were found as compared to the control. This phenomenon shows that the addition of Cu and Zn at those rates is not related with the release of the two metals.

Table 1 – Exchangeable cations in $H_2O + CO_2$ percolates (meq 100g^{-1})

Treat.	1st L	2nd L	3rd L	4th L	5th L	6th L	ΣK_{lav}
C	0,76	1,77	1,10	0,51	0,71	0,69	5,54 a
C1	3,99	0,61	0,93	1,16	1,21	0,50	8,40 ab
C2	1,72	2,09	0,69	0,73	0,52	0,44	6,19 ab
C3	8,80	2,75	0,50	0,42	0,68	0,73	13,89 b
C4	4,35	1,88	1,03	0,48	0,51	0,95	9,22 c

L – leaching - † Letters indicates differences determined by Duncan's test (p<0.05)

Table 2-.Barley biomass analyses and Cu and Zn uptake

Treat.	culms (n°)	h (mean) cm	h (max) cm	Yield dry weight (g)	Cu mg kg^{-1} d.w.	Zn mg kg^{-1} d.w.
Csv	61,0	26,3	39,3	8,35 b	15,0 a	49,2 a
S1	57,8	26,3	35,3	6,57 a	37,3 a	454,2 c
V1	58,3	20,2	33,7	7,17 ab	114,8 b	330,0 b
S2	64,3	23,3	34,3	7,30 ab	17,0 a	440,7 c
V2	60,3	30,3	31,3	7,90 ab	15,3 a	297,3 b
S3	61,0	25,0	30,7	7,17 ab	18,8 a	45,3 a
V3	60,0	23,3	34,2	7,45 ab	104,7 b	46,5 a

† Letters indicates differences determined by Duncan's test ($p<0.05$)

Table 3 – Soil Characteristic of plots at the beginning (B) and at the end of the experiment

Treat.	B	Csv	S1	V1	S2	V2	S3	V3
pH	8,4	8,4	8,0	8,0	8,1	8,1	8,2	8,2
Ca meq100g^{-1}	32,9	33,5	31,6	30,4	30,6	26,2	27,0	27,7
Mg meq100g^{-}	2,5	1,2	1,1	1,2	1,2	1,1	1,2	1,2
Na meq100g^{-}	0,2	0,2	0,2	0,2	0,2	0,3	0,2	0,2
K meq100g^{-}	0,6	0,3	0,3	0,2	0,2	0,2	0,3	0,3
CEC meq100g^{-}	36,1	35,0 d	32,9 c,d	32,1 a,b,c	332,3 b,c,d	27,9 a	28,8 a,b	29,3 a,b
Avail. Cu mg kg^{-1}		1,5 a	47,6 b	27,2 a,b	2,3 a	1,7 a	82,0 c	46,4 b
Tot. Cu mg kg^{-1}		26,2	182,9	111,0	29,1	27,7	216,2	146,8
Avail. Zn mg kg^{-1}		0,4 a	71,8 c	50,0 b,c	42,1 b	34,6 b	1,1 a	0,6 a
Tot. Zn mg kg^{-1}		31,4	256,2	192,2	242,5	194,6	32,7	31,2

† Letters indicate differences determined by Duncan's test ($p<0.01$)

In the treatments with simultaneous application of Cu + Zn (C1 and C2), an appreciable decrease of exchangeable cations in the percolates was noticed, as compared the treatments in which the two metals had been added singularly. Moreover, exchangeable cations released by the C1 and C2 treatments presented concentrations more similar to the control (Table 1).

The separate application of Cu (C3) and Zn (C4) unexpectedly induced a greater release of exchangeable cations in the percolates and, according to the metal species added and its concentration, the amount of the release ranged from 2.5 to 3.3 times that of the control. It was also found that Cu is more active than Zn in its ability of removing cations.

As regards the Mitcherlich pot experiment, the barley biomass yield showed to be influenced by the Cu and Zn additions to the soil (Table 2). In fact, even though no flattening phenomenon took place, the added metals caused a depressing effect, the letting production vary between treatments. The lowest production was found in S1, in which a double amount of Cu + Zn had been applied at sowing. This phenomenon leads again to assume that a synergism between the two metals results in a lesser amount of exchangeable cations available to the plant (Biondi, et al. 1995).

As regards heavy metals plant uptake (Table 2), it was found that Zn uptake is higher in the treatments with Cu and Zn application at sowing.

The highest concentration of available Cu in the soil was found in S3 with 82 mg kg^{-1}, while the highest concentration of available Zn was found in S1 with 72 mg kg^{-1}; in this last treatment was also found the maximum plant uptake of the two metals. At the end of the experience, all the treatments showed a decrease of soil CEC, even though sometimes slight, as compared to the control. This phenomenon evidences the impoverishing effect of the exchangeable cations (Table 3).

4. Conclusion

On the whole, from the experiment carried out, it can be assumed what follows: i) a stronger activity of the Cu ion compared to the Zn ion in their ability of removing cations in the open laboratory system; ii) a detrimental synergism of Cu+Zn in relation to the leacheability of Ca, Mg, Na and K in the open laboratory system; iii) a possible relation between the trend of the biomass yield in the pot experience and the reported phenomena in the open laboratory system.

5. References

Biondi F.A., Figliolia A., Di Dio C., Sequi P., Socciarelli S. (1995). Influenza del Cu e dello Zn sui Cationi del Complesso di Scambio. I Congresso Nazionale di Chimica Ambientale. Roma

Chapman H.D. (1966). Zinc. In Diagnostic Criteria for Plants and Soils. Ed. Chapman. pp 489-499. Univ. of California

FAO (1971-81) FAO/UNESCO. Soil map of the World. Vol.1-10, UNESCO, Paris

Lindsay W.L. (1979). Chemical Equilibrial in Soils. J. Wiley, New York

Martin T.D., Martin E.R. and G.D. Mckee (1987). Inductively Coupled Plasma Atomic Emission Spectrometric Method for Trace Element Analysis. EPA Method 200.7

Mi.R.A.A.F - Ministero delle risorse Agricole, Alimentari e Forestali (1994). Metodi ufficiali di analisi chimica del suolo. -Osservatorio Nazionale Pedologico e per la Qualità del Suolo. Roma

Ottavi C., Ottaviani M. e Figliolia A. (1995). Aspetti tecnico-economici, agronomici, pedologici, igienico-sanitari e normativi dei fanghi di depurazione civile. I.S.S.. pp.103-114-. Roma

Robson A.D. (1993). Zinc in Soils Plants. Development in Plant and Soil Science. Kluwer Acad. Publish. Vol.55. Australia

Ross M.S. (1994). Toxic Metals in Soil-Plant System. Wiley Publishers

Vismara R., Genevini P.L., Mezzanotte V. (1983). Utilizzo Agricolo dei Fanghi di Depurazione. Ing. Amb. 12 (9) pp.359-493

Chapter 64

Suitability of Mehlich III method for assessing the plant nutrients in Erzurum plain and acid soils

N. Yıldız, O. Aydemir, A. Aydın, F. Ulusu
University Of Atatürk, Faculty Of Agriculture, Department of Soil Science, Erzurum- Turkey

Key words: soil tests, availability index, multielement extractant.

Abstract: The purpose of this investigation was to find out suitability of Mehlich III. extraction method in determining some of the plant available nutrients (Ca, Mg, Na, K, Fe, Cu, Zn, Mn and P) in Erzurum plain and some acid soils. Dry matter yield ,nutrient content and total nutrient uptake of the test plant were taken as biological index. The result of the statistical analysis indicate that, Mehlich III method was not interrelated with biological index for the plant available nutrients investigated in Erzurum plain and acid soils.

1. Introduction

The Mehlich III method was developed by Mehlich (1984) as a multielement soil extractant which is composed of 0.2 M CH_3COOH, 0.25 M NH_4NO_3 , 0.015 M NH_4F, 0.013 M HNO_3 and 0.001 M EDTA. In the Mehlich III procedure, phosphorus is extracted by reaction with acetic acid and fluoride compounds, exchangeable K, Ca, Mg and Na are extracted by the action of ammonium nitrate and nitric acid and finally, the micronutrients (Fe, Cu, Mn, Zn) are extracted by NH_4^+ and chelating agent EDTA. The amount of Mehlich III extractable P was found to be closely related to that determined by other conventional extractants such as Bray 1, Bary 2, Olsen, Mehlich II and bicarbonate resin (Mehlich 1984; Wolf and Baker, 1985; Michaelson et al., 1987; Ness et al., 1988; Tran et al., 1990). The Mehlich III extractant is neutralized less by carbonate compounds than the double acid and Bray 2 extractants. The amount of Mehlich III extractable P is highly correlated with plant P uptake in a wide range of soils in Quebec (Tran and Giroux, 1985; 1989).

The amount of exchangeable bases (K, Ca, Mg, Na) determined by the Mehlich III method are nearly identical to those obtained by ammonium acetate method (Hollond and Johnson, 1984; Michaelson et al., 1987; Tran and Giroux 1989). Moreover, the Mehlich III extractable amount of Cu, Zn, Mn and Fe are also closely related to those obtained by the double acid, diethylene triamine pentaacetic acid triethanolamine (DTPA-TEA) or 0.1 M HCl methods and can be used in soil testing for micronutrients (Mehlich 1984; Makarim and Cox 1983; Mascagni and Cox 1985; Tran 1989).

2. Materials and Methods

Twenty surface soils (depth of 0-20) representing Erzurum region and 14 surface soils representing acid soils were chosen for the greenhouse experiments. The pH, organic matter, $CaCO_3$ content, CEC and the texture of Erzurum plain soils were 6.1-8.5, 1.31-5.26%, 0.05-19.21%, 14.39-46.2 cmol kg^{-1} and C, CL, L, SCL respectively. In acid soils, these values were 3.2-5, 1-6.4%, 8.4-33.2 cmol kg^{-1} and C-L respectively.

Corn plants were grown in pots under greenhouse conditions in randomised block design with two replicates (Özbek, 1969). Plants were harvested 8 weeks after germination and dry matter yield, Ca, Mg, Na, K, Fe, Cu, Zn, Mn and P content and uptake of the corn plant were determined (Kacar, 1972).

In determining plant available Ca, Mg, Na, K, Fe, Cu, Zn, Mn and P contents of soils, Mehlich III chemical extraction method was used. In the Mehlic III method precedure; 30 ml Mehlich extraction solution added onto 3 g air dry soil, and the content (soil: solution ratio 1:10) was shaken for 5 minutes and filtered. In clear extracts, Ca, Mg, Fe, Cu, Zn and Mn concentrations were determined by Atomic Absorption Spectrophotometer, Na and K were by flame emission and P by colorimetric method (Tran and Simard, 1993).

The correlations between Mehlich III method extracted soil nutrients and biological index were calculated (Düzgünes, 1963).

3. Results and Discussion

Dry matter yield, nutrient contents (Ca, Mg, Na, K, P, Fe, Cu, Zn, Mn and P) and total uptake of corn plants were determined. The mean values are shown in Table 1. In determining the available Ca, Mg, Na, K, P, Fe, Cu, Zn and Mn contents of Erzurum plain and acid soils, Mehlich III extraction method was used .Results are given in Table 2a and 2b.

The result of this study showed that plant available nutrients (Ca, Mg, Na, K, P, Fe, Cu, Zn and Mn) obtained with Mehlich III extraction method were not interrelated with biological index in the Erzurum plain and acid soils (Table 3a and 3b). Results also showed that the Mehlich III method might not be used for plant available soil nutrients at least in the soils investigated.

Although it has orginally been developed as a multielement extraction method and suitability for extracting plant available nutrients was proved under various experimental conditions elsewhere, it has not been found as succesfull in our study. The reason for the failure may be the completely different soil and climatic conditions in this area.

Before concluding that the method is completely unsuitable in evaluating plant available nutrients under semiarid conditions, the method must be tried extensively in other experimental conditions.

Table.1.The mean values of biological index.

Dry matter g p^{-1}	Mineral content of corn plants grown in Erzurum soils								
	Ca	Mg	K	P	Fe	Cu	Zn	Mn	Na
	%				mg kg^{-1}				
0.75-1.42	0.21-1.00	1.25-1.93	2.1-9.4	0.11-0.35	14-234	12-68	75-129	11-255	13-462
	Total nutrient uptake of corn plants grown in Erzurum soils								
	mg kg^{-1}				µg p^{-1}				
	2.1-21.8	9.9-47.3	19.5-99	1.18-4.4	17.9-234	9-68	35-169	12-113	17-210
Dry matter g p^{-1}	Mineral content of corn plants grown in acid soils								
	Ca	Mg	K	P	Fe	Cu	Zn	Mn	Na
	%				mg kg^{-1}				
0.60-2.25	0.77-1.32	1.45-1.96	2.7-8.5	0.21-0.61	87-402	15-153	110-215	11-107	88-970
	Total nutrient uptake of corn plants grown in acid soils								
	mg kg^{-1}				µg p^{-1}				
	6.3-28	10.3-40	19-175	1.9-9.2	76-743	14-344	91-371	25-133	61-260

Table 2a. Extractable Ca, Mg, Na, K, P, Fe, Cu, Zn and Mn obtained by Mehlich III method in Erzurum plain soils.

Soil Sample	Extractable cations				Microelements				
	Ca	Mg	Na	K	Fe	Cu	Zn	Mn	P
	cmol.kg.$^{-1}$				mg kg^{-1}				
1	19.8	15	1.5	1.5	105	11	4.2	50	137
2	6.0	8.5	1.9	0.4	80	6.0	4.6	70	122
3	27	13	2.0	1.2	58	25	3.7	100	185
4	42	21	2.5	0.6	70	17	4.3	60	140
5	34	15	2.8	3.0	65	10	7.7	120	590
6	39	18	2.0	2.0	80	9.0	4.0	60	118
7	9.0	7.5	1.9	1.7	105	5.5	6.5	300	760
8	30	15	4.0	1.2	45	24	3.0	140	176
9	68	33	5.5	2.2	60	5.0	1.9	100	420
10	48	13	2.0	1.8	44	20	2.3	150	98
11	68	25	5.0	2.5	45	7.0	2.0	140	154
12	88	23	2.1	1.1	25	7.5	3.5	75	273
13	30	10	2.4	4.5	75	16	8.0	180	587
14	38	15	3.0	6.0	78	15	7.5	200	803
15	92	27	1.9	2.4	18	7.0	3.6	70	175
16	10	8.5	1.7	1.4	88	8.0	3.4	180	244
17	60	10	2.0	1.0	33	14	3.3	140	503
18	33	13	1.8	0.9	90	18	3.0	190	97
19	9.0	10	2.5	0.5	110	14	3.0	70	47
20	36	18	2.0	1.4	70	4.0	3.1	180	78

Table 2b. Extractable Ca, Mg, Na, K, P, Fe, Cu, Zn and Mn obtained by Mehlich III method in acid soils.

Soil Sample	Extractable cations				Microelements				
	Ca	Mg	Na	K	P	Fe	Cu	Zn	Mn
	cmol kg^{-1}				mg kg^{-1}				
1	7.0	2.3	2.00	0.87	38	165	3.2	5.3	90
2	16.6	13	1.87	1.55	74	116	5.6	3.5	260
3	7.0	5.5	1.92	1.59	23	148	4.8	6.4	120
4	6.5	6.2	1.92	0.82	37	160	8.0	4.0	200
5	19	13.8	2.6	1.13	17	168	19	5.9	360
6	9.0	12.3	2.0	1.41	22	152	6.4	4.4	150
7	6.6	9.2	1.96	2.54	99	204	52	32	142
8	7.35	6.6	1.83	0.71	17	180	5.6	4.0	150
9	5.5	7.0	1.94	0.63	57	132	17.6	22	180
10	5.45	8.0	2.08	3.5	95	228	43	45	172
11	7.6	7.3	1.87	0.42	68	140	25	23	200
12	7.0	14.1	1.87	0.45	85	156	7.2	22	148
13	1.3	6.3	1.82	0.42	36	148	14.4	23	162
14	0.80	2.83	1.89	0.49	19	120	5.6	5.7	141

284

Table 3a The linear correlation cofficients between Mehlich III method and biological index for Erzurum plain soils.

Biological Index	The result of Mehlich III extraction method in Erzurum soils								
	Ca	Mg	Na	K	Fe	Cu	Zn	Mn	P
	cmol kg^{-1}				mg kg^{-1}				
Dry matter (g p^{-1})	-0.29	-0.20	-0.08	-0.11	0.28	-0.23	0.31	0.06	0.20
Mineral content (%, mg kg^{-1})	-0.36	0.16	0.38	0.79	-0.13	0.06	0.08	0.05	0.32
Total nutrient uptake (mg p^{-1}, µg p^{-1})	-0.36	0.07	0.34	0.78	-0.08	-0.06	0.33	0.06	0.37

Table 3b The linear correlation cofficients between Mehlich III method and biological index for acid soils.

Biological Index	The result of Mehlich III extraction method in acid soils								
	Ca	Mg	Na	K	Fe	Cu	Zn	Mn	P
	cmol kg^{-1}				mgkg^{-1}				
Dry matter (g p^{-1})	0.54*	0.41	0.15	-0.16	-0.26	-0.35	-0.51	0.35	-0.17
Mineral content (%, mg kg^{-1})	-0.23	0.47	-0.25	0.68*	0.09	-0.18	0.40	-0.31	0.12
Total nutrient uptake (mg p^{-1}, µgp^{-1})	0.49	0.42	-0.11	0.25	0.33	-0.24	-0.31	-0.05	0.18

4. References

Düzgüneş O.(1963). Bilimsel Araştırmalarda İstatistik Prensipleri ve Metodları. Ege Üniv. Matbaası. İzmir.

Hanlond E.A and Johnson G.V. (1984). Bray/Kurtz Mehlich II. AB/D and acetate extractions of P, K and Mg in four Oklohoma soils. Comm. Soil Sci. Plant Anal. 15:227-294.

Kacar B (1972) Bitki ve Toprağın Kimyasal Analizleri; II. Bitki Analizleri A.Ü.Z.F. Yayınları.453.Uygulama Klavuzu:155, Ank. Üni. Basımevi, Ankara.

Makarim AK and Cox FR (1983) Evaluation of the need for copper with several soil extractant Agron J. 75: 493-496.

Mascagni HJ and Cox FR (1985) Calibration of a manganase availability index for soybean soil test data. Soil Sci. Soc. Am. J. 49: 382-386.

Mehlich A (1984) Mehlich-3 soil test extractant: a modification of Mehllich-2 extractant. Comm. Soil Sci. Plant Anal. 15:1409-1416.

Michaelson GJ Ping CL and Mitchell CA (1987) Correlation of Mehlich –3, Bray 1 and ammonium acetate extractable P, K, Ca and Mg for Alaska agricultural soils. Soil Sci. Plant Anal. 18:1003-1015.

Ness P. Grave J and Bloom PR (1988) Correlation of several tests for phosphorus with resin extractable phosphorus for 30 alkaline soils. Common. Soil Sci. Plant Anal. 19:675-689.

Özbek N (1969) Deneme Tekniği. I. Sera Deneme Tekniği ve Metodları. A.Ü.Z.F. Yayınları: 525 Ders Kitabı: 170, A.Ü. Basımevi, Ankara.

Tran T Sen and Giroux M. (1985) Disponibilite du phosphore dans les sols neutres et calcaires du Quebec en relation evec les les proprietes chimiques et physiques. Can. J. Soil Sci. 67: 1-16.

Tran T Sen and Giroux M (1989)Evaluation de la methode Mehlich III pour determiner les element nutritifs (P, K, Ca, Mg, Na) des sols du Quebec. Agrosol. 2:27-33.

Tran T Sen. 1989. Determination des mineraux et oligo-elements parla methode Mehlich III. Methods d' analyse des sols, des fumiers, et des tissus vegetoux. Conseil des productions vegetables du Quebec. Ministere de I ' Agriculture, des Pecheries et de I' Alimentation du Quebec. Agdex.533.

Tran T Sen Giroux M Guilbeault J and Audesse P (1990) Evaluation of Mehlich III extractant to estimate the available P'ın Quebec soils. Common. Soil Sci. Plant Anal. 21:1-28.

Tran T Sen and Simard RR (1993) Mehlich III extractable elements. In Soil Sampling and Methods of Analysis Canadian Society of soil Sciency.

Wolf AM and Baker DE (1985) Comparisons of soil test phosphorus by Olsen. Bary 1, Mehlich –1 and Mehlich –III methods. Common. Sci. Plant Anal. 16: 467-484.

Chapter 65

Microbial biomass and activity in soils under different cropping systems

N. Okur, M. Çengel and H.S. Uçkan

Ege University, Faculty of Agriculture, Department of Soil Science, 35100 Bornova İzmir/Turkey

Key words: microbial biomass, different cropping systems, C and N-mineralization

Abstract: In agricultural soils, the microbial biomass and activity act as a source-sink for labile nutrients. Practices which alter the size of the biomass or its rate of N and C turnover can affect crop growth. Fifteen soil samples were analyzed for microbial biomass, mineralizable C and N contents. These soils were covered with different crops such as fruit, cereal, vegetable, industry crop and pasture for long years. Microbial biomass ranged from 57,39 to 105,30 mg C 100 g^{-1} and varied significantly with different cropping systems. The highest microbial biomass values were determined in soils which were covered by vegetables. Soils which were under cereal had the lowest microbial biomass, mineralizable C, and N. Inorganic fertilization and minimum tillage increased mineralizable C and N contents.

1. Introduction

In agricultural soils, the microbial biomass and activity act as a source-sink for labile nutrients and practices which alter the size of the biomass or its rate of N and C turnover can affect crop growth. Microbial biomass carbon (C $_{micr}$) reflects the long-term amount of C input into a soil (McGill et al., 1986; Anderson and Domsch, 1986). Apart from substrate quantity, also its quality and distribution determine the amounts of C_{micr} (Insam et al., 1989). In addition, the microbial biomass is sensitive to tillage-induced changes in crop residue incorporation, root activity, cropping systems, and environmental factors such as differences in soil moisture regimes (Carter, 1986; Doran et al., 1987).

Since microbial biomass is generally correlated with the level of soil organic matter and is influenced by climatic variables, its absolute value alone measured once can not show whether a specific cropping or tillage system is gaining or losing organic matter. In order to elucidate changes in organic matter equilibrium, emphasis has been placed on the proportion of total organic C within the microbial biomass (Wu and Brookes, 1988; Anderson and Domsch, 1989). Using this concept, organic matter equilibrium functions have been provided for the ratio of biomass C to total organic C in a wide range of crop rotations (Anderson and Domsch, 1989) and macroclimates (Insam et al., 1989).

Soil microbial biomass also plays an important role in soil N cycling both as a transformation agent and a source-sink of N (Bonde et al., 1988; Duxbury et al., 1991).Biological cycling of N in soil is strongly linked to the dynamics of C. It has been suggested that each soil has an inherent potential to mineralize N under standart conditions (Stanford and Smith, 1972).

The objective of this study was to determine the amount of microbial biomass and mineralizable C and N in soils which were cultivated with different crops such as fruit, cereal, vegetable, industry crop and pasture for long years.

Table 1. The crops and some chemical characteristics of soils

Site No	Crop	Crop Group	Rotation	Fertilization	PH (H$_2$O)	N$_t$ (%)	C$_{org}$ (%)
1	Plum	Fruit	No	Inorganic	8.27	0.080	0.72
2	Sugarbeet	Ind. Crop	Yes	Inorganic	8.26	0.110	0.97
3	Sorghum	Pasture	No	Inorganic	7.48	0.110	1.00
4	Maize	Cereal	Yes	Inorganic	7.58	0.110	1.07
5	Apple	Fruit	No	Inorganic	7.89	0.160	1.35
6	Wheat	Cereal	Yes	Inorganic	6.61	0.080	0.88
7	Pepper	Vegetable	Yes	Organic	7.84	0.150	1.35
8	Tomato	Vegetable	Yes	Organic	7.95	0.180	1.26
9	Vineyard	Fruit	No	Inorganic	8.24	0.100	0.88
10	Bean	Vegetable	Yes	Organic	7.84	0.130	1.21
11	Alfaalfa	Pasture	No	Inorganic	7.97	0.100	0.83
12	Sunflower	Ind. Crop	Yes	Inorganic	7.61	0.110	1.20
13	Sunflower	Ind. Crop	Yes	Inorganic	7.41	0.090	0.82
14	Nectarine	Fruit	No	Inorganic	8.28	0.090	0.94
15	Wheat	Cereal	Yes	Inorganic	7.19	0.070	0.77

2. Materials and Methods

Sampling sites of soils. Fifteen Alluvial soil samples in the northwest Anatolia-Uludağ University-Experimental Station, Bursa were sampled in May 1997 after long years of cultivation with different crops. Table 1 shows the crops and the characteristics of soils.

Soil analysis. The collected soils were sieved (4 mm) and stored at 4°C until needed. Subsamples for the determination of physicochemical parameters were air dried and sieved (2 mm) before analysis. Organic C was determined as described by Walkley and Black, 1934; total N by Bremner, 1960. Other physicochemical analysis were determined according to standard methods.

Measurement of microbial biomass. Microbial biomass C was determined by substrate-induced respiration method (Anderson and Domsch, 1978). Moist soil samples (100 g) were amended with glucose (400 mg), and the pattern of respiration response was recorded for 4 h. By a conversion factor, values were converted to mg biomass-C.

Mineralizable C and N. Mineralizable C and N were estimated from the quantity of CO$_2$-

C and NH$_4$-N+NO$_3$-N , respectively, mineralized from soil sample during a 10-d incubation at 25°C (Campell et al., 1991).

3. Results and Discussion

Microbial Biomass

Microbial biomass ranged from 57.39 to 105.30 mgC 100 g^{-1}, and varied significantly with different cropping systems (Table 2).

Table 2. Mean values of microbial biomass and the ratio of C$_{mic}$ to total C of soils for different cropping systems*

Cropping System	Mic.Biomass C (mg C 100g^{-1} soil)	Organic C in Biomass (%)
Vegetable	105.30 a	8.27
Industry crop	71.47 bc	7.03
Fruit	80.08 b	8.52
Cereal	57.39 c	6.30
Pasture	83.43 ab	9.35

*Means associated with the same letter are not significantly different (p< 0.01, Duncan's multiple range test)

The highest microbial biomass values were determined in soils which were cultivated by vegetables, and the lowest in soils which were cultivated cereals. In most soils, the microbial biomass comprises about 1-3 % of the soil organic matter and there is a reasonably close linear relationship between biomass and soil organic matter (Jenkinson and Ladd, 1981; Anderson and Domsch, 1989). The same relationship was also determined in the research soils (r = 0.655**).

The proportion of organic C in the biomass C for different cropping systems is given in Table 3. The highest proportion was obtained in pasture soils and subsequently in fruit and vegetable soils. The lowest proportion was appeared in soils cultivated by cereals. These differences showed that the organic matter residues in pasture soils allows a more efficient organic matter utilization per unit biomass as compared to the cereal residues. Insam et al. (1989) have shown that soils under permanent monoculture had significantly lower amounts of microbial C per unit soil C than soils under continuous crop rotations.

Mineralizable C and N

The effects of different cropping systems on mineralizable C and N were significant at 1 % level (Table 3). Mineralizable C was highest in the soils of fruit orchards and vegetables, possibly due to intensive fertilization. The fertilizers increased the biomass which could be attribute to better plant growth resulting in greater roots. Soils sown by cereals had lowest mineralizable C. The fact that inputs of the C to soil is more less than the soils planted to the other crops, the mineralizable C amount decreased. C input from crop roots, rhizosphere products (i.e. root exudates, mucilage, sloughed cells, etc.), and crop residues can have a large effect on mineralizable C and N in soils (Ross, 1987), which in turn, affect the ability of soil to supply nutrients to plants through soil organic turnover (Bonde and Roswall, 1987).

The highest mineralizable N was found in soils planted by industry crops and lowest in the soils of fruit orchards. Intensive nitrogen fertilization applied to the soils planted by industry crops may have caused an increase in mineralizable N. Furthermore, it also may have been an indirect result of greater mineralizable

C due to increased deposition of rhizosphere products in the surface layer (Recous et al., 1988). Low mineralizable N in the soils of fruit

Table 3 Mean values of mineralizable C and N for different cropping systems

Cropping System	Mineralizable C (mg CO_2-C g^{-1})	Mineralizable N (mg inorg N g^{-1})
Vegetable	6.35 a	5.40 b
Industry crop	5.75 b	7.09 a
Fruit	7.00 a	2.31 d
Cereal	2.36 d	3.19 c
Pasture	3.66 c	3.00 c

*Means associated with the same letter are not significantly different (p< 0.01, Duncan's multiple range test)

orchards resulted from the decreased substrate quality. In these kind of soils, roots and rhizosphere products can be low in N concentration and high in mineralizable C, leading to significant N immobilization (Mary et al., 1993).

3. References

Anderson,J.P.E. and Domsch,K.H. 1978. A physilogical method for the quantitative measurement of microbial biomass in soils. Soil Biol. Biochem. 10 : 215-221.

Anderson,T.H. and Domsch,K.H. 1986 . Carbon link between microbial biomass and soil organic matter. In proceeding of the Fourth International Symposium on Microbial Ecology (F.Megusar and M.Gantar, Eds.), pp. 476-471. Slovene Society for Microbiology, Ljubljana, Yugoslavia.

Anderson,T.H.and Domsch ,K.H. 1989 . Ratios of microbial biomass carbon to total organic carbon in arable soils. Soil Biol. Biochem. 21: 417-479.

Bonde,T.A and Rosswall,T. 1987 . Seasonal variation of potentially mineralizable nitrogen in four cropping systems. Soil Science Society of America Journal 51, 1508-1514.

Bonde,T.A., Schnüber,J. and Rosswall,T. 1988 . Microbial biomass as a fraction of potentially mineralizable nitrogen in soils from long-term field experiments. Soil Biology & Biochemistry 20, 447-452.

Bremner,J.M. 1960. Determination of nitrogen in soil by the Kjeldahl method. Journal of Agricultural Sciences 55, 11-13.

Campell,C.A., Biederbeck,V.O., Zentner, R.P. and Lafond,G.P. 1991. Effect of crop rotations and cultural practices on soil organic matter , microbial biomass and respiration in a thin Black Chernozem. Canadian Journal of Soil Science 71, 363-376.

Carter,M.R. 1986. Microbial biomass as an index for tillage induced changes in soil biological properties. Soil Tillage Res. 7 : 29-40

288

Doran,J.W., Fraser,D.G., Culik,M.N. and Liebhardt,W.C. 1987. Influence of alternative and conventional agricultural management on soil microbial processes and nitrogen availability. Am. J. Alternative Agric. 2:99-106.

Duxbury,J.M., Lauren,J.G. and Fruci,J.R 1991. Measurement of the biologically active soil nitrogen fraction by a ^{15}N technique. Agriculture, Ecosystems and Enviroment 34, 121-129.

Insam,H., Parkinson,D. and Domsch,K.H. 1989. Influence of macroclimate on soil microbial biomass. Soil Biol. Biochem. Vol. 21. No. 2, 211-221.

Jenkinson, D.S. and Ladd,J.N. 1981. Microbial biomass in soil measurement and turnover. In Soil Biochemistry (E.A. Paul and J.N. Ladd, Eds.), Vol 5, 415-471. Marcell Decker, New York.

Mary ,B., Frensneau,C., Morel,J.L. and Marriotti,A. 1993. C and N cycling during decomposition of root musilage, roots and glucose in soil. Soil Biology & Biochemistry 25, 1005-1014.

McGill,W.B., Cannon,K.R., Robertson, J.A. and Cook,F.D. 1986. Dynamics of soil microbial biomass and water soluble organic C in Breton L after 50 years of cropping the two rotations. Canadian Journal of Soil Science 66, 1-19.

Recous,S., Frensneau,C., Faurie,G. and Mary,B. 1988. The fate of labelled ^{15}N urea and ammonium nitrate applied to a winter wheat crop. Nitrogen transformations in the soil. Plant and Soil 112, 205-214.

Ross,D.J. 1987. Soil microbial biomass estimated by the fumigation-incubation procedure : seasonal fluctuation and influence of soil moisture content. Soil Biology & Biochemistry 19, 397-404.

Stanford,G. and Smith,S.J. 1972. Nitrogen mineralization potentials of soils. Soil Science Society of America Proceedings 36, 465-472.

Walkley,A. and Black,I.B. 1934. An examination of the Degtjareff method for determining soil organic matter a proposed modification of the chromic acid titration method. Soil Science 37, 29-38.

Wu, J. and Brookes,P.C. 1988. Microbial biomass and organic matter relationships in arable soils. J. Sci. Food Agric. 45 : 138-139.

Chapter 66

Effects of different parent materials on some plant nutrients and heavy metals in the arid regions of Turkey

S. Irmak[1] and A. K. Sürücü [2]

1 Dept. Soil Sciences, Fac. Agriculture, University of Harran, Şanlıurfa, Turkey.
2 Dept. Soil Sciences, Fac Agriculture, University of Ondokuz Mayıs, Samsun, Turkey.

Key words: parent material, heavy metals, plant nutrient elements

Abstract: In this study, seven soils which were developed on lime stone, basalt, marine and alluvium parent materials were studied to understand the effects of different parent materials on plant nutrient levels and heavy metal contents in the arid regions. Results showed that different parent materials have influenced the plant nutrients of soils. Available K_2O content of soils developed on basalt rocks are higher than other soils and change between 112.5 and 133.9 kg da^{-1}. Available K_2O contents of soils developed on marine parent material is the lowest and range from 27.0 to 36.7 kg da^{-1}. The high available K_2O content of soils, developed on the basalt material, may be attributed to mica content of basalt rocks. Available K_2O content of unweathered basalt rock is 127.1 kg da^{-1}. Also, available P_2O_5 content of soils, developed on basalt parent material are higher than other soils and range between 2.9 and 6.1 kg da^{-1}. The available P_2O_5 content of soils, developed on marine is very low and range between 0.7 and 1.0 kg da^{-1}. The high P_2O_5 content of soils, developed on basalt parent material, may be related with the chemical composition of basalt rocks. Also, the P_2O_5 content of unweathered basalt rock is 2.9 kg da^{-1}. Nitrogen content of all soils, depending on organic matter, are very low and range between 0.035 and 0.13 %.

1. Introduction

Soils form as a result of weathering and disintegration of parent rocks and materials. Jenny (1949) included biota as one of the five soil-forming factors, along with climate, topography, parent material, and time, in his formal expressions of soil-forming functions. Jenny's works inspired much of soil genesis research, utilizing the factor functions, or sequence approach. In such studies, situations in nature are sought where all factors are constant, except the one under investigation so that the effect of that individual factor on soil formation can be assessed (Buol et al., 1980; Graham and Wood, 1991). The soils exhibit similar characteristics where these soil-forming factors are the same. Parent material affects soil characteristics especially in arid and semiarid

regions (Buringh, 1979). It is of great importance to study the soils by various methods according to their characteristics in order to establish a correlation between soils and plants and to classify them accordingly. Different parent materials affect the morphology and chemistry of soils under the same conditions, such as topography and vegetation, especially in arid and semiarid regions. Differences in physical, chemical and mineralogical properties of soils are related primarily to parent material (Washer and Collins, 1988). A soil landscape pattern generally reflects the original parent material; however, saprolite is highly weathered prior to soil formation (Wysocki et al., 1988). The original separation of soils was based on the type of parent rock and on morphological properties. The objective of this study was to examine the effects of different parent materials on some plant nutrients and heavy metals of the soils in the arid regions of Southeast Anatolia Region, Turkey.

2. Materials and Methods

The study area is characterised by arid climate and lies between 37° 46' and 36° 43' N latitudes and 37° 59' and 39° 46' E longitudes in the Southeast Anatolia Region of Turkey. The average amount of annual rainfall is 400 mm and total evaporation is 2047.7 mm. The mean annual air temperature is 17.8 °C.

Seven soil profiles on four different parent materials were described according to Soil Survey Staff (1996). Soils on different parent materials were sampled from each horizon and air dried to pass a 2 mm siever for laboratory analysis. The organic matter, available P_2O_5, K_2O, Cu, Mn, Mo, Zn analysis were carried out following methods (Black, 1965; Olsen et al., 1954; Kick et al., 1980).

3. Results and Discussion

The available nutrients

Some plant nutrients (N, P2O5, K2O) of the studied soils are presented in Table 1. The nitrogen content depending on the organic matter of all soils are very low. The low N content of soils may be attributed to low organic matter and indirectly to poor vegetation.

The available phosphorus of soils, developed on basalt rocks and $CaCO_3$ rocks, is higher than other soils, forming on marines. The available phosphorus content of profiles PL3 and PL4, developed on marines, is very low and changes between 1.0 and 0.7 kg da^{-1}. The P_2O_5 content of PL6 and PL7, developed on basalts, ranges between 2.1 and 5.7 kg da^{-1}. In the R horizon of profile PL7 and in the C_2k_2 horizon of profile PL6 the available P_2O_5 contents are 2.9 and 2.1 kg da^{-1}. These data show that the P_2O_5 content of parent rock of basalt and limestone is rich and phosphorus mixed into soil by weathering.

The available potassium content of soils, developed on basalt rocks, are relatively higher than other soils and changes between 112.5 and

133.9 kgda^{-1}. The available K_2O content of profile PL4, developed on the marine, is relatively lower than other soils and ranges between 27.0 and 36.7 kg da^{-1}. In the R horizon (basalt rock) of profile PL7 the K_2O content is higher than A horizon (Table 1). The K_2O content in the Ck horizon of profile PL2, on $CaCO_3$ rock, is 120.9 kg da^{-1}. These data show that the K_2O content of parent rock of basalt and limestone is rich and K_2O mixed into soil by weathering. Sandy soils generally have lower K buffering capacities than loamy soils (Beegle and Baker, 1987; Uribe and Cox, 1988) and that is why such soils require higher soil solution K concentration to satisfy the K - accumulation- rate requirement (Classen and Barber, 1977). Some researchers report that fine textured soils generally have higher K contents than sandy soils and loamy soils (Heckman and Kamprath, 1992).

Heavy metal contents of soils

Heavy metal contents of the studied soils are presented in Table 1. The total Zn contents of the soils PL6 and PL7, developed on basalt parent materials, are relatively higher than other soils and change between 54.0 and 110.2 mg kg^{-1}. However, total Zn contents of soils PL3 and PL4, developed on the marine, are low and is between 35 and 57.2 mg kg^{-1}.

Total Mn contents of soils, developed on basalt and alluvium materials, are higher than other soils. PL3 and PL4 soils, developed on the marines, have a relatively lower total Mn content also and change between 105 and 144 mg kg^{-1}. The high total Zn and Mn content of soils formed on the basalts may be attributed to chemical composition of these rocks. Analysis of the unweathered basalt rock sample of profile PL7, also shows similar mineralogical composition.

The total Cu content of profiles PL6 and PL7, on the basalt parent material, seems relatively higher than other soils and range from 11.4 to 18.9 mg kg^{-1}. On the other hand, the total Cu contents of profiles PL1 and PL2 of the $CaCO_3$ rocks change between 2.7 and 10.5 mgkg^{-1}. Of the profiles PL3 and PL4 formed on marines, the total Cu contents change between

3.7 and 11.7 mg kg^{-1} and of profile PL5, on the alluvium parent material, between 4.5 and 20.5 mg kg^{-1}.

The total Co content of profiles PL6 and PL7, on the basalts, are higher than other soils and range from 17.8 to 27.3 mg kg^{-1}; of profiles PL1 and PL2, on the CaCO$_3$ rocks, from 8.6 to 18.6 mg kg^{-1}; of profiles PL3 and PL4, on the marines, from 9.4 to 18.5 mg kg^{-1}; and of profile PL5, on the alluvium parent material, from 9.2 to 21.4 mg kg^{-1} (Table 1).

4. References

Black, C.A. 1965. Methods of soil analysis. Part. 2. Am. Soc. Agron. No: 9. Madison, WI.

Beegle D.B., and Baker, D.E. 1987. Differential K buffer behavior of individual soils related to K corrective treatments. Soil Sci. Plant anal. 18: 371-385.

Buol, S.W., Hole, F.D., and Mc Cracken, R.J. (1980). Soil Genesis and Classification. The Iowa State. University Press, Ames. 360 pp.

Buringh, H.P. (1979). Introduction of the Study of Soil in Tropical and Subtropical Regions, Center for Agricultural Publishing and Documentation. Wageningen.

Classen N., and S.A. Barber. 1977. Potassium influx characteristics of corn roots and interaction with N, P, Ca, and Mg influx. Agron. J. 69: 860-864.

Graham, R.C. and Wood., H.B. (1991). Morphologic development an clay redistribution in lysimeter soils under chaparral and pine. Soil Sci. Soc. Am. J. 55: 1638-1646.

Heckman J.R., and E.J. Kamprath. 1992. Potassium accumulation and corn yield related to potassium fertilizer rate and placement. Soil Sci. Soc. Am. J. 56: 141-148.

Kick, H., H. Burger, K. Sommer. 1980. Gesamtgehalte on Pb, Zn, Sa, As, Hg, Cu, Ni, Cr und Co in landwirtschaftlich und gardnerisch genutzten B,den. Nordrhein-westfaleus. Land. Forschung. 33 (1): 12-22.

Olsen S.R., V. Cole, F.S. Watanabe and L.A. Dean. 1954. Estimation of available phosphorus in soils by exraction with with sodium bicarbonate. U.S. Dept. of Agric. Cir. 939, Washington: D.C.

Soil Survey Staff. 1996. Keys to Soil Taxonomy. SMSS Tec. Monog. No: 19, Pocahontas, Inc. Virginia.

Uribe, E., and F.R. Cox. 1988. Soil properties affecting the availability of potassium in highly weathered soils. Soil Sci. Soc. Am. J. 52: 148-152.

Washer, N.E., and Collins, M.E.1988. Genesis of adjacent morphologically distinct soils in Northwest Florida. Soil Sci. Soc. Am. J. 52: 191-196.

Table 1 Some plant nutrient and heavy metal contents of the studied soil

Horizon	Depth (cm)	P_2O_5 kg da^{-1}	K_2O kg da^{-1}	N (%)	Zn mgkg^{-1}	Mn mgkg^{-1}	Cu mgkg^{-1}	Co mgkg^{-1}
			Profile PL1 (on the CaCO$_3$ rocks)					
Ap	0-18	3.8	120.9	0.092	47.8	284.4	8.5	22.8
A	18-30	3.2	118.8	0.090	54.9	218.4	10.5	14.8
Bw1	30-55	3.0	108.0	0.079	57.6	306.1	9.5	17.9
Bw2	55-70	2.7	112.5	0.077	52.8	152.0	2.7	12.7
C1k1	70-95	2.9	123.1	0.077	53.7	172.2	8.5	16.1
C2k2	95-110	1.5	25.9	0.052	38.3	160.2	3.2	8.6
			Profile PL2 (on the CaCO$_3$ rocks)					
A1	0-15	6.1	118.8	0.148	52.0	213.8	8.8	10.5
Bw	15-30	3.8	123.1	0.116	53.1	215.5	3.2	15.8
Ck	30-48	3.2	120.9	0.088	55.2	176.8	4.6	18.6
			Profile PL3 (on the marine)					
Ap	0-15	1.0	69.1	0.118	50.2	137.5	111.5	16.1
A	15-30	0.9	64.8	0.101	47.1	144.6	4.8	18.5
Bw	30-50	0.9	66.9	0.098	35.4	124.1	3.7	9.4
C1	50-70	0.7	41.0	0.055	41.9	105.4	11.7	10.9
			Profile PL4 (on the marine)					
Ap	0-15	1.0	29.1	0.034	54.7	241.7	4.3	13.6
C1	15-40	1.0	27.0	0.046	57.2	241.5	7.4	12.4
C2	40-60	0.7	36.7	0.052	53.4	253.7	5.5	14.4
			Profile PL5 (on the alluvium material)					
Ap	0-28	4.5	47.9	0.100	56.6	254.6	4.5	10.3
A1	28-45	4.0	49.6	0.114	34.3	259.9	7.7	9.2
Bw1ss	45-56	5.1	45.3	0.122	35.3	183.6	7.3	13.2
Bw2ss	56-75	3.8	41.0	0.122	69.8	334.3	8.5	16.6
Bkss	75-90	3.6	51.8	0.114	61.4	277.3	20.5	28.1
Ck	90-125	2.0	51.3	0.123	58.0	301.6	13.3	21.4
			Profile PL6 (on the basalt rocks)					
Ap	0-22	5.7	112.5	---	64.1	159.0	13.2	20.4
A1	22-38	5.0	129.6	0.120	74.5	265.1	17.5	23.0
A2	38-50	4.3	133.9	0.122	74.5	232.7	12.4	21.9
Bw1ss	50-76	5.1	118.8	0.104	57.6	303.4	11.9	27.3
Bw2ss	76-90	4.7	127.4	0.092	58.9	244.4	18.9	25.8
C1k1	90-120	3.0	129.6	0.122	56.5	243.8	13.5	25.1
C2k2	120-150-	2.1	112.5	0.088	110.2	317.2	13.9	22.9
			Profile PL7 (on the basalt rocks)					
Ap	0-18	4.8	118.8	0.092	57.5	252.3	13.3	27.2
A1	18-30	4.2	120.9	0.088	69.2	292.5	18.2	22.7
A2	30-40	4.6	120.9	0.094	54.0	273.6	13.8	21.2
Bw1ss	40-55	3.7	123.1	0.122	61.4	299.5	18.3	26.6
Bw2ss	55-64	2.1	129.6	0.052	63.3	270.2	14.1	17.8
Ck	64-75	3.1	118.8	0.055	60.7	224.5	11.4	24.5
R	75+	2.9	127.1	0.060	49.5	197.4	12.7	20.1

Chapter 67

Roots distribution, yield and protein content of broad bean (*Vicia faba* L. *major*) treated with different tillage methods in southern Italy.

D. De Giorgio, G. Convertini, D. Ferri, L. Giglio, P. La Cava
Istituto Sperimentale Agronomico (Ministry of Agriculture)-BARI, Italy

Key words: broad bean, tillage, root, protein

Abstract: This research was carried out at Foggia (southern Italy) on a silty-clay soil (Typic Chromoxererts) in an experimental field characterized by hot-arid summers, cold winters and scanty and sporadic rains.Roots distribution, agronomic parameters and nitrogen balance were recorded at harvest of broad beans after durum wheat in a two-year rotation. Four different tillage methods (A - traditional mouldboard ploughing; B - ripper subsoiling; C - skim ploughing; D - minimum tillage) and three N rates (0, 50, 100 kg N ha^{-1} only to wheat) were used. Nitrogen balance in selected "treatment" was done by comparing soil mineral nitrogen variations (N-NO$_3$ + exchangeable N-NH$_4$), in all experimental plots, with nitrogen distribution within broad bean plant organs. The findings show a strong interaction between tillage methods and N rates as affected by the modifications of water infiltration, temperature gradients, organic carbon turnover and soil nitrogen transport and transformations.

1. Introduction

The main interest in growing broad beans (*Vicia faba* L. *major*) is the peculiar chemical composition of the seeds (B1 vitamin, niacin, Fe and Ca contents etc.) and their agronomic role as a symbiotic N-fixing crop. In the last thirty years Italian and European farmers have preferred crops other than broad bean for their greater income. However, the availability of the new agronomic techniques have recently increased interest in broad beans mainly for food. To improve the performance of this crop it's necessary to apply field specific agronomic inputs with high precision and grow cultivars which are more productive and stable in their annual yield. (Pandey, 1981 Stringi, 1994; De Pace, 1994; Costa et. al, 1997; De Giorgio et al., 1997).

In southern Italy some aspects are being investigated: they include the effects of differing tillage methods on the broad bean root and on the changes in quality components of broad bean seeds. Some results are presented here.

2. Materials and Methods

In a typical area of southern Italy ("Apulian Tavoliere") broad bean (cv Supersimone) followed by durum wheat was cropped from 1995 with four different tillage methods.

A – Conventional (double share ploughing at 45 cm depth, two rotary tillage at 20 cm with disc plough, 10 cm rotary tillage); B – two- layer (combined equipment – 60 cm subsoiling and 10 cm rotary tillage); C – surface (25 cm five-share ploughing, 10 cm rotary tillage); D minimum (10 cm rotary tillage). The experimental design was a "split-plot" but three replicates in which the "tillage methods" were arranged in "plots" and N in

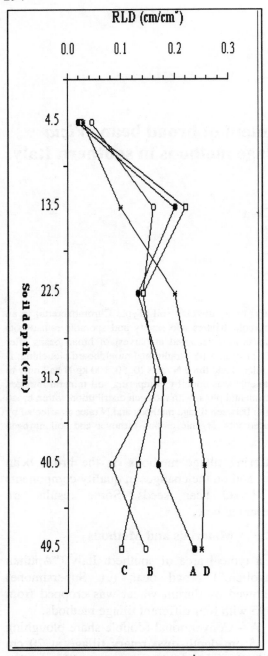

Figure 1. Trend of the mean RLD vs. soil depth in broad bean (Vicia faba L. major) in relation to 4 tillage treatment. (A conventional ——•——, B ripper subsoiling ——o——, C surface disharrowing ——□——, D minimum tillage ——*——).

"subplots" only for durum wheat. Crop residues of previous crops were removed from the field.

Soil tillage was done in the period from the end of August and the beginning of October.

In the first week of November, before the sowing, the minirhizotrons (length one meter) were inserted into the soil at an angle of 45°. The soil of experimental site is silty-loam classified as Typic Chromoxerert by the USDA "Soil Taxonomy". The rainfall in the period August '94-July '95 was of 436 mm (maximum values of 70 mm in January and March; 50 mm at the end of May).

Differing plant organs (seeds, crop residues and roots) were oven-dried at 70°C for 24 hours, weighed, ground, sieved at 2 mm and analyzed, using standard procedures to determine N, P_2O_5, and K_2O contents. Throughout the experiment soil nitrate and ammonium concentrations were determined. Images of the roots were recorded by a camera inserted in the minirhizotrons to a depth of 50 cm. The numbers were stored and transformed in RLD (Root Length Density) as Upchurch and Ritchie (1983) and evaluated by statistical analysis.

3. Results and Discussion

Minimum tilled soil did not affect yield and other characteristics of broad bean cropped after durum wheat. However during the period between emergence and tillering, the broad bean plants grown on minimum tilled soil were less developed than on other tilled soils but successively the differences were decreased. These results were confirmed by root development (fig. 1). In the 0-15 cm layer with minimum tillage RLD was lower than in differently tilled soil layers.

The maximum value of RLD was observed with A and C, treatments, because with these treatments the first soil layer was treated with a great number of tillage treatment. In the deeper layers were observed: i) a decrease of RLD with A, B and C treatments, ii) at the depth of 40 cm RLD increased until 0,25 cm/cm^3 with minimum tillage (d); iii) in the deepest layer, RLD below the conventionally tilled soil reached a value very similar to the RLD value of minimum tilled soils.

The trend of root development help to explain the differences observed in the first stage of epigeal crop development. The broad bean shows good possibilities to grow in minimum tilled plots contrary to other findings which showed that crop performances improved with the intensity of soil tillage .

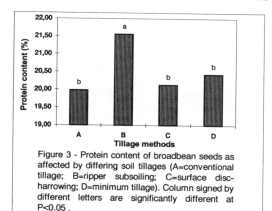

Figure 3 - Protein content of broadbean seeds as affected by differing soil tillages (A=conventional tillage; B=ripper subsoiling; C=surface disc-harrowing; D=minimum tillage). Column signed by different letters are significantly different at P<0.05 .

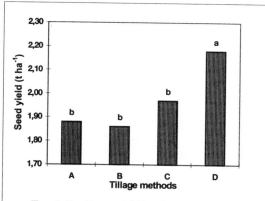

Figure 2 . Broad bean seed yield as affected by differing soil tillages (A=conventional tillage; B=ripper subsoiling; C=surface disc-harrowing; D=minimum tillage). Column signed by different letters are significantly different at P<0.05.

The different soil tillage methods affected broad bean yield and its protein content. In Figure 2 it can be observed that the highest broad bean seed yield was with minimum tillage; probably the soil characteristic and broad bean root system improved seed d.m. accumulation. In fact longer root system were observed with A, B, C tillage treatments that reduced epigeal crop development and consequently seed d.m. accumulation. On the contrary the highest protein content was observed (Fig. 3) with ripper subsoiling (treatment B), because proteic N accumulation was determined mainly by the size and shape of the root system that assimilates inorganic N from the soil and starts the transformations in NH_4^+ and aminoacids.

The differences in the mechanism of dry matter and protein accumulation appear evident when both are compared for treatment D and B. Obviously protein yield of broad bean (Fig. 4) appears more strictly dependent from broad bean seed yield (i.e.: the maximum value was observed in "D" treated plots). The investigation on broad bean (*Vicia faba major*) responses (yield and protein content) as affected by differing soil tillages shows that

in semiarid condition of southern Italy broad bean d.m. and protein yields can be increased by using m.inimum soil tillage.

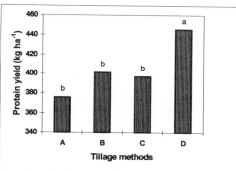

Figure 4 - Broad bean protein yield as affected by differing soil tillages (A=conventional tillage; B=ripper subsoiling; C=surface disc-harrowing; D= minimum tillage). Column signed by different letters are significantly different at P< 0.05.

4. References

Bryan J.A., Berlyn G. P., Cordon, C. (1996) Toward a new concept of the evolution of symbiotic nitrogen fixation in the Leguminosae. Plant and Soil, 186: 151-159.

Caballero R., Haj Ayed, M., Galvez J. F., Hernaiz P.J. (1995) Yield components and chemical composition of some annual legumes and oat under continental Mediterranean conditions. Agr. Med., 125: 222-230.

Carnovale E., Cappelloni M., Zaza G. (1979) Problemi connessi con l'utilizzazione per l'alimentazione umana dei preparati proteici di fava (*Vicia faba* **L.** *major*). Riv. di Agron., 1: 50. 54

De Costa W.A.J.M., Dennet M.D., Ratnaweera U., Nyalemegbe k. (1997) Effects of different water regimes on field-grown determinate and indeterminate broad bean (Vicia faba L.). I, canopy growth and biomass production. Field Crops Research, 49: 83-93.

De Giorgio D., Stelluti M., Rizzo V. (1997) Effects of four tillage methods on growth analysis and on the main biometrics parameters of broad bean (vicia faba l.,). Proc. 14th ISTRO Conference. July 27-August 1, Pulawy, Poland, 243-246.

De Pace C. (1994) V. faba: costituzione di nuovi tipi varietali per la produzione di granella secca. Agricoltura Ricerca, 155: 53-94.

Stringi, L. (1994) Acquisizioni e prospettive di sviluppo nelle tecnica colturale della fava da granella (Vicia faba L.). Agricoltura. Ricerca, 155: 31-46.

Upchurch, D.R., Ritchie, J. T. (1983) Root observation using a video recording system in mini-rhizotrons. Agron.J.,76:1015-10

Chapter 68

Effects of simulated erosion and amendments on grain yield and quality of wheat

S.S. Malhi[1], E.D. Solberg[2], R.C. Izaurralde[3] and M. Nyborg[3]

[1] *Agriculture and Agri-Food Canada, P.O Box 1240, Melfort, Saskatchewan, Canada, SOE 1AO*

[2] *Alberta Agriculture, Food and Rural Development, Edmonton, Alberta, Canada*

[3] *Department of Renewable Resources, University of Alberta, Edmonton, Alberta, Canada*

Key words: erosion, wheat

Abstract: Field experiments were conducted in 1991 at Josephburg (Black Chernozem soil) and Cooking Lake (Gray Luvisol soil), Alberta, Canada, to determine the influence of depth of erosion (Simulated-erosion approach by artificially removing the topsoil) and amendments on grain yield, protein concentration and thousand kernel weight (TKW) of hard-red spring wheat (*Triticum aestivum* L. Cv. Roblin).There were five erosion levels (0, 5, 10, 15 and 20 cm) and four amendments (control, addition of 5 cm of topsoil, addition of fertilizer at 100 kg N ha^{-1} and 20 kg P ha^{-1}, and cattle manure at 75 Mg ha^{-1} on a dry basis). Yield reduction due to topsoil removal was a function of nutrient loss. The applications of fertilizer N and P, and cattle manure improved yield substantially and reduced the impact of yield loss due to erosion. Return of 5 cm of topsoil to the plots made the rate of yield loss an independent function of erosion. At both sites, protein concentration in wheat grain was influenced by the erosion level imposed and the kind of amendment used. The protein concentration at Josephburg decreased from 141 g kg^{-1} in the noneroded treatment to 124 g kg^{-1} in the 20 cm erosion treatment. In the same order the values for the Cooking lake site were 132 and 123 g kg^{-1}, respectively. The application of commercial fertilizer at Josephburg resulted in the highest protein concentration (153 g kg^{-1}). At both sites, TKW decreased with increasing depth of soil erosion. At Cooking Lake, all amendments produced heavier seeds than the control plots. In conclusion, the productivity of artificially-eroded soil was mostly a function of nutrient removal. The productivity and quality of wheat was partially restored by using either fertilizer or organic amendments.

1. Introduction

Many agricultural landscapes in western Canada exhibit features of past erosion by wind and water. Erosion reduces the thickness of topsoil and alters other soil properties. Eroded soils can adversely affect crop yield and quality through increased bulk density, reduced organic matter, low availability of N and other nutrients, and reduced water holding capacity (Eck et al. 1965, Frye et al. 1982, Dormaar et al. 1986, Tanaka and Aase 1989, Malhi et al. 1996, Fowler 1998). Since eroded and non-eroded areas in cultivated fields often occupy different topographic position, it is very difficult to assess the effect of soil erosion on crop production and quality under natural field conditions. One practical way to determine loss in crop yield and quality due to erosion is to erode the soil artificially by physically removing the surface soil layers. The objective of the study was to determine the influence of depth of erosion (simulated-erosion approach by artificially removing the topsoil) and amendments on grain yield, protein concentration (PC) and thousand kernel weight (TKW) of wheat.

2. Material and Methods

Field experiments were conducted in 1991 at Josephburg (Black Chernozem soil) and Cooking Lake (Gray Luvisol soil), Alberta, Canada. The mean annual precipitation of the area is about 500 mm, with 60 % occurring in the growing season from May to August. There were five erosion levels (0, 5, 10, 15 and 20 cm) randomly arranged as main-plot treatments; and four amendments (control, addition of 5 cm of topsoil, addition of fertilizer at 100 kg N plus 20 kg P ha^{-1}, and cattle manure at 75 Mg ha^{-1} on a dry basis) as sub-plot treatments (10 m x 3.65 m).The test crop was hard-red spring wheat (*Triticum aestivum* L. Cv. Roblin).

All plots were soil sampled prior to sowing in 1991 and the samples were analyzed for texture, moisture limits, bulk density and nutrients. At maturity, wheat was harvested and the dried plant samples were threshed for grain yield. A grain subsample from each plot was then used to determine TKW and total N. The protein content in grain was estimated by multiplying total N with 5.87.

3. Results and Discussion

Soil Properties:

At both sites, soil physical, chemical and biochemical properties changed with depth of simulated erosion (Table 1). Total C, NO_3-N, total N, extractable P, total P and N mineralization potential of soil markedly decreased as depth of topsoil removal increased. Soil bulk density increased with 0 cm topsoil removal to 20 cm topsoil removal from 1.17 to 1.31 Mg m^{-3} at Josephburg and from 1.25 to 1.38 Mg m^{-3} at Cooking Lake.

Grain Yield:

Grain yields at both sites markedly decreased with increased erosion depth (Table 2). Mean grain yields (averaged across all amendments) decreased in non-eroded soil to 20 cm erosion treatment from 4314 to 2769 kg ha^{-1} at Josephburg and 4228 to 1730 kg ha^{-1} at Cooking Lake. The marked decreased in grain yield due to erosion depth was associated with a drastic decline in soil available N. In this study,

there was a close relations between nutrient loss. These results are in agreement with those reported by other researchers (Dormaar et al. 1986, Shafig et al. 1998, Tanaka and Aase 1989, Verity and Anderson 1990).

Grain yields improved with all amendments, but the magnitude of yield increase varied with the kind of amendment (Table 2). Commercial fertilizer was most effective to restore productivity of eroded soil, followed by manure and topsoil addition. Return of 5 cm topsoil to the plots made the rate of yield loss an independent function of erosion. The superiority of fertilizer N and P over the other amendments suggest that eroded soil became extremely deficient in plant-available N and P.

Grain Quality:

Protein concentration in wheat grain at both sites was influenced by the erosion level imposed and the kind of amendment used, but the kind of amendment used had a more pronounced effect on grain protein concentration than did the erosion level imposed (Table 2). The protein concentration in wheat grain at Josephburg decrease from 141 g kg^{-1} in the non-eroded soil to 124 g kg^{-1} in the 20 cm erosion treatment. The values at Cooking Lake were 132 and 123 g kg^{-1}, respectively, in the 0 cm and 15 cm erosion treatments. The application of fertilizer N and P at Josephburg resulted in the highest grain protein concentration (i.e., 153 g kg^{-1}). The protein concentration in the manure and topsoil treated wheat was lower with respect to wheat in the control plots. At Cooking Lake , wheat from plots receiving fertilizer and other amendments had less protein concentration than the control plots.

At both sites, TKW decrease with increasing level erosion (Table 2). The mean TKW (averaged across amendments) in non-eroded soil to 20 cm eroded soil decreased from 35.1 to 35.1 to 32.6 g at Josephburg and from 35.6 to 31.2 g at Cooking Lake. At Cooking Lake, All amendments produced heavier seed than 35.6 to 31.2 g. at Cooking Lake. At Cooking Lake, all amendments produced heavier seed than the control plots, but at Josephburg only manure produced the heaviest seeds.

Plant-available N has a significant effect on grain quality because N is required for the formation of grain storage proteins that have a direct influence on most grain quality measurements. There is a positive relationship between protein concentration in wheat and level of plant-available N (Fowler 1998). Erosion decreases soil-N fertility not only by physically removing organic and inorganic N but also by reducing the mineralization potential of the remaining soil N. This can have an adverse impact on protein concentration. In the present study, the lower protein concentration in wheat on eroded soil treatments compared to non-eroded soil can be attributed to lower amounts of NO_3-N in soil in spring and the reduction in N-supplying capacity of soil during growing season, particularly at grain filling. The protein concentration in grain is determined by the ratio of protein yield to grain yield. When grain yield is relatively increased more than protein yield, then a drop in protein concentration is expected from an amendment compared to the unfertilized treatment. In this study, a decrease in protein concentration in wheat, particularly with manure, happened because of a dilution effect from increased grain yield.

4. Conclusion

Crop productivity on artificially-eroded soils was a function of nutrient removal. The lost productivity and quality of wheat was partially restored by using either fertilizer or organic amendments.

5. References

Dormaar, J.F., Lindwall, C.W. and Kozub, G.C. 1986. Restoring productivity to an artificially eroded Dark Brown Chernozemic soil under dryland conditions. Can. J. Soil Sci. 66: 273-285.

ECK, H.V., Hauser, V.L. and Ford, R.H. 1965. Fertilizer needs for restoring productivity on Pullman silt loam after various degrees of soil removal. Soil Sci. Soc. Am. Proc. 29-213.

Fowler, D.B. 1998. Grain protein concentration response to plant-available nitrogen. *In* D.B. Fowler et al. (ed), Proc. Wheat Protein Symposium, 9-10 March, 1998, Saskatoon, Saskatchewan, Canada. pp. 281-284.

Frye, W.W., Ebelhar, S.A., Murdock, L.W. and Blevins, R.L. 1982. Soil erosion effects on properties and productivity of two Kentucky soils. Soil Sci. Soc. Am. J. 46: 1051-1055.

Power, J.F., Sandoval, F.M., Ries, R.E. and Merrill, S.D. 1981. Effects of topsoil and subsoil thickness on soil water content and crop production on a disturbed soil. Soil Sci. Soc. Am. J. 45: 124-129.,

Shafiq, M., Zafar, M.I., Ikram, M.J. and Ranjha, A.Y. 1988. The influence of simulated soil erosion and restorative fertilization on maize and wheat production. Pakistan J. Sci. Ind. Res. 31: 502-505.

Tanaka, D.L. and Aase, J.K. 1989. Influence of topsoil removal and fertilizer application on spring wheat yields. Soil Sci. Soc. Am. J. 53:228-232.

Verity, G.E. and Anderson, D.W. 1990. Soil erosion effects on soil quality yield. Can. J. Soil Sci. 70: 471-484.

Table 1. Some characteristics of soil (10 cm deep) after various depths of topsoil removal.

Site	Topsoil removal (cm)	Bulk density (Mg m^{-3})	Total C (g kg^{-1})	NO$_3$-N (mg kg^{-1})	Total P (mg kg^{-1})	Ext. P (mg kg^{-1})	Total P (mg kg^{-1})	N min. potential (mg N kg^{-1})
Josephburg	0	1.17	40	11	3600	23	800	188
	5	1.21	40	11	3110	20	790	161
	10	1.24	34	10	2850	14	720	134
	15	1.28	29	8	2440	12	640	99
	20	1.31	25	6	2050	12	640	74
Cooking Lake	0	1.25	34	16	3000	40	840	328
	5	1.28	29	12	2580	32	720	251
	10	1.30	18	8	1700	12	540	94
	15	1.34	12	7	1300	13	550	57
	20	1.38	8	6	930	2	780	38

Table 2. Effects of erosion level and amendments on grain yield, protein concentration (PC) and thousand kernel weight (TKW) of wheat

Treatment	Josephburg	Cooking Lake		Josephburg	Cooking Lake
		Yield (kg ha^{-1})			
Erosion (cm)[a]			Amendment [b]		
0	4314a	4228a	Control	2137d	1514d
5	4184a	3624b	Fertilizer	4854a	4169a
10	3693b	3133c	Manure	4477b	3655b
15	3090c	2256d	Topsoil	2972c	3640c
20	2769c	1730e			
		PC (g kg^{-1})			
Erosion (cm)[a]			Amendment [b]		
0	141a	132a	Control	140b	153a
5	136ab	133a	Fertilizer	153a	137b
10	131bc	128ab	Manure	122c	113c
15	125c	123b	Topsoil	110d	111c
20	124c	127ab			
		TKW (g)			
Erosion (cm)[a]			Amendment [b]		
0	35.1a	35.6a	Control	33.5bc	32.4b
5	34.8a	34.4ab	Fertilizer	33.9b	33.5a
10	34.3ab	33.4bc	Manure	35.4a	34.1a
15	33.3bc	32.0cd	Topsoil	33.2c	33.4a
20	32.6c	31.2d			

[a] Averaged across all amendments.

[b] Averaged across all erosion levels.

Chapter 69

Effects of town waste on some plant nutrient levels, physical and chemical properties of soils in Harran plain

S Irmak[1], Y. Kasap[1] and A.K. Sürücü[2]

[1]Dept. Soil Science., Fac. Agriculture, University of Harran, Şanlıurfa, Turkey.
[2]Dept. Soil Science, Fac. of Agriculture, University of 19 Mayıs, Samsun, Turkey

Key words: town waste, soil properties

Abstract: In this study, the effects of town waste on plant nutrients, physical and chemical properties of soils taken from three different soil series (Urfa, Akabe, Yenice) in Harran Plain were researched. It was observed that Şanlıurfa town wastes have affected largely plant nutirent levels and physical, chemical properties of the soils. While the available K_2O in untreated soils of Urfa series is 108 kg da^{-1}, the same available K_2O amount for exposed soils from sewage reaches upto 485 kg da^{-1}. For Yenice series.it is 89 kg da^{-1} for clean soils and 263 kg da^{-1} for soils irrigated by waste water. The available K_2O for Akabe series of clean sample is 119 kg da^{-1} and the sample exposed from town waste (garbage) is 402 kg da^{-1}. In the same way, the available P_2O_5 amount in untreated soils of Urfa series is 4.9 kg da^{-1} while the available P_2O_5 soil exposed from town waste (sewage) reach upto 16.8 kg da^{-1}. The increase in P_2O_5 may be attributed to domestic water waste mixed to general town waste. The P_2O_5 amount for Yenice series of unexposed soil is 3.9 kg da^{-1} while the P_2O_5 amount of exposed soil from town waste is 13.4 kg da^{-1}. The same measurement is 4.6 kg da^{-1} for Akabe series of unexposed soil from town waste and 11.4 kg da^{-1} for the treated soils. Town wastes have affected the Cl^-, and SO_4^{-2} of contents soils.

1. Introduction

Huge volume of sewage water is being produced in cities due to ever increasing population. Also the cost of scientifically treated sewage water for recycling is too high to be generally feasible in developing countries like Turkey. However, this water and town wastes contain valuable nutrients (Dahatonde et al., 1995) and thus its use as irrigation water to certain crops may lead to increased agricultural production. However, systematic studies on the effect of irrigation of soil with sewage water on soil properties and plant growth are practically lacking (Essington and Mattigod, 1991) Sewage sludge is continually used as a source of nutrients and as an organic amendment for the improvement of soil physical properties to increase food production (Bansal and Viniti,

1998). Land application of sewage water is one of the methods of disposal of waste products, while recycling the elements contained in the sludge. However, one constraint with this approach is the possible contamination of the human food chain with toxic substances such as heavy metals, when food crops are grown on the sewage-treated soils (Darmody et al., 1983; Adhikari et al., 1998). The concentration of trace metals in sewage sludge is a potential hazard in many of the disposal methods currently being employed. A suggested method for sewage-sludge disposal is application to agricultural land. Sludge amendments substitute for or supplement N and P fertilization and have a beneficial effect on soil physical condition. It is inherent in the nature of most sewage sludges, however, to have elevated levels of potentially phytotoxic trace elements

(Essington and Mattigod, 1990). Repeated applications of sewage sludge to soils has been shown to result in a substantial accumulation of trace-metal elements in the soil surface horizons and in crops grown on them (Kirkham, 1975; Essington and Mattigod, 1991). The objective of this study was to examine the effects of town waste on plant nutrient elements and physical, chemical properties of soils .

2. Materials and Methods

Six soil profiles (exposed and unexposed from town waste) from three different soil series were researched. Soil samples were taken from 0-30, 30-60, 60-90, and 90-120 cm depths for laboratory analysis. Exchangeable cations and anions, $CaCO_3$ content, available P_2O_5, K_2O, N, and heavy metal (Cu, Zn, Fe, Mn) analysis were carried out.

3. Results and Discussions

Effects of town waste on some plant nutrient elements and heavy metals

The major plant nutrient elements of the studied soils have been presented in Table 1. Town wastes have considerably affected some of plant nutrient contents. Results showed that available P_2O_5 content of the clean soils of Urfa series unexposed from town wastes ranged from 4.9 to 1.0 kg da^{-1} and the available P_2O_5 content of the same soil series irrigated by waste water increased 4-fold changing between 13.8 and 10.7 kg da^{-1}. The increase in the P_2O_5 content may be associated with the phosphorous compounds mixing into domestic wastewater. As the available K_2O content of Urfa series unexposed from town wastes varied between 108.0 and 324.0 kg da^{-1} available K_2O content of the same soil series irrigated by wastewater changed from 480.0 to 192.2 kg da^{-1}. N percentage of the soils of the same series unexposed from town wastes was found between 0.048 and 0.094%. It ranged from 0.0198 to 0.049% for the soils exposed from the town wastes.

The available P_2O_5 content of the Yenice series clean soils was assessed between 5.1 and 3.0 kg da^{-1}, and the available P_2O_5 content of the same soil series exposed from town waste

between 13.9 and 12.6 kg da^{-1}. As the available K_2O content of the same series unexposed from town waste changed from 110.1 to 168.1 kgda^{-1}, the available K_2O content of the irrigated soil series was determined between 263.5 and 157.6 kg da^{-1}.

The available P_2O_5 content of the soils of Akabe series unexposed from town waste ranged from 4.6 to 3.9 kg da^{-1} and the available P_2O_5 content of the same soil series exposed from town waste from 10.2 to 11.4 kg da^{-1}. As the available K_2O content of this soil series unexposed from town waste changed between 118.8 and 79.9 kg da^{-1}, the available K_2O content of the same soil series exposed from town waste ranged from 324.0 to 402.4 kg da^{-1}. Likewise, N content of the Akabe treated soils was determined between 0.082 and 0.063% and N content of Akabe soils exposed from town wastes between 0.154% to 0.175 (Table 1).

Cu content of the clean soils of Yenice was measured between 29.3 and 65.9 mg kg^{-1} and the Cu content of the same soil series exposed from town waste varied between 69.1 and 503.5 mgkg^{-1}. Of the Akabe soil series unexposed from town waste, Cu content ranged from 43.2 to 45.9 mg kg^{-1} and Cu content of the same soil series irrigated by waste water was between 312.8 and 362.2 mg kg^{-1}. In the case of Zn, the clean soils of Yenice series unexposed from town waste showed a range of 199.5 and 212.9 mg kg^{-1} while the Zn content of the same soil series exposed from town waste changing between 362.5 and 1550.0 mg kg^{-1}. The Zn content of the soils of Akabe series unexposed from town waste was found between 179.4 and 641.5 mg kg^{-1} and Zn content of the same soil series exposed from town waste was measured between 1716.9 and 1838.2 mg kg^{-1}.

The effects of town waste on the chemical properties of the soils

The major chemical properties of the studied soils have been presented in Table 2. The town wastes have greatly affected some chemical properties of soils in the Harran Plain. The town waste water affecting $CaCO_3$ content increased the lime up to a certain level in the soil profile.

The available $CaCO_3$ content of the soils of Urfa series unexposed from town wastes changed between 23.3 and 28.8 % and $CaCO_3$ content of the same soil series irrigated by waste water decreased and ranged from 9.1 to 12.5 % (Table 1). The decrease in the $CaCO_3$ content may be associated with the wastewater dissolving $CaCO_3$ in the soil and leaching it in the profile. The chemical solvents mixing with domestic wastewater increases the solubility of $CaCO_3$ further. The exchangeable Na content of the Urfa soil series unexposed from town waste was present uniformly in the soil profile and was measured between 0.012 and 0.016 $meq100^{-1}g$. On the other hand, the exchangeable Na content of the same soil series irrigated by town wastewater increased and reached 0.106 $meq100g^{-1}$. The increase in Na may be associated with the soluble NaCl mixing into the wastewater. While the exchangeable SO_4 content of the Urfa soils unexposed from town wastes change between 0.006 and 0.016 $meq100\ g^{-1}$, the exchangeable SO_4 content of the Urfa soils irrigated by wastewater increased and reached the value of 1.103 $meq100g^{-1}$. The exchangeable SO_4 content of the Yenice soils unexposed from town wastes varied from 0.004 to 0.042 $meq100g^{-1}$ and the exchangeable SO_4 content of the same but treated soil series increased and reached a value of 3.689 $meq100^{-1}\ gr$. Of the Akabe clean soil series the exchangeable Na content changed from 0.018 to 0.020 $meq100g^{-1}$ and the exchangeable Na content of the same soil series treated with town wastes ranged between 0.612 and 0.266 $meq100g^{-1}$. While the exchangeable K content of the same soil series unexposed from town wastes between 0.001

and 0.003 $meq100g^{-1}$, K content of the soils exposed from town wastes ranged from 0.694 to 0.903 $meq100g^{-1}$. The exchangeable Cl content of the same soil series unexposed from town wastes was determined between 0.318 and 0.337 $meq100g^{-1}$. And the exchangeable Cl content of the Akabe treated soils changed between 1.353 and 0.728 $meq100g^{-1}$. While the exchangeable SO_4 content of the Akabe soils unexposed from town wastes ranged from 0.202 to 0.439 $meq100g^{-1}$., exchangeable SO_4 content of the same soil series irrigated by waste water increased considerably and changed between 4.150 and 6.924 $meq100g^{-1}$.

4. References

Adhikari, S., S.K. Gupta and S.K. Banerjee. 1998. Pollutant Metal Contents of Vegetables Irrigated with Sewage Water. J. of Ind. Soc. Soil Sci. Vol. 46: 153-155.

Bansal, O.P. and Viniti Gupta. 1998. Effect of Sewage Sludge and Nitrogen Fertilizers on Adsorption, Persistence Mobility and Degradation of Oxamyl in Soils. J. Ind. Soc. Soil Sci. Vol. 46: 36-42.

Dahatonde, B.N., K.S. Chaudhary and R.V. Nalamwar. 1995. International Conference Sustainable Agric. Hisar, p. 27.

Darmody, R.G., J.E. Foss, M. McIntosh and D.C. Wolf. 1983. J. environ , Qual. Vol. 12, 231.

Essington, M.E. and S.V. Mattigod. 1990. Element Partitioning in Size and Density-Fractionated Sewage Sludge and Sludge Amended Soil. Soil Sci. Soc. Am. J. 54: 385.

Essington, M.E. and S.V. Mattigod. 1991. Trace Element Solid-Phase Associations in Sewage Sludge and Sludge-Amended Soil. Soil Sci. Soc. Am. J. 55: 350-356.

Kirkham, M.B. 1975. Trace elements in corn grown on long-term sludge disposal site Environ Sci. 9: 765-768.

Sommers, L.E. 1977. Chemical composition of sewage sludges and analysis. J. Environ Qual. 6: 225-232.

304

Table 1 Plant nutrients and heavy metal contents of the studied soils

Depth (cm)	Cu (mg kg^{-1})	Zn (mg kg^{-1})	Mn (mg kg^{-1})	Fe (%)	CaCO$_3$ (%)	Ava P$_2$O$_5$ (kg da^{-1})	Ava K$_2$O (kg da^{-1})	N
			Urfa Series (unexposed from sewerage)					
0-30	79.0	179.2	2282	2.20	28.5	4.9	324.0	0.0
30-60	65.4	160.5	2163	2.46	28.1	3.1	245.0	0.0
			Urfa Series (exposed from sewerage)					
0-30	97.5	210.1	981	0.12	12.5	13.8	480.0	0.1
30-60	179.8	245.0	974	2.12	11.7	12.2	316.0	0.0
			Yenice Series (unexposed from town waste)					
0-30	29.3	199.5	2516	2.16	9.5	5.1	168.1	0.0
30-60	65.9	212.9	2375	2.47	10.2	3.6	141.5	0.0
			Yenice Series (exposed from town waste)					
0-30	503.5	1496.1	1260	2.11	19.0	13.4	263.5	0.0
30-60	340.1	1550.0	1120	2.50	17.4	12.6	192.2	0.0
			Akabe Series (unexposed from town waste)					
0-30	45.9	641.5	1114	0.06	19.2	4.6	118.8	0.0
30-60	43.2	179.4	160	2.41	17.6	3.9	79.9	0.0
			Akabe Series (exposed from town waste)					
0-30	362.2	1838.2	3398	2.03	26.6	11.4	402.4	0.1
60-90	312.8	1716.9	1984	2.10	20.2	10.2	324.0	0.1

Table 2 Exchangeable cations and anions of soils

Depth (cm)	Exchangeable cations (meq 100gr^{-1})				Exchangeable anions (meq 100g^{-1})		
	Na$^+$	K$^+$	Ca^{++}	Mg	Cl$^-$	HCO^{-3}	SO$_4^{-2}$
		Urfa Series (unexposed from sewerage)					
0-30	0.094	0.058	0.568	0.135	0.408	0.143	0.304
30-60	0.106	0.030	1.031	0.589	0.509	0.144	1.103
		Urfa Series (exposed from sewerage)					
0-30	0.016	0.009	0.370	0.998	0.337	0.219	0.009
30-60	0.014	0.005	0.348	0.183	0.337	0.206	0.006
		Yenice Series (unexposed from town waste)					
0-30	0.036	0.031	0.436	0.373	0.312	0.201	0.363
30-60	0.088	0.046	1.806	1.267	0.325	0.151	2.732
		Yenice Series (exposed from town waste)					
0-30	0.011	0.008	0.224	0.336	0.399	0.138	0.042
30-60	0.009	0.004	0.175	0.272	0.338	0.118	0.004
		Akabe Series (unexposed from town waste)					
0-30	0.612	0.694	0.160	0.496	1.353	0.250	4.150
30-60	0.266	0.903	3.588	2.421	0.728	0.238	6.924
		Akabe Series (exposed from town waste)					
0-30	0.018	0.003	0.228	0.729	0.318	0.222	0.439
60-90	0.020	0.001	0.154	0.539	0.337	0.176	0.202

Chapter 70

Relationship of available nutrients with organic matter and microbial biomass in MSW compost amended soil

Mondini C.[1], Cantone P.[1], Marchiol L.[2], Franco I.[2], Figliolia A.[3] and Leita L.[1]

[1]*Istituto Sperimentale per la Nutrizione delle Piante, via Trieste 23, I-34170 Gorizia, Italy*

[2]*Dipartimento di Produzione Vegetale e Tecnologie Agrarie, via delle Scienze 208, I-33100 Udine, Italy*

[3]*Istituto Sperimentale per la Nutrizione delle Piante, via della Navicella 2-4, I-00184 Rome, Italy*

Key words: MSW compost Element availability Organic matter Soil microbial biomass Soil biological fertility

Abstract: The importance of quantity and quality of organic matter (OM) added to the soil on nutrients availability is well established. Nevertheless the role of soil microbial biomass (SMB), the more active fraction of soil organic matter (SOM), in the determination of availability of elements has not been fully investigated. The aims of the present work were to investigate the influence of quantity and quality of OM and the role of soil microbial biomass in determining element availability. The experiment was carried out in drainage lysimeters filled to a depth of 0.5 m with the following treatments: 1. 100% soil (control), 2. mixture soil-30 t dry matter per hectare of compost, 3. mixture soil-150 t dry matter per hectare of compost. Samples were taken at the beginning of the experiment, at 7, 14, 21, 35 and 280 days and analyzed for the following properties: total organic carbon (TOC), humification index (HI), microbial biomass C (B_C), total and DTPA-extractable content of Cu, Fe, K, Mn, Ni, P and Zn. Soil amendment caused a significant increase of B_C, TOC, B_C/TOC ratio and percentage, with respect their total content, of DTPA-extractable elements. The increase in elements availability following compost application is probably related, among the other things to the more available form of the elements present in organic substrates respect to that of soil native elements. Dynamics of B_C and percentage of DTPA-extractable elements in amended soil showed a maximum value after 7 days of compost application and a following decrease towards lower values, suggesting a possible relation between B_C and elements mobility. The decreasing trend of elements availability may be also related with the depletion of the non humified organic substances capable to chelate elements. Simple linear correlation coefficients of percentage of DTPA-extractable element with B_C and B_C/TOC ratio were highly significant. Results showed clearly the relationship existing between SOM, as a whole, and the mobilization equilibria of soil elements. More difficult is to enucleate the role that a fraction of the soil organic matter, the microbial biomass, exerts on this aspect of soil fertility, due to the close relation existing between B_C and TOC.

1. Introduction

The achievement and maintenance of a suitable degree of soil fertility is of fundamental importance to improve crop quality. Among the causes that nowadays endanger soil fertility, the decrease of organic matter (OM) content in agricultural soil of warmer countries is particularly worrying. To counteract this trend is not easy due to the scarcity of organic substrates suitable for soil application. Composting of different wastes is a well known system for a rapid organic matter stabilization and humification and therefore could make suitable as organic fertilizers huge quantity of organic wastes that were previously disposed. The importance of quantity and quality of OM added to the soil on nutrients availability is well established. Nevertheless the role of soil microbial biomass (SMB), the more active fraction of soil organic matter (SOM), in determining elements availability has not been fully investigated (Leita et al., 1995). SMB could influence the availability of elements with different process such as: mineralization of OM, production of chelating agents, element immobilization, adsorption and transformation. Application of MSW compost, increasing the SOM content and promoting microbial growth, would be expected to change the mobility of elements, but to predict their availability is difficult due to the many factors involved.

The aims of the present work were to investigate the influence of quantity and quality of the added OM and the role of the soil microbial biomass in determining elements availability.

2. Materials and Methods

The experiment was carried out in drainage lysimeters (1 x 0.9 x 0.7 m) filled to a depth of 0.5 m with the following treatments: 1. 100% soil (control), 2. mixture soil-30 t d.m. ha^{-1} compost (maximum triennial load permitted by the current Italian law) 3. mixture soil-150 t d.m. ha^{-1} compost. The soil used in the mixtures was a sandy clay (28% clay, 18% silt and 54% sand, total organic C 0.7%, total N 0.08%). MSW compost (pH 7.25, organic C 34.9%, total N 1.37%) was obtained from a urban composting facility. Samples were taken at the beginning of the experiment and at 7, 14, 21, 35 and 280 days. Total organic carbon (TOC) was measured by an elemental analyser (CHN). Humification index (HI) was determined according with Sequi et al. (1986). Analysis on soil microbial biomass C (B$_C$) were carried out according with Wu et al. (1990). Total elements content were determined by ICP-AES (Perkin Elmer, Optima 3000) after mineralization of samples with concentrate HCl-HNO$_3$ (Leita et al., 1995). Available fractions of elements were extracted following Lindsay-Norvell method (1978) and measured by ICP-AES.

3. Results and Discussion

Mean values of TOC content for the whole experiment were 0.98% in control soil, 1.32% and 1.88% in soils amended with 30 t and 150 t ha^{-1} of compost, respectively (Table 1). Soil amendment caused a significant increase of B$_C$ content of about three-fold and seven-fold, respectively for the 30 and 150 t ha^{-1} compost treatments, showing that microbial biomass is a more sensitive indicator of changing soil conditions respect to the TOC content (Powlson, 1994). The ratio between B$_C$ and TOC is a further reliable parameter to early assess changes in the equilibrium of soil system (Brookes, 1995). Soil amendment caused a significant increase of this parameter related with the application rates. Increment of this ratio showed as the addition of compost to a depleted soil caused a specific improvement of soil biological fertility, through the microbiological component. Addition of compost caused an increase in the percentage of DTPA-extractable elements with respect their total content, with some difference among the elements (table 1). The increase in the availability of elements following compost application is probably related, among the other things, to the more available form of the elements present in organic substrates respect to that of soil native elements (Chander and Brookes, 1993).

Table 1 - Mean values of TOC, B$_C$, B$_C$/TOC ratio and percentage of DTPA extractable elements in treatments 1 (Control), 2 (mixture soil - 30 t ha^{-1} compost) and 3 (mixture soil - 150 t ha^{-1} compost)

Treat.	TOC (%)	B$_C$ (µg/g)	B$_C$/TOC	Cu	Fe	K	Mn (%)	Ni	P	Zn
1	0.98 c†	87.0 c	0.90 c	2.40 c	0.022 b	0.05 c	0.60 b	0.14 b	0.06 b	2.98 c
2	1.32 b	318.9 b	2.41 b	4.21 b	0.034 b	0.33 b	3.63 b	0.26 a	0.09 b	4.14 b
3	1.88 a	686.4 a	3.64 a	7.29 a	0.081 a	1.56 a	12.8 a	0.30 a	0.15 a	6.66 a

† Letters indicates differences determined by Duncan's test (P<0.05)

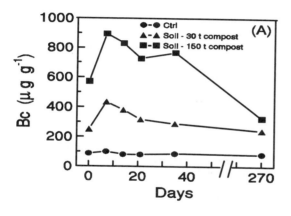 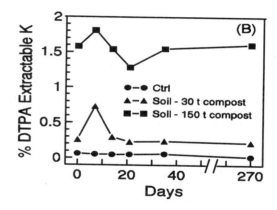

Figure A: Dynamics of soil microbial biomass C (B$_C$) (A) and percentage of DTPA-extractable K (B) during the experiment

In both amended soil B$_C$ reached a maximum values 7 days after the addition of compost. Afterwards the B$_C$ content decreased, probably for the depletion of the more available substrate suitable for microbial growth (Figure 1A).

A similar trend was also recorded for the dynamics of B$_C$/TOC ratio. HI (ratio between not humified and humified fraction of extractable SOM) in control soil decreased from 0.7 at the beginning to about 0.5 at the end of the experiment. The amended soils displayed a more pronounced variation of this parameter, i.e. from 0.8 to 0.4 and from 1 to 0.15 respectively for the 30 and 150 t ha^{-1} compost treatment. Results showed that the OM of the not well-stabilized compost, underwent in the soil to a marked process of humification. Time course of the percentage of DTPA-extractable Cu Fe K Mn and Zn showed a relative peak after 7 days of compost addition. Afterwards the percentage of extractable elements tended to stabilize to lower values. On Figure 1B has been reported the dynamics of the percentage of DTPA-extractable K, as an example of the general behaviour of elements. The similarity of dynamics of the percentage of available elements with B$_C$ suggests a possible implication of microbial biomass on the mobilization of elements. In addition, the compost used in the experiment was not

properly stabilized (HI 1.0). The not humified fraction is mainly represented by chelating substances such as aminoacids, low molecular weight organic acids and polysaccharides that could be expected to increase the availability of elements. With the ongoing of the humification process (as showed by HI values at the end of the experiment in the amended soils) the non humified fraction subsided while increased the content of humic substances. There are controversial criticism about the specific implication of OM in the mobilization dynamics of elements, but generally humic substances are thought to immobilize elements through binding and/or complexation.

Total element content is referred as a poor indicator of element availability in soils receiving biosolids (Rogers, 1997). The increase in the percentage of element availability following compost application could be expected to be more related with TOC and B$_C$. Table 2 reports simple correlation coefficients among the percentage of DTPA-extractable elements with TOC and B$_C$/TOC ratio. The B$_C$/TOC ratio allows a normalization of B$_C$ values with respect to TOC content and may be a more reliable parameter of B$_C$ content, when comparing treatments with different levels of TOC. All the elements displayed significant correlation

Table 2 - Simple correlation coefficients among percentage of DTPA-extractable elements, TOC and B_C/TOC ratio

	Percentage of DTPA-extractable elements						
	Cu	Fe	K	Mn	Ni	P	Zn
TOC (%)	0.88***	0.84***	0.93***	0.86***	0.86***	0.68**	0.90***
B_C/TOC (%)	0.76***	0.83***	0.82***	0.90***	0.86***	ns	0.81***

***significant for P< 0.001   ** significant for P< 0.01   ns not significant

(P<0.001) with both TOC and microbial biomass. The only exception was represented by P, those availability seems to be related only with TOC content

Results showed clearly the relationship existing between SOM, as a whole, and the mobilization equilibria of soil elements. More difficult is to enucleate the role that SMB, a fraction of the soil organic matter, exerts on this aspect of soil fertility. The problem to discriminate between TOC and B_C, with respect their implications on elements availability is of difficult solutions, for the reason that these parameters are not independent variables. It is widely demonstrated that there is a close relationship between these two soil properties (Powlson, 1994) and also in the present work a significant relation was found (r=0.91 P<0.001). Further research is need to better understand the role of SMB in determining element mobility, for example studying variations in element availability following treatment of soil with little amounts of molecules (i.e. water soil extracts) stimulating significant increasing of microbial size and activity without modification of the OM content.

4. References

Brookes P.C. (1995). The use of microbial parameters in monitoring soil pollution by heavy metals. Biol. Fertil. Soil 19: 269-279.

Chander K. and Brookes P.C. (1993). Residual effects of zinc, copper and nickel in sewage sludge on microbial biomass in a sandy loam. Soil Biol. Biochem. 25:1231-1239.

Leita L., De Nobili M., Muhlbachova G., Mondini C., Marchiol L., Zerbi G. (1995). Bioavalability and effects of heavy metals on soil microbial biomass survival during laboratory incubation. Biol. Fertil. Soil 19: 103-108.

Lindsay W.L. and Norwell W.A. (1978). Development of a DTPA soil test for zinc, iron, manganese, and copper. Soil Sci. Soc. Am. J. 42: 421-428

Powlson D.S. (1994). The soil microbial biomass: Before, beyond and back. In : Ritz K, Dighton J, Giller K.E. (Eds) "Beyond the biomass". John Wiley & Sons, Chichester, pp. 3-20.

Rogers S. (1997). Measurements of total heavy metals in biosolids and soil receiving biosolids is a poor indicator of metal availability. In: Proceedings of Fourth International Conference on the Biogeochemistry of Trace Elements, Berkeley, pp. 147-148.

Sequi P., De Nobili M., Leita L. and Cercignani G. (1986). A new index of humification. Agrochimica 30: 175-179.

Wu J., Joergensen R.G., Pommering B., Chaussod R., Brookes P.C. (1990). Measurement of soil microbial biomass C by fumigation-extraction. An automated procedure. Soil Biol. Biochem. 22: 1167-1169.

AUTHOR INDEX